PROCEEDINGS OF SYMPOSIA IN PURE MATHEMATICS
Volume XVII

APPLICATIONS OF CATEGORICAL ALGEBRA

AMERICAN MATHEMATICAL SOCIETY
Providence, Rhode Island
1970

Proceedings of the Symposium in Pure Mathematics
of the American Mathematical Society
Held in New York, New York
April 10–11, 1968

EDITED BY
ALEX HELLER

Standard Book Number 821-81417-6
Library of Congress Catalog Number 72-89866

AMS 1968 Subject Classifications
Primary: 1801
Secondary: 0201, 1810, 1815, 1820, 1680, 1390, 2010, 5550, 5540, 5701, 5705, 1455

Preface

This Symposium, sponsored by the American Mathematical Society, took place in New York on April 10th and 11th, 1968. The organizing committee consisted of Hyman Bass, John C. Moore, and the editor.

If categorical algebra may be said to have been born with the article of Eilenberg and MacLane on the "General theory of natural equivalences," published in 1945, it seems reasonable to assume that it has by now attained its majority. Nowadays in fact it is widely studied in its own right; a number of conferences and colloquia in recent years have adequately shown this to be so. The organizing committee for the present symposium felt that it was now appropriate to devote attention to the *applications* of categorical algebra rather than to its autonomous development. It was of course explicit problems in topology and algebra which led to the engendering of category theory, which in turn has continued throughout its existence to serve in such applications. That they continue to be numerous and lively, we hope that this symposium has helped to show.

Indeed, the applications are now so widespread that it seemed impossible to cover all of them. In particular, the manifold uses of category theory in algebraic geometry are almost unrepresented in this symposium. Topology and algebra are the staples here, as they traditionally have been. But even within these fields there is some indication of the diversity which now exists.

With two exceptions the papers appear below in the order in which they were to be presented at the symposium. That of Eilenberg, on categorical methods in the theory of computation, does not appear here. Mazur's paper "Finite flat structures," which he was unable to read in person, has been placed at the end of the list.

ALEX HELLER
New York, April 1969

Contents

Equality in hyperdoctrines and comprehension schema as an adjoint functor

F. William Lawvere

0. The notion of hyperdoctrine was introduced (*Adjointness in Foundations*, to appear in Dialectica) in an initial study of systems of categories connected by specific kinds of adjoints of a kind that arise in formal logic, proof theory, sheaf theory, and group-representation theory. It appears that abstract structures of this kind are also intimately related to Gödel's proof of the consistency of number theory (Dialectica 1958) and to Läuchli's complete semantics for intuitionistic logic (to appear in Proceedings of the Buffalo Conference on Intuitionism and Proof Theory), although the precise relationship is yet to be worked out. Since then the author has noticed that yet another "logical operation", namely that which assigns to every formula φ its "extension" $\{x:\varphi(x)\}$ is characterized by adjointness, and that the "same" adjoint in a different hyperdoctrine leads to the notion of fibered category (or in particular the covering groupoid of a permutation group). The second part of this article is devoted to a preliminary discussion of this sort of adjoint, which we call tentatively the Comprehension Schema. The first part of the article concerns two kinds of identities which a hyperdoctrine may satisfy, and which lead in particular to a more or less satisfactory theory of the attribute "equality". One of these kinds of identities is formally similar to, and reduces in particular to, the Frobenius reciprocity formula for permutation representations of groups. Actually our definition of "equality" is *not* satisfactory when these identities do not hold, though from examples one surmises that a satisfactory theory could be developed by introducing still more structure into the already rather rich notion of hyperdoctrine.

We recall the basic ingredients of a hyperdoctrine: there is to be a category **T** of "types", whose morphisms are called "terms", and which is assumed to be cartesian closed. For each type X there is a cartesian closed category $P(X)$ of "attributes of type X", whose morphisms are called "deductions over X", and for every term $f:X \to Y$ there is a functor $f \cdot (\ \ ) \ P(Y) \to P(X)$ called "substitution of f in $(\ \ )$" for which it is assumed that $f \cdot (g \cdot \varphi) = (fg) \cdot \varphi$ for $g: Y \to Z$ a term and φ an attribute of type Z (or a deduction over Z). Actually we should in principle only give natural isomorphisms $f \cdot (g \cdot (\ \ )) \cong (fg) \cdot (\ \ )$ and assume that these are coherent, but actual equality holds in the examples which we consider here. Finally there are given, for each term $f:X \to Y$, two functors $(\ \ )\,\Sigma f$ and $(\ \ )\,\Pi f$ respectively left and right adjoint to substitution, called "existential, respectively universal, quantification along f". By general properties of adjoints we have then canonical natural isomorphisms

$$(\Phi\Sigma f)\Sigma g \cong \varphi\Sigma(fg), \qquad (\varphi\Pi f)\Pi g \cong \varphi\Pi(fg)$$

for any attribute φ of type X.

All the adjointness relations involved in a hyperdoctrine are supposed to involve given front and back adjunction maps, so that the theory of hyperdoctrines is a purely equational calculus. Nevertheless, we shall mostly use only the hom-set bijections induced by the adjunction morphisms, and in fact we will indicate these bijections in the manner usually used for rules of inference. Thus the cartesian closed structure of $\mathbf{T}$, for example, involves three adjoints: First there is the terminal object 1, right adjoint to $\mathbf{T} \to \mathbf{1}$, whose characteristic property is

$$\frac{X \to 1 \text{ (in } \mathbf{T})}{\circ \quad \text{(in } \mathbf{1})}$$

where the horizontal line indicates the canonical bijection of the morphisms of the sort above the line with those of sort below the line, and the dot denotes the unique morphism of the category $\mathbf{1}$. Secondly there is the cartesian product, right adjoint to the diagonal functor $\mathbf{T} \to \mathbf{T} \times \mathbf{T}$, whose adjunction morphisms are the diagonal $X\delta: X \to X \times X$ and the projections $\langle Y_1, Y_2\rangle\pi_i: Y_1 \times Y_2 \to Y_i$, and whose characteristic property is expressed by the bijection

$$\frac{X \to Y_1 \times Y_2}{X \to Y_1,\ X \to Y_2}$$

where the ordered pair below the line may be thought of as a morphism in $\mathbf{T} \times \mathbf{T}$. Finally, for each type A, we have the right adjoint to $A \times (\ \ )$, called exponentation by A, whose adjunction natural transformations λ_A and ϵ_A can be "deduced" from the basic bijection

$$\frac{X \to Y^A}{A \times X \to Y}$$

by setting $Y = A \times X$ and considering the identity term below the line, respectively by setting $X = Y^A$ and considering the identity term above the line.

In the cartesian closed category $P(X)$ of attributes of type X, we call the terminal object 1_X the "identically true attribute of type X" (deductions over X with domain 1_X will sometimes be called "proofs over X"), and we denote product and exponentation as conjunction and implication, respectively. Thus the "evaluation" natural transformation ϵ could instead be called "modus ponens", and the adjointness relations become bijections of deductions over X as follows.

$$\frac{\varphi \to \psi_1 \wedge \psi_2}{\varphi \to \psi_1,\ \varphi \to \psi_2}, \qquad \frac{\varphi \to (\alpha \Rightarrow \psi)}{\alpha \wedge \varphi \to \psi}.$$

Finally the adjointness property for existential (and dually for universal) quantification along $f: X \to Y$ is expressed by the bijection

$$\frac{\varphi \Sigma f \to \psi}{\varphi \to f \cdot \psi}$$

between deductions over Y above and deductions over X below for each attribute φ of type X and attribute ψ of type Y. Here we have not bothered to give names to the adjunction transformations. This neglect, and our use of the "rule of inference" notation, indicates in particular that we are ignoring coherence questions; that is, in our assertions below in which we assert the existence of a canonical natural deduction $\varphi_1 \to \varphi_2$, we have not verified that there do not exist several such. Lambek, in the Proceedings of the Batelle Conference on Categorical Algebra and Homology Theory, has made a healthy start on the coherence problem by establishing Cut-Elimination for certain categories closely related to cartesian closed categories. In the same place, Gray, by introducing the appropriate notion of 2-dimensional adjointness, has shown that all the features of a hyperdoctrine, including our comprehension schema, can be obtained by defining a type to be an arbitrary category and an attribute of type **B** to be any fibration over **B**.

As pointed out in our Dialectica article, terms corresponding to all higher-type primitive recursive functions can be guaranteed by assuming a left adjoint to the forgetful functor $\tilde{T} \to \mathbf{T}$ (the domain being the usual category whose objects are endo-terms). However we have not here included this adjoint in our general definition as it plays no role in this paper.

We mention now some examples of hyperdoctrines. Given any theory (several sorted, institutionistic or classical) formulated in the language of finite types, define **T** to have as objects all type symbols $1, V_0, V_1, V_2, \ldots$ (one V_i for each sort), $V_i \times V_j, V_i^{V_j}, (V_i \times V_j)^{V_k} \times V_e^{(V_i^{V_n})}, \ldots$ (i.e. all expressions obtained by closing the V_i with respect to product and exponentation) and as morphisms suitable equivalence classes of (tuples of) terms from the theory. The adjunction equations force certain identifications of terms, and additional identifications may be forced by axioms of the theory if there are terms provided by the theory in addition to those guaranteed by the requirement that **T** be cartesian closed (for instance, in higher-order number theory, the recursion-adjoint F of the preceding paragraph exists, and the natural numbers $1 \xrightarrow{n} 1F$, the successor map, $1F \xrightarrow{\sigma} 1F$, etc. are such additional terms, while the distributive law is such an additional identification). As objects in $P(X)$ take all formulas of the theory whose free variables correspond to the type X. For deductions over X, one may take provable entailments (so that the category $P(X)$ reduces to a preordered set) or one may take suitable "homotopy classes" of deductions in the usual sense. One can write down an inductive definition of the "homotopy" relation, but the author does not understand well what results (some light is shed on this question by the work of Läuchli and Lambek cited above). Thus, although such syntactically presented hyperdoctrines are quite important, it is fortunate for the intuition that there are also semantically-defined examples, as below.

There are two basic examples in which $\mathbf{T} = \mathcal{S}$ the category of all (small) sets and mappings. One has $P(X) = 2^X =$ the partially-ordered set of all propositional functions defined on X; if we confuse propositional functions with the corresponding subsets, we then must have that $\varphi_1 \wedge \varphi_2 = \varphi_1 \cap \varphi_2$ and that $\varphi\Sigma f$ is the direct image of φ along f (understanding that substitution is defined by composition, so that, under the confusion, $f \cdot \psi$ is the inverse image by f of ψ). Every model of a higher-order theory induces a morphism from the corresponding hyperdoctrine to this set-hyperdoctrine, and conversely. The other example has $P(X) = \mathcal{S}^X$, so that an attribute φ of type X is any family $x \cdot \varphi$ of sets indexed by $x \in X$ and a deduction $\varphi_1 \xrightarrow{d} \varphi_2$ over X is any family $x \cdot \varphi_1 \xrightarrow{x \cdot d} x \cdot \varphi_2$ of mappings. Thus $P(1) = \mathcal{S}$ is the "category of truth-values" for this hyperdoctrine. The relations

$$x \cdot (\alpha \Rightarrow \psi) = (x \cdot \psi)^{(x \cdot \alpha)}, \qquad y \cdot (\varphi \Pi f) = \prod_{xf=y} x \cdot \varphi,$$

$$y \cdot (\varphi \Sigma f) = \sum_{xf=y} x \cdot \varphi \text{ (disjoint sum)}$$

follow (from the definition of substitution as composition). By our general definition of "proof over X" it follows that the proofs (over 1) of $x \cdot \varphi$ for $1 \xrightarrow{x} X$ are precisely the elements of the set $x \cdot \varphi$. Thus, this hyperdoctrine may be viewed as a kind of set-theoretical surrogate of proof theory (honest proof theory would presumably also yield a hyperdoctrine with nontrivial $P(X)$, but a syntactically-presented one). For example, by the above equations, a proof over X of $\alpha \Rightarrow \psi$ is a function which, for each $1 \xrightarrow{x} X$ assigns to every proof that $x \cdot \alpha$ a corresponding proof that $x \cdot \psi$, while a proof over Y of $\varphi\Sigma f$ is a function assigning to every $1 \xrightarrow{y} Y$ an ordered pair consisting of an x such that $xf = y$ and a proof that x has the attribute φ.

The functor $\mathcal{S} \to \mathbf{2}$ taking the empty set to 0 and every other set to 1 induces a functor from the "proof" hyperdoctrine on $\mathbf{T} =$ sets to the "propositional-function" hyperdoctrine on $\mathbf{T} =$ sets which commutes with all the mentioned logical operations. The fact that it commutes with universal quantification is equivalent to the axiom of choice, or in the language of proofs, to a strong form of ω-completeness.

We will consider three examples in which types are small categories and terms are all functors between them. Here of course exponentiation of types must be the usual functor-category construction. One has $P(\mathbf{B}) = \mathbf{2}^{\mathbf{B}} =$ the category of all functors from $\mathbf{B}$ into the arrow category $\simeq$ the Brouwerian lattice of all sets φ of objects of $\mathbf{B}$ with the property that if $B \xrightarrow{b} B'$ in $\mathbf{B}$ and $B \in \varphi$ then $B' \in \varphi$; we leave as an exercise the computation of implication and quantification. The second example has $P(\mathbf{B}) = \mathscr{S}^{\mathbf{B}}$. Hence one has

$$\alpha \Rightarrow \psi : B \rightsquigarrow \text{nat}\,(H^B \times \alpha, \psi) = \mathscr{S}^{\mathbf{B}}(H^B \wedge \alpha, \psi),$$

$$\varphi\Sigma f : C \rightsquigarrow \varinjlim\,[(f, C) \to \mathbf{B} \xrightarrow{\varphi} \mathscr{S}]$$

for $\mathbf{B} \xrightarrow{f} \mathbf{C}$ a functor and $\alpha, \varphi, \psi : \mathbf{B} \to \mathscr{S}$. The third example also has $P(\mathbf{B}) = \mathscr{S}^{\mathbf{B}}$, but we restrict the category $\mathbf{T}$ of types to consist only of those $\mathbf{B}$ which are groupoids i.e. categories in which all morphisms are isomorphisms. $\mathbf{T}$ is still cartesian closed since in fact $\mathbf{B}^{\mathbf{A}}$ is a groupoid for any category $\mathbf{A}$ if $\mathbf{B}$ is. If $\mathbf{B}$ and $\mathbf{C}$ have one object (i.e. are groups) then $P(\mathbf{B})$ is the category of all permutation representations of $\mathbf{B}$ and $\varphi\Sigma f$ is the so-called induced representation of $\mathbf{C}$. (Actually, there are two induced representations, the other being $\varphi\Pi f$, calculated roughly as the fixed point set of $\varphi^{\mathbf{C}}$ rather than the orbit set of $\varphi \times \mathbf{C}$. If f is of finite index the analogous constructions for *linear* representations yield isomorphic results, which is perhaps why there seems to be no established name for "universal quantification" in representation theory.)

Since we have not taken recursion as part of the definition, hyperdoctrines are also obtained if in the last five examples we replace small set, category, functor, etc. by finite set, category, functor.

Finally we remark that although our discussion below of comprehension hinges on the operation Σ, there is at least one structure, namely with types = Kelly spaces and attributes = set-valued sheaves in which all features of hyperdoctrines except Σ exist ($f \cdot (\ \ )$ is only exact, not continuous in general) but in which there is clearly a kind of "extension", namely the espace etalé.

1. We define, for each type X, an attribute of type $X \times X$ as follows

$$\Theta_X = 1_X\Sigma(X\delta).$$

The adjunction then provides a canonical deduction $1_X \to (X\delta) \cdot \Theta_X$ which we interpret to mean that "reflexivity" holds for "equality" so defined. We wish to investigate what other expected properties of equality hold, and more generally to study the interaction of existential quantification of attributes and cartesian products of types.

There are other expected properties of equality which we have not investigated; for example, considering the projections p, π_1, π_2, and the evaluation adjunction ϵ in

$$\begin{array}{ccc} X \times Y^X \times Y^X & \xrightarrow{\ p\ } & Y^X \times Y^X \\ \Big\downarrow{\scriptstyle \pi_i} & & \\ X \times Y^X & \xrightarrow{\ \epsilon\ } & Y \end{array}$$

one might expect $\Theta_{Y^X} = ((\pi_1\epsilon)\Theta_Y(\pi_2\epsilon))\Pi p$ to hold. The intuitive interpretation of this equation, $f_1 = f_2 \Leftrightarrow \forall x[xf_1 = xf_2]$ does not quite reflect it adequately, for it does not necessarily mean that 1 is a generator for $\mathbf{T}$; for example, the equation holds in the hyperdoctrine derived from a higher-order theory, even though there may be no morphisms $x: 1 \to X$ in $\mathbf{T}$. However for what we are able to prove in this paper neither exponentiation of types nor universal quantification of attributes plays any role. Thus we only assume that we work in an arbitrary eed (elementary existential doctrine, defined like a hyperdoctrine except that Y^X and $\varphi\Pi f$ are not necessarily assumed to exist).

Reasonable relationships in an eed between products and equality as we have defined it turn partly on implication being strictly preserved by substitution.

PROPOSITION (SUBSTITUTIVITY OF EQUALITY). *In any eed in which, for every term $f: X \to Y$ and any two attributes α, ψ of type Y, the canonical deduction*

$$f \cdot (\alpha \Rightarrow \psi) \to f \cdot \alpha \Rightarrow f \cdot \psi$$

over X is an isomorphism, one also has, for any attribute φ of type X, a canonical deduction

$$\Theta_X \to \pi_1 \cdot \varphi \Rightarrow \pi_2 \cdot \varphi$$

over $X \times X$.

PROOF. The identity deduction $\varphi \to \varphi$ yields a canonical

$$1_X \to \varphi \Rightarrow \varphi = (\delta\pi_i)\varphi \Rightarrow (\delta\pi_2) \cdot \varphi \xleftarrow{\sim} \delta \cdot (\pi_1 \cdot \varphi \Rightarrow \pi_2 \cdot \varphi)$$

which by the adjointness of existential quantification along the diagonal used to define equality yields the conclusion: Thus in fact we only used the assumption for the case $f = \delta$.

DEFINITION-THEOREM. *In any eed, the following are equivalent:*

(1) *Frobenius Reciprocity holds.*

(2) *For any $f: X \to Y$, α, ψ in $P(Y)$* $f \cdot (\alpha \Rightarrow \psi) \xrightarrow{\approx} f \cdot \alpha \Rightarrow f \cdot \psi$.

(3) *For any $f: X \to Y$, $\varphi \in P(X)$, $\alpha \in P(Y)$* $((f \cdot \alpha) \wedge \varphi)\Sigma f \xrightarrow{\approx} \alpha \wedge (\varphi \Sigma f)$.

PROOF. The second condition means that the diagram of functors

$$\begin{array}{ccc} P(Y) & \xrightarrow{\alpha \Rightarrow (\ )} & P(Y) \\ {\scriptstyle f\cdot(\ )}\big\downarrow & & \big\downarrow{\scriptstyle f\cdot(\ )} \\ P(X) & \xrightarrow[(f\cdot\alpha) \Rightarrow (\ )]{} & P(X) \end{array}$$

commutes up to canonical natural equivalence. Hence replacing each functor by its left adjoint also yields a diagram which commutes up to canonical natural equivalence:

$$\begin{array}{ccc} P(Y) & \xleftarrow{\alpha \wedge (\ )} & P(Y) \\ {\scriptstyle (\ )\Sigma f}\big\uparrow & & \big\uparrow{\scriptstyle (\ )\Sigma f} \\ P(X) & \xleftarrow[(f\cdot\alpha) \wedge (\ )]{} & P(X) \end{array}$$

But the latter is just the third condition. Conversely if the third condition holds, we can replace the functors in the latter diagram by their right adjoints, yielding the second condition.

It is clear that Frobenius Reciprocity holds in both the 2-valued and set-valued hyperdoctrines with sets as types. However it does not hold in the set-valued hyperdoctrines with small categories as types. We provide a

COUNTEREXAMPLE. Let $f:\mathbf{1}\to\mathbf{2}$ be 1 and consider $\alpha, \psi:\mathbf{2}\to\mathcal{S}$ represented as $A\xrightarrow{\alpha}B$, $U\xrightarrow{\psi}V$ in $\mathcal{S}$. Then in general $f\cdot(\alpha\Rightarrow\psi)\to f\cdot\alpha\Rightarrow f\cdot\psi$ is not an isomorphism in $\mathcal{S}$.

PROOF. $f\cdot\zeta$ is just the value of ζ at 1 for any $\zeta\in\mathcal{S}^2$. We do have

$$(\alpha\Rightarrow\psi)_0 = U^A = \alpha_0\Rightarrow\psi_0$$

but, since $H^1 = 1_2$,

$$(\alpha\Rightarrow\psi)_1 = \mathcal{S}^2(\alpha,\psi) = U^A \underset{V^A}{\times} V^B$$

while

$$a_1\Rightarrow\psi_1 = V^B.$$

Nevertheless, group theory is simpler than category theory.

PROPOSITION. *In the groupoid-permutation hyperdoctrine, Frobenius Reciprocity holds.*

PROOF. We need only show that substitution preserves implication. But in fact we have for any *groupoid* $\mathbf{C}$ and object $c\in\mathbf{C}$ and any two functors α, $\psi:\mathbf{C}\to\mathcal{S}$ that

$$\mathcal{S}^{\mathbf{C}}(H^C\times\alpha,\psi)\to(C\psi)^{(C\alpha)}$$

defined by evaluating a natural transformation at the identity in $(C)H^C = \mathbf{C}(C, C)$, is a bijection. The inverse sends any mapping: $g:C\alpha\to C\psi$ into the natural transformation $\bar{g}$ for which

$$D\bar{g}:\langle u, x\rangle\to x(u^{-1}\alpha)g(u\psi)$$

for any $C\xrightarrow{u}D$ in DH^C and $x\in D\alpha$. This actually shows that for $\mathbf{1}\xrightarrow{C}\mathbf{C}$, $C\cdot(\ ):\mathcal{S}^{\mathbf{C}}\to\mathcal{S}^{\mathbf{1}}=\mathcal{S}$ preserves implication, that implication is defined objectwise. ($C\rightsquigarrow(C\psi)^{(C\alpha)}$ becomes a functor by means of $u\rightsquigarrow(u\psi)^{(u^{-1}\alpha)}$). Thus for any $f:\mathbf{B}\to\mathbf{C}$ the sets involved in an implication-representation are preserved, and it is clear that the action is also preserved.

In order to prove the theorems we are aiming at in this section, we need to consider another condition, which first came to the writer's attention in unpublished work of Jon Beck on Descent Theory but which was surely considered earlier in topology.

DEFINITION. An eed satisfies the Beck condition iff for every diagram

$$\begin{array}{ccc} X & \xrightarrow{x} & X' \\ {\scriptstyle f}\downarrow & & \downarrow{\scriptstyle f'} \\ Y & \xrightarrow[y]{} & Y' \end{array}$$

of types and terms which is a *meet* (pullback, fibered product) diagram and for any attribute ψ of type Y, the canonical deduction $(f \cdot \psi)\Sigma x \to f' \cdot (\psi\Sigma y)$ (induced by the identity deduction on $\psi\Sigma y$) is an isomorphism. (We should require the same for Π if it exists.) Since we have not assumed that **T** has meets in general, we are led to ask

QUESTION. What is the form of the diagrams which must be meet diagrams in any category with products? Only two forms enter into our theorems; we do not know whether there are essentially different forms.

PROPOSITION. *For any morphism* (*term*) $f : X \to Y$

$$\text{(a)} \qquad \begin{array}{ccc} X & \xrightarrow{\langle X, f\rangle} & X \times Y \\ {\scriptstyle f}\downarrow & & \downarrow{\scriptstyle f \times Y} \\ Y & \xrightarrow[Y\delta]{} & Y \times Y \end{array}$$

is a meet diagram.

PROOF. Clear.

Case (a) of the Beck condition enables us to settle the following, which may have puzzled some readers. Our notion of quantification along an arbitrary term seems a considerable generalization of the usual quantification with respect to a variable x, which corresponds to the case when the term f quantified along is a projection $\pi_Y : X \times Y \to Y$. The greater generality was used in defining equality, since there we quantified along a diagonal term, which is not reducible to quantification along a projection. But perhaps that is the only essential case gained by the generalization: that is, perhaps the general case of $\varphi\Sigma f$ can be expressed in terms of Θ_Y and $(\ \ )\Sigma\pi_Y$. In fact, that is true in the basic set-propositional function hyperdoctrine where $y \cdot (\varphi\Sigma f) = 1$ iff $\exists x[xf = y \wedge x \cdot \varphi = 1]$. More generally, this relation (suitably translated into our variable-free language) holds in many eeds, as asserted below. First we introduce a slight abbreviation of notation: if $f_i : X \to Y$, $i = 1, 2$, denote by $f_1 \Theta f_2 = \langle f_1, f_2\rangle \cdot \Theta_Y$ the attribute of type X obtained by substituting $\langle f_1. f_2\rangle : X \to Y \times Y$ into the equality attribute of type $Y \times Y$. Then

THEOREM. *In any eed in which Frobenius Reciprocity and case* (a) *of the Beck condition holds,*

$$\varphi\Sigma f \xrightarrow{\approx} (\pi_X \cdot \varphi \wedge (\pi_X f \Theta \pi_Y))\Sigma\pi_Y.$$

PROOF. We show first that Frobenius Reciprocity implies

$$\varphi\Sigma f \xrightarrow{\approx} (\pi_X \cdot \varphi \wedge (1_X\Sigma\langle X,f\rangle))\Sigma\pi_Y.$$

Indeed,

$$\varphi\Sigma f = \varphi\Sigma(\langle X,f\rangle\pi_Y) \cong (\varphi\Sigma\langle X,f\rangle)\Sigma\pi_Y$$

so we are reduced to showing that

$$\varphi\Sigma\langle X,f\rangle \xrightarrow{\approx} \pi_X \cdot \varphi \wedge (1_X\Sigma\langle X,f\rangle),$$

which is equivalent to

$$(\langle X,f\rangle \cdot (\pi_X \cdot \varphi) \wedge 1_X)\Sigma\langle X,f\rangle \xrightarrow{\approx} \pi_X \cdot \varphi \wedge (1_X\Sigma\,\langle X,f\rangle);$$

but the latter follows from our statement of Frobenius Reciprocity by making the substitutions $a \rightsquigarrow \pi_X \cdot \varphi$, $\varphi \rightsquigarrow 1_X$, $f \rightsquigarrow \langle X,f\rangle$.

To complete the proof we show that Beck's condition applied to diagrams of form (a) yields a canonical isomorphism $1_X\Sigma\langle X,f\rangle \xrightarrow{\approx} \pi_X f\Theta\pi_Y$ (note that both of the expressions intuitively express the attribute of type $X \times Y$ which corresponds to the graph of f). In fact Beck (a) is explicitly $(f \cdot \psi)\Sigma\langle X,f\rangle \xrightarrow{\sim} (f \times Y)\cdot(\psi\Sigma(Y\delta))$; noting that $f \times Y = \langle \pi_X f, \pi_Y\rangle$ and that $f \cdot 1_Y \xrightarrow{\sim} 1_X$, the stated isomorphism follows by setting $\psi = 1_Y$ and using our definition of equality.

PROPOSITION. *For any type A and term $f{:}X \to Y$, the following is a meet diagram*

$$\text{(b)}\qquad \begin{array}{ccc} A \times X & \xrightarrow{A\times f} & A \times Y \\ \downarrow & & \downarrow \\ X & \xrightarrow[f]{} & Y \end{array}$$

PROOF. Special case of the following, whose proof is clear.

PROPOSITION. *For any pair of terms $f_i{:}X_i \to Y_i$, $i = 1, 2$ the following is a meet diagram*

$$\text{(c)}\qquad \begin{array}{ccc} X_1 \times X_2 & \xrightarrow{X_1\times f_2} & X_1 \times Y_2 \\ {\scriptstyle f_1\times X_2}\downarrow & & \downarrow{\scriptstyle f_1\times Y_2} \\ Y_1 \times X_2 & \xrightarrow[Y_1\times f_2]{} & Y_1 \times Y_2 \end{array}$$

Our other theorem concerning the interaction of products and quantifications will have a corollary concerning equality of vectors, and will be based on Beck's condition applied to diagrams of form (b). The theorem itself states in effect that, though conjunction and existential quantification do not usually commute, they do in a certain sense if the quantified variables are "independent" of each other inside the matrix. First, to make the notation more readable, we define the functor

$$P(X_1) \times P(X_2) \xrightarrow{\otimes} P(X_1 \times X_2)$$

by $\varphi_1 \otimes \varphi_2 = \pi_1 \cdot \varphi_1 \wedge \pi_2 \cdot \varphi_2$, the conjunction being of course the product in $P(X_1 \times X_2)$.

THEOREM. *In an eed in which Frobenius Reciprocity and Beck* (b) *hold, one has for any term $f: X \to Y$ and type A, and for any attributes φ and α of types X and A respectively, a canonical natural isomorphism $(\alpha \otimes \varphi)\Sigma(A \times f) \xrightarrow{\approx} \alpha \otimes (\varphi\Sigma f)$ of attributes of type $A \times Y$.*

PROOF. Let $\pi_X : A \times X \to X$, $\pi_Y : A \times Y \to Y$ denote the projections. Then Beck (b) yields explicitly

$$(\pi_X \cdot \varphi)\Sigma(A \times f) \xrightarrow{\approx} \pi_Y \cdot (\varphi\Sigma f).$$

Thus

$$\begin{aligned} \alpha \otimes (\varphi\Sigma f) &\xleftarrow{\sim} \pi_A \cdot \alpha \wedge (\pi_X \cdot \varphi)\Sigma(A \times f) \\ &\xleftarrow{\sim} ((A \times f) \cdot (\pi_A \cdot \alpha) \wedge \pi_X \cdot \varphi)\Sigma(A \times f) \quad \text{by Frobenius} \\ &= (\alpha \otimes \varphi)\Sigma(A \times f) \end{aligned}$$

since $(A \times f)\pi_A = \pi_A$.

COROLLARY. *If $g: B \to A$, $f: X \to Y$ are any two terms, β and φ attributes of types B and X respectively, then under the hypotheses of the foregoing theorem, one has a canonical natural isomorphism*

$$(\beta \otimes \varphi)\Sigma(g \times f) \xrightarrow{\sim} (\beta\Sigma g) \otimes (\varphi\Sigma f).$$

PROOF. Set $\alpha = \beta\Sigma g$ in the foregoing theorem, use also the symmetrized form

$$(\beta \otimes \varphi)\Sigma(g \times B) \xrightarrow{\approx} (\beta\Sigma g) \otimes \varphi$$

of the theorem and the fact that $g \times f = (g \times B)(A \times f)$.

COROLLARY. *Under the hypotheses of the theorem, one has for any two types X_1, X_2 an isomorphism*

$$\Theta_{X_1 \times X_2} \xrightarrow{\approx} \theta \cdot (\Theta_{X_1} \otimes \Theta_{X_2})$$

where θ is the term "middle four exchange isomorphism":

$$\theta : (X_1 \times X_2)^2 \xrightarrow{\approx} X_1^2 \times X_2^2.$$

Thus our culminating result states that two ordered pairs are equal iff their first components are equal and their second components are equal.

PROOF. Setting $\beta = 1_{X_1}, \varphi = 1_{X_2}, g = X_1\delta, f = X_2\delta$ in the previous corollary, one obtains

$$(1_{X_1} \otimes 1_{X_2})\Sigma(X_1\delta \times X_2\delta) \xrightarrow{\sim} \Theta_{X_1} \otimes \Theta_{X_2}.$$

But $1_{X_1} \otimes 1_{X_2} = 1_{X_1 \times X_2}$, since both conjuncts are $\pi_i \cdot 1_{X_i} = 1_{X_1 \times X_2}$. Finally

$$(X_1 \times X_2)\delta = (X_1\delta \times X_2\delta)\theta^{-1}$$

but since θ is an isomorphism

$$\xi\Sigma\theta^{-1} \xrightarrow{\sim} \theta \cdot \xi,$$

so the statement follows.

Even these meager theorems apparently do not hold in the doctrines whose attributes are set-valued functors on small categories or groupoids. *Counterexample* (albeit to the hypotheses, not the conclusion, of the theorems). Let $\mathbf{G}$ be the groupoid

$$A \underset{f^{-1}}{\overset{f}{\rightleftarrows}} B$$

with only four morphisms and consider the two constant endofunctors ∂_A, ∂_B of $\mathbf{G}$. Then

$$\begin{array}{ccc} \mathbf{O} & \longrightarrow & \mathbf{G} \\ \downarrow & & \downarrow{\scriptstyle \partial_B} \\ \mathbf{G} & \underset{\partial_A}{\longrightarrow} & \mathbf{G} \end{array}$$

is a meet diagram (where $\mathbf{O}$ is the empty category) and yet the Beck condition applied to this diagram does not hold at any nonempty attribute φ.

PROOF. Obviously $((\mathbf{O} \to \mathbf{G}) \cdot \varphi)\Sigma(\mathbf{O} \to \mathbf{G}) = 0$ and yet, since

$$\begin{array}{ccc} \mathcal{S}^{\mathbf{G}}(\partial_B \cdot \phi, \psi) & \approx & \mathcal{S}^{\mathbf{G}}(\phi, \partial_A \cdot \psi) \\ \wr\!\!\searrow & & \swarrow\!\!\wr \\ & \mathcal{S}(B \cdot \varphi, A \cdot \psi), & \end{array}$$

we have that $\partial_B \cdot (\varphi\Sigma\partial_A) = \partial_B \cdot (\partial_B \cdot \varphi) = \partial_B^2 \cdot \varphi = \partial_B \cdot \varphi \neq 0$.

This should not be taken as indicative of a lack of vitality of $\mathcal{S}^{\mathbf{B}}$, $\mathbf{B} \in \mathrm{Cat}$ as a hyperdoctrine, or even of a lack of a satisfactory theory of equality for it. Rather, it indicates that we have probably been too naive in defining equality in a manner too closely suggested by the classical conception. Equality should be the "graph" of the identity term. But present categorical conceptions indicate that, in the context of set-valued attributes, the graph of a functor $f : \mathbf{B} \to \mathbf{C}$ should be, not $1_{\mathbf{B}}\Sigma\langle \mathbf{B}, f\rangle$, but rather the corresponding "profunctor", a binary attribute of *mixed* variance in $P(\mathbf{B}^{\mathrm{op}} \times \mathbf{C})$. Thus in particular "equality" should be the functor $\hom_{\mathbf{B}}$ (rather than the rather uninformative attribute $\Theta_{\mathbf{B}}$ in $P(\mathbf{B} \times \mathbf{B})$, given by our present definition). The term which would take the place of δ in such a more enlightened theory of equality would then be the forgetful functor

$$\tilde{\mathbf{B}} \to \mathbf{B}^{\mathrm{op}} \times \mathbf{B}$$

from the "twisted morphism category", as follows from the "extensional" considerations of the following section. Of course to abstract from this example would require at least the addition of a functor $\mathbf{T} \xrightarrow{\text{op}} \mathbf{T}$ to the structure of an eed.

2. In any elementary existential doctrine we have a functor

$$(\mathbf{T}, B) \xrightarrow{1(\,)\Sigma(\,)} P(B)$$

for each type B, defined on objects by

$$\begin{array}{c} E \\ p \Big\downarrow \quad \rightsquigarrow 1_E \Sigma p \\ B \end{array}$$

The morphisms in the category $(\mathbf{T}, B)$ are of course arbitrary commutative triangles

$$\begin{array}{ccc} E & \xrightarrow{f} & E' \\ & {\scriptstyle p} \searrow \quad \swarrow {\scriptstyle p'} & \\ & B & \end{array}$$

of terms, and it is easy to verify that the above definition can be canonically extended to these morphisms to become a functor. For example, in the hyperdoctrine with $\mathbf{T} = \mathscr{S}$, $P(X) = 2^X$, our functor

$$(\mathscr{S}, B) \to 2^B$$

assigns to any mapping p with codomain B the propositional function $\bar{p}$ defined on B such that $b\bar{p} = 1$ iff $b \in \text{image}\,(p)$ or in the example $P(X) = \mathscr{S}^X$, our functor takes $p: E \to B$ into the family E_b, $b \in B$ of sets in which E_b is the fiber of p over b.

When the functor defined in the previous paragraph is equipped with a right adjoint

$$P(B) \to (\mathbf{T}, B)$$

we say that the eed satisfies the *Comprehension Schema* and denote the adjoint by

$$\begin{array}{c} \{B:\psi\} \\ \psi \rightsquigarrow \quad \Big\downarrow p_\psi \\ B \end{array}$$

The new rule of inference is then expressed by the adjointness bijection

$$\frac{\begin{array}{ccc} E & \longrightarrow & \{B:\psi\} \\ & {\scriptstyle p} \searrow \quad \swarrow & \\ & B & \end{array}}{1_E \Sigma p \to \psi}$$

between terms $E \xrightarrow{f} \{B:\psi\}$ for which $fp_\psi = p$ and deductions $1_E \Sigma p \to \psi$ over B. We may call $p_\psi:\{B:\psi\} \to B$ the "extension" of ψ, justified by the fact that in the hyperdoctrine 2^X, $X \in \mathscr{S}$, p_ψ reduces to the inclusion $\tilde{p}_\psi$ of that part of B whose characteristic function is the propositional function ψ. For since $\tilde{p}_\psi$ is then monomorphic, there is for any p at most one f such that $f\tilde{p}_\psi = p$; there is such an f iff the image of p is contained in the part of B in question, which holds iff there is a "deduction" $1_E \Sigma p \to \psi$ in 2^B; hence $p_\psi \approx \tilde{p}_\psi$ in $(\mathscr{S}, B)$.

Similarly the set-valued hyperdoctrine on $\mathbf{T} = \mathscr{S}$ satisfies the comprehension schema; for a family ψ of sets indexed by the elements of B, $\{B:\psi\} = \sum_{b \in B} b \cdot \psi$, the disjoint sum, with p_ψ the obvious projection. Thus in this case the Comprehension Schema is more nearly the Replacement Schema.

Given $f_i:X \to Y$, $i = 1, 2$ one would expect that the extension $\{X:f_1\Theta f_2\}$ of the attribute of type X expressing that f_1 and f_2 are equal should in fact give the equalizer in the category $\mathbf{T}$ of f_1, f_2. This is true under certain conditions.

THEOREM. *Suppose that in a given eed in which the Comprehension Schema holds, we have further the following conditions for any two terms $h_i:E \to Y$:*

(i) *There is at most one proof* $1_E \to h_1 \Theta h_2$
(ii) *If there is such a proof, then* $h_1 = h_2$.

Then if $f_i:X \to Y$ are any two given terms and we set $\varphi = f_1\Theta f_2$, it follows that

$$\{X:\varphi\} \xrightarrow[p_\varphi]{} X \overset{f_1}{\underset{f_2}{\rightrightarrows}} Y$$

is an equalizer diagram.

NOTE. It would be too restrictive to replace (i) by the assumption that all attributes have at most one proof. An equality statement tends to be a very special sort of attribute; consider for example $P(X) = \mathscr{S}^X$, $X \in \mathscr{S}$, where (i) holds but most attributes have many distinct proofs. Condition (ii) seems difficult to guarantee by other kinds of assumptions.

PROOF. Consider any "test" term $E \xrightarrow{p} X$ as an object in $(\mathbf{T}, X)$. We must show that there is at most one term $E \to \{X:f_1\Theta f_2\}$ which when composed with p_φ gives p, and that there is such a term iff $pf_1 = pf_2$. By adjointness

$$\begin{array}{c} E \to \{X:f_1\Theta f_2\} \\ {}_{p}\searrow \quad \swarrow_{p_\varphi} \\ X \\ \hline 1_E \Sigma p \to f_1 \Theta f_2 \\ \hline 1_E \to p \cdot (f_1 \Theta f_2) \end{array}$$

But $p \cdot (f_1\Theta f_2) = p\langle f_1, f_2\rangle \cdot \Theta = \langle pf_1, pf_2\rangle \cdot \Theta = pf_1 \Theta pf_2$ so the result follows by setting $h_i = pf_i$.

The notation of "extension" surely belongs to logic, yet its own extension is considerably broader than the case traditionally considered by logicians. For example

THEOREM. *The hyperdoctrine with* $\mathbf{T} = \mathrm{Cat}$, $P(\mathbf{B}) = \mathscr{S}^{\mathbf{B}}$ *satisfies the Comprehension Schema. Indeed, if* $\varphi : \mathbf{B} \to \mathscr{S}$ *is any functor, its extension* $p_\varphi : \{\mathbf{B} : \varphi\} \to \mathbf{B}$ *is the op-fibration with discrete fibers associated to* φ.

PROOF. We need only show that the op-fibration $\tilde{\varphi} \to \mathbf{B}$ in question, has the required universal property. Recall that $\tilde{\varphi}$ has as objects pairs $\langle B, x\rangle$ with $1 \xrightarrow{x} B\varphi$ in $\mathscr{S}$, and as morphisms $\langle B, x\rangle \to \langle B', x'\rangle$ the morphisms $B \to B'$ in $\mathbf{B}$ which under the action φ take $x \rightsquigarrow x'$. For any $p : \mathbf{E} \to \mathbf{B}$ one has clearly that the commuting diagrams

$$\begin{array}{ccc} \mathbf{E} & \longrightarrow & \tilde{\varphi} \\ & {\scriptstyle p}\searrow \quad \swarrow & \\ & \mathbf{B} & \end{array}$$

correspond to the elements of proj lim $(p \cdot \varphi)$. But on the other hand for deductions (i.e. natural transformations) one has

$$\frac{1_{\mathbf{E}}\Sigma p \to \varphi}{1_{\mathbf{E}} \to p \cdot \varphi}$$

and the deductions of the sort below the line also correspond canonically to the elements of proj lim $(p \cdot \varphi)$ since the terminal object represents the inverse limit functor on $\mathscr{S}^{\mathbf{E}}$. Thus

$$\frac{\begin{array}{ccc} \mathbf{E} & \longrightarrow & \tilde{\varphi} \\ & {\scriptstyle p}\searrow \quad \swarrow & \\ & \mathbf{B} & \end{array}}{1_{\mathbf{E}}\Sigma p \to \varphi}$$

canonically for all $\mathbf{E}, p$, and hence

$$\tilde{\varphi} \approxeq \{\mathbf{B} : \varphi\}.$$

It is clear that if $\varphi : \mathbf{B} \to \mathscr{S}$ is a functor whose domain $\mathbf{B}$ is a groupoid, then the corresponding cofibered category $\tilde{\varphi}$ is also a groupoid; it is in fact the "covering groupoid" used by Higgins in his proof of the subgroup theorem and, in a measure-theoretic context, by Mackey in his theory of virtual subgroups. Thus

COROLLARY. *The hyperdoctrine with* $\mathbf{T} =$ *groupoids*, $P(\mathbf{B}) = \mathscr{S}^{\mathbf{B}}$ *satisfies the Comprehension Schema, with* $\{\mathbf{B} : \varphi\} =$ *the covering groupoid of* φ *for any permutation representation* φ *of the groupoid* $\mathbf{B}$.

CUNY GRADUATE CENTER
FORSCHUNGSINSTITUT FÜR MATHEMATIK
ETH ZÜRICH, SWITZERLAND

Homology of simplicial objects

Michel André

The aim of the following pages is to give a naive approach to the part of homological algebra involving simplicial objects. This is an expository paper having much to do with **[1]** and with **[5]**, and I hope it is general enough as background for most of the applications in algebra. This approach is naive in the sense that I use classical homological algebra **[2]** as much as possible and combinatorial homotopy theory **[4]** as little as possible. First we define the framework of the exposition, the notion of a simplicial category; a simplicial category is a set of categories, one in each dimension $n \geq 0$, glued together in a simplicial way. Then we consider simplicial objects in a simplicial category (a simplicial module over a simplicial ring is an example) and models that are simplicial objects with a good behaviour with respect to the simplicial structure of the simplicial category (a simplicial free module over a simplicial ring is an example). Then if a functor from the simplicial category to an abelian category is given, homology objects are defined for any simplicial object. The definition involves a double complex; thus there are two spectral sequences. The first one shows that a first approximation of the homology at the level of the simplicial category is the homology at the level of the different categories which are components of the simplicial category. The second one shows that, sometimes, a morphism of simplicial objects gives an isomorphism of the corresponding homology objects. Then the two preceding facts show that the homology objects can be computed by means of so-called simplicial resolutions, when they exist. Actually they exist in most cases appearing in algebra and the existence is proved by attaching cells: it is the only point where homotopy theory is needed.

1. **Introduction.** A brief survey of the beginning of **[1]** will play the role of an introduction. We consider a model category **N**, a good abelian category **A** and a covariant functor F from **N** to **A**. A *model category* is a category plus a set of objects in this category, named the models: M^λ, $\lambda \in \Lambda$. A *good abelian category*

is an abelian category for which all direct sums exist and are exact. Then *derived functors* L_nF or *homology objects*

$$H_n(N, F) = L_nF(N)$$

are defined. For each $n \geq 0$ we get a functor from **N** to **A**. Let us do the construction for a fixed object N in **N**, the naturality being clear.

First we consider a set of indices $K_n(N)$. An element of this set is a set of $n + 1$ models plus a set of $n + 1$ morphisms

$$M^{\lambda^0}, \ldots, M^{\lambda^n} \quad \text{and} \quad \alpha^0, \ldots, \alpha^n$$

satisfying the following conditions:

(i) the composition $\alpha^0 \cdots \alpha^n$ exists,

(ii) the domain of α^i is equal to M^{λ^i},

(iii) the codomain of α^0 is equal to N,

$$M^{\lambda^n} \xrightarrow{\alpha^n} M^{\lambda^{n-1}} \xrightarrow{\alpha^{n-1}} \cdots M^{\lambda^0} \xrightarrow{\alpha^0} N.$$

Such an element is denoted by

$$(\alpha^0, \lambda^0, \alpha^1, \lambda^1, \ldots, \alpha^n, \lambda^n).$$

Then we consider a set of objects in **A** indexed by $K_n(N)$ and depending on F:

$$F[\alpha^0, \ldots, \lambda^n] = F(M^{\lambda^n}).$$

Finally we consider the direct sum of the objects just defined:

$$C_n(N, F) = \sum_{(\alpha^0, \ldots, \lambda^n) \in K_n(N)} F[\alpha^0, \ldots, \lambda^n].$$

We denote by $[\alpha^0, \ldots, \lambda^n]$ the natural morphism

$$F[\alpha^0, \ldots, \lambda^n] \to C_n(N, F).$$

Furthermore we define a differential d of degree -1 by the following equalities:

$$d_n = \sum_{i=0}^{n} (-1)^i s_n^i : C_n(N, F) \to C_{n-1}(N, F)$$

$$s_n^i \circ [\alpha^0, \ldots, \lambda^n] = [\alpha^0, \ldots, \lambda^{i-1}, \alpha^i\alpha^{i+1}, \lambda^{i+1}, \ldots, \lambda^n] \quad \text{if} \quad 0 \leqslant i < n$$

$$[\alpha^0, \ldots, \lambda^{n-1}] \circ F(\alpha^n) \quad \text{if} \quad i = n.$$

Now we have a *complex* $C_*(N, F)$ and we may consider its *homology*:

$$H_n(N, F) = H_n[C_*(N, F)].$$

2. **Simplicial category.** The preceding definition of derived functors/homology objects is simple, but we have to make the following two remarks. On the one hand, simplicial objects are involved, as soon as we try to compare those homology objects with other ones (singular homology of a topological space, derived functors of an additive functor, for example). On the other hand, the background of a

model category does not suffice in general (simplicial modules over a simplicial ring, for example). Thus we have to go a little further and use simplicial theory.

We denote by **S** the *basic category* of simplicial theory: an object $[m]$ is an ordered set $(0, 1, \ldots, m)$, and a morphism $\omega: [p] \to [q]$ is an order preserving map $(0, 1, \ldots, q) \to (0, 1, \ldots, p)$. A simplicial object N_* in a category **N** is a covariant functor from **S** to **N**; in other words there is an object N_m for each $m \geq 0$, and there is a morphism $\bar{\omega}: N_p \to N_q$ for each morphism $\omega: [p] \to [q]$.

Now let us introduce the notions of a simplicial category and of a simplicial object in a simplicial category. A *simplicial category* is a pair $(\mathbf{N}, T)$ where **N** is a category and T a covariant functor from **N** to **S**. A *simplicial object* in this simplicial category is a covariant functor N_* from **S** to **N** such that $T \circ N_*$ is equal to the identity functor of **S**. In other words it is a simplicial object N_* in the category **N**, in the usual sense, fulfilling the following condition:

$$T(N_m) = [m] \quad \text{and} \quad T(\bar{\omega}) = \omega.$$

For a simplicial category $(\mathbf{N}, T)$ it is convenient to introduce the following notation: if $\omega: [p] \to [q]$ is a morphism in **S** and if P and Q are two objects in **N** with $T(P) = [p]$ and $T(Q) = [q]$, then by definition $\mathrm{Hom}_\omega(P, Q)$ is the set of morphisms $\alpha: P \to Q$ with $T(\alpha) = \omega$.

A *simplicial model category* is a simplicial category $(\mathbf{N}, T)$ plus a set of simplicial objects in this simplicial category, named the *models*: M_*^λ, $\lambda \in \Lambda$. We suppose that any model M_* satisfies the following *condition*: if $\omega: [p] \to [q]$ is a morphism in **S** and if N is an object in **N** with $T(N) = [q]$, then the morphism $\bar{\omega}: M_p \to M_q$ gives a bijection:

$$\mathrm{Hom}_{1_q}(M_q, N) \to \mathrm{Hom}_\omega(M_p, N).$$

A model category gives a trivial example of a simplicial model category. The model category consists of the category **N** and of the set of objects M^λ, $\lambda \in \Lambda$. Then the category of the simplicial model category is the product category $\mathbf{S} \times \mathbf{N}$, the functor of the simplicial model category is the projection functor $\mathbf{S} \times \mathbf{N} \to \mathbf{S}$, and the models M_*^λ of the simplicial model category correspond to the models M^λ of the model category in the following way:

$$M_n^\lambda = ([n], M^\lambda) \quad \text{and} \quad \bar{\omega} = (\omega, 1_M\lambda).$$

A simplicial ring R_* gives another example of a simplicial model category. An object of the category is a pair $([m], N)$ where N is an R_m-module and a morphism of the category is a pair (ω, α):

$$\omega: [p] \to [q] \quad \text{and} \quad \alpha: P \to Q$$

where P is an R_p-module, where Q is an R_q-module and consequently an R_p-module through $\bar{\omega}: R_p \to R_q$, and where α is an R_p-homomorphism. With the functor T mapping $([m], N)$ onto $[m]$ we get a simplicial category. Then a simplicial object in this simplicial category is nothing but a simplicial module over R_*. The set of models is a set (large enough) of direct sums of copies of the simplicial module R_*.

3. **Homology.** We consider a simplicial category (N, T) with models M_*^λ, $\lambda \in \Lambda$, a good abelian category $\mathbf{A}$ and a covariant functor F from $\mathbf{N}$ to $\mathbf{A}$. Then homology objects $\bar{H}_n(N_*, F)$ are defined. For each $n \geq 0$ we get a functor from the category of simplicial objects (in the simplicial category) to the abelian category. Let us do the construction for a fixed simplicial object N_* in $(\mathbf{N}, T)$.

First we consider a set of indices $\bar{K}_{p,q}(N_*)$ with $p \geq 0$ and $q \geq 0$. An element of this set is a set of $p + 1$ models plus a set of $p + 1$ morphisms

$$M_*^{\lambda^0}, \ldots, M_*^{\lambda^p} \quad \text{and} \quad \alpha^0, \ldots, \alpha^p$$

satisfying the following conditions:

(i) the composition $\alpha^0 \cdots \alpha^p$ exists,

(ii) the domain of α^i is equal to $M_q^{\lambda^i}$,

(iii) the codomain of α^0 is equal to N_q,

(iv) the morphism $T(\alpha^i)$ is equal to 1_q,

$$M_q^{\lambda^p} \xrightarrow{\alpha^p} M_q^{\lambda^{p-1}} \xrightarrow{\alpha^{p-1}} \cdots M_q^{\lambda^0} \xrightarrow{\alpha^0} N_q$$

$$[q] \xrightarrow{1_q} [q] \xrightarrow{1_q} \cdots [q] \xrightarrow{1_q} [q].$$

Such an element is denoted by

$$(\alpha^0, \lambda^0, \alpha^1, \lambda^1, \ldots, \alpha^p, \lambda^p).$$

Then we consider a set of objects in $\mathbf{A}$ indexed by $\bar{K}_{p,q}(N_*)$ and depending on F

$$\bar{F}[\alpha^0, \ldots, \lambda^p] = F(M_q^{\lambda^p}).$$

Finally we consider the direct sum of the objects just defined

$$\bar{C}_{p,q}(N_*, F) = \sum_{(\alpha^0,\ldots,\lambda^p)\in\bar{K}_{p,q}(N_*)} \bar{F}[\alpha^0, \ldots, \lambda^p].$$

We denote by $[\alpha^0, \ldots, \lambda^p]$ the natural morphism

$$\bar{F}[\alpha^0, \ldots, \lambda^p] \rightarrow \bar{C}_{p,q}(N_*, F).$$

Then we define two differentials, d' of degree -1 in p and of degree 0 in q, and d'' of degree 0 in p and of degree -1 in q. We define the *first differential* by the following equalities:

$$d'_{p,q} = \sum_{i=0}^{p} (-1)^i s'^i_{p,q} : \bar{C}_{p,q}(N_*, F) \rightarrow \bar{C}_{p-1,q}(N_*, F)$$

$$s'^i_{p,q} \circ [\alpha^0, \ldots, \lambda^p] = [\alpha^0, \ldots, \lambda^{i-1}, \alpha^i\alpha^{i+1}, \lambda^{i+1}, \ldots, \lambda^p] \quad \text{if} \quad 0 \leq i < p$$

$$[\alpha^0, \ldots, \lambda^{p-1}] \circ F(\alpha^p) \quad \text{if} \quad i = \text{p}.$$

We define the *second differential* by the following equalities:

$$d''_{p,q} = \sum_{j=0}^{q} (-1)^j s''^j_{p,q} : \bar{C}_{p,q}(N_*, F) \rightarrow \bar{C}_{p,q-1}(N_*, F)$$

$$s''^j_{p,q} \circ [\alpha^0, \lambda^0, \alpha^1, \lambda^1, \ldots, \alpha^p, \lambda^p] = [\tilde{\epsilon}^j_p\alpha^0, \lambda^0, \tilde{\epsilon}^j_q\alpha^1, \lambda^1, \ldots, \tilde{\epsilon}^j_q\alpha^p, \lambda^p] \circ F(\tilde{\epsilon}^j_q)$$

where the symbols $\bar{\epsilon}_q^j$ and $\tilde{\epsilon}_q^j$ are described below. We denote by ϵ_q^j the morphism of the basic category **S** given by the injection of $(0, 1, \ldots, q-1)$ into $(0, 1, \ldots, q)$ avoiding the element j (the jth face in dimension q). Thus a morphism $\bar{\epsilon}_q^j$ is defined for any simplicial object, for example N_* or M_*^λ. Then the symbol $\tilde{\epsilon}_q^j$ is completely described by the following equalities:

$$T(\tilde{\epsilon}_q^j\alpha^i) = 1_{q-1} \quad \text{and} \quad (\tilde{\epsilon}_q^j\alpha^i)\circ\bar{\epsilon}_q^j = \bar{\epsilon}_q^j\circ\alpha^i$$

(remember that a model satisfies a nice condition). In order to have a global view of the component $s''^{j}_{p,q}$ of the differential $d''_{p,q}$, let us consider the following diagrams:

$$\begin{array}{ccccccc}
M_q^{\lambda^p} & \xrightarrow{\alpha^p} & M_q^{\lambda^{p-1}} & \xrightarrow{\alpha^{p-1}} \cdots & M_q^{\lambda^0} & \xrightarrow{\alpha^0} & N_q \\
\downarrow{\scriptstyle \bar{\epsilon}_q^j} & & \downarrow{\scriptstyle \bar{\epsilon}_q^i} & & \downarrow{\scriptstyle \bar{\epsilon}_q^j} & & \downarrow{\scriptstyle \bar{\epsilon}_q^j} \\
M_{q-1}^{\lambda^p} & \xrightarrow{\tilde{\epsilon}_q^j\alpha^p} & M_{q-1}^{\lambda^{p-1}} & \xrightarrow{\tilde{\epsilon}_q^j\alpha^{p-1}} \cdots & M_{q-1}^{\lambda^0} & \xrightarrow{\tilde{\epsilon}_q^j\alpha^0} & N_{q-1}
\end{array}$$

$$\begin{array}{ccccccc}
[q] & \xrightarrow{1_q} & [q] & \xrightarrow{1_q} \cdots & [q] & \xrightarrow{1_q} & [q] \\
\downarrow{\scriptstyle \epsilon_q^j} & & \downarrow{\scriptstyle \epsilon_q^j} & & \downarrow{\scriptstyle \epsilon_q^j} & & \downarrow{\scriptstyle \epsilon_q^j} \\
[q-1] & \xrightarrow{1_{q-1}} & [q-1] & \xrightarrow{1_{q-1}} \cdots & [q-1] & \xrightarrow{1_{q-1}} & [q-1]
\end{array}$$

$$F(\bar{\epsilon}_q^j): F(M_q^{\lambda^p}) \to F(M_{q-1}^{\lambda^p}).$$

It is long and easy to verify that we get a *double complex* $\bar{C}_{*,*}(N_*, F)$.

Now we denote by $\bar{C}_*(N_*, F)$ the complex associated to this double complex and we consider its *homology*:

$$\bar{H}_n(N_*, F) = H_n[\bar{C}_*(N_*, F)].$$

The purpose of the following pages is to learn a little more of those homology objects.

4. **Comparison theorem.** We consider two simplicial objects N_* and N'_* in a simplicial category $(\mathbf{N}, T)$ and a morphism $\nu_*: N_* \to N'_*$ between them, that is, a collection of morphisms

$$\nu_m: N_m \to N'_m, \qquad m = 0, 1, \ldots$$

satisfying the following conditions: $T(\nu_m) = 1_m$ and for any $\omega: [p] \to [q]$ the morphisms $\bar{\omega}\circ\nu_p$ and $\nu_q\circ\bar{\omega}$ are equal. Out of the definition of the homology objects it is clear how to define a morphism

$$\bar{H}(\nu_*, F): \bar{H}_n(N_*, F) \to \bar{H}_n(N'_*, F)$$

as soon as we have a simplicial model category and a functor with values in a good abelian category: in dimension (p, q) replace the index $(\alpha^0, \lambda^0, \alpha^1, \ldots, \lambda^p)$ by the index $(\nu_q\circ\alpha^0, \lambda^0, \alpha^1, \ldots, \lambda^p)$. Our problem is to show that sometimes the morphism $\bar{H}_n(\nu_*, F)$ is an isomorphism.

Now we consider two simplicial objects M_* and N_* in the simplicial category $(\mathbf{N}, T)$ and we suppose that M_* satisfies the condition used in the definition of a model: if $\omega: [p] \to [q]$ is a morphism in $\mathbf{S}$ and if N is an object in $\mathbf{N}$ with $T(N) = [q]$, then the morphism $\bar{\omega}: M_p \to M_q$ gives a bijection:

$$\mathrm{Hom}_{1_q}(M_q, N) \to \mathrm{Hom}_{\omega}(M_p, N).$$

We can define a useful *simplicial set* $\langle M_*, N_* \rangle_*$ in the following way:

$$\langle M_*, N_* \rangle_p = \mathrm{Hom}_{1_p}(M_p, N_p);$$

in other words, a p-simplex is a morphism $\alpha: M_p \to N_p$ with $T(\alpha) = 1_p$, and for the morphism $\omega: [p] \to [q]$ the map

$$\bar{\omega}: \langle M_*, N_* \rangle_p \to \langle M_*, N_* \rangle_q$$

is defined by the equality $\bar{\omega}(\alpha) = \tilde{\omega}\alpha$ where $\tilde{\omega}\alpha$ is uniquely determined by the following commutative diagram:

$$\begin{array}{ccc} M_p & \xrightarrow{\alpha} & N_p \\ {\scriptstyle\bar{\omega}}\downarrow & & \downarrow{\scriptstyle\bar{\omega}} \\ M_q & \xrightarrow{\tilde{\omega}\alpha} & N_q \end{array} .$$

We are specially interested by the *singular homology*, of this simplicial set with the rational integers as coefficients:

$$H_*(M_*, N_*) = H_*^{\mathrm{sing}}(\langle M_*, N_* \rangle_*, \mathbf{Z}).$$

Now we consider a simplicial category $(\mathbf{N}, T)$ with models M_*^{λ}, $\lambda \in \Lambda$ and a morphism $\nu_*: N_* \to N'_*$ of simplicial objects in the simplicial category. We say that ν_* is an *equivalence* if the following condition is satisfied: for any model M_*^{λ},the homomorphism

$$H_*(M_*^{\lambda}, \nu_*): H_*(M_*^{\lambda}, N_*) \to H_*(M_*^{\lambda}, N'_*)$$

is an isomorphism.

THEOREM. *Let $\nu_*: N_* \to N'_*$ be an equivalence in a simplicial model category and let F be a functor from this category to a good abelian category. Then the morphisms*

$$\bar{H}_n(\nu_*, F): \bar{H}_n(N_*, F) \to \bar{H}_n(N'_*, F), \qquad n \geq 0$$

are isomorphisms.

PROOF. If C is a double complex with differentials d' and d'', we know that the three different types of homology (H' for the first differential, H'' for the second differential, and H for the total differential) are related by a spectral sequence

$$H'_p[H''_q[C]] \underset{p}{\Longrightarrow} H_n[C]$$

having the best properties we can imagine. In particular a morphism of double complexes giving an isomorphism for the homology with respect to the second differential gives an isomorphism for the homology with respect to the total differential. Thus we can prove that the morphisms

$$\bar{H}_n(\nu_*, F): \bar{H}_n(N_*, F) \to \bar{H}_n(N'_*, F), \qquad n \geq 0$$

are isomorphisms in proving that the morphisms

$$H''_q[\bar{C}_{p,*}(N_*, F)] \to H''_q[\bar{C}_{p,*}(N'_*, F)], \qquad p, q \geq 0$$

are isomorphisms.

Let us have a look at the definition of the double complex we are using:

$$\bar{C}_{p,q}(N_*, F) = \sum_{(\alpha^0, \ldots, \lambda^p) \in \bar{K}_{p,q}(N_*)} \bar{F}[\alpha^0, \ldots, \lambda^p].$$

We can rewrite the index:

$$(\alpha^0, \lambda^0, \ldots, \alpha^p, \lambda^p) = (\lambda^0, \ldots, \lambda^p)(\alpha^0)(\alpha^1, \ldots, \alpha^p)$$

and remark the following: $(\lambda^0, \ldots, \lambda^p)$ is not involved directly in the definition of the second differential and does not depend on N_*, $(\alpha^1, \ldots, \alpha^p)$ is involved directly in the definition of the second differential and does not depend on N_*, (α^0) is involved directly in the definition of the second differential and depends on N_*, and $\bar{F}[\alpha^0, \ldots, \lambda^p]$ is involved directly in the definition of the second differential and does not depend on N_*:

$$\bar{C}_{p,q}(N_*, F) = \sum_{(\lambda^0, \ldots, \lambda^p)} \sum_{(\alpha^0)} \sum_{(\alpha^1, \ldots, \alpha^p)} \bar{F}[\alpha^0, \ldots, \lambda^p].$$

Consequently, in the proof of the isomorphism

$$H''_q[\bar{C}_{p,*}(N_*, F)] \to H''_q[\bar{C}_{p,*}(N'_*, F)]$$

the index $(\lambda^0, \ldots, \lambda^p)$ can be fixed, the morphism $\nu_*: N_* \to N'_*$ appears at the level (α^0), and the part

$$\sum_{(\alpha^1, \ldots, \alpha^p)} \bar{F}[\alpha^0, \ldots, \lambda^p]$$

plays the role of coefficients. To be more precise, let us remark the following.

Let us consider a simplicial set E_* and a simplicial object A_* in the abelian category **A**. Then singular homology objects $H_n^{\text{sing}}(E_*, A_*)$ are defined in the following way (usually the simplicial object A_* is trivial $A_k \equiv A$) the object $H_n^{\text{sing}}(E_*, A_*)$ is the nth homology object of the complex $C_*^{\text{sing}}(E_*, A_*)$ defined as follows:

$$C_n^{\text{sing}}(E_*, A_*) = \sum_{E_n} A_n,$$

the direct sum of as many copies of A_n as there are elements in E_n, and

$$d_n = \sum_{i=0}^{n} (-1)^i s_n^i : C_n^{\text{sing}}(E_*, A_*) \to C_{n-1}^{\text{sing}}(E_*, A_*)$$

where s_n^i can be described componentwise by the following commutative diagrams, one for each n-simplex $e \in E_n$:

$$\begin{array}{ccc} A_n \text{ of } e & \xrightarrow{\bar{\epsilon}_n^i} & A_{n-1} \text{ of } \bar{\epsilon}_n^i(e) \\ \downarrow & & \downarrow \\ \sum_{E_n} A_n & \xrightarrow{s_n^i} & \sum_{E_{n-1}} A_{n-1} \end{array}$$

as usual $\bar{\epsilon}_n^i$ denotes the ith face morphism of any simplicial object. Now we have the following result: let $\omega_*: E_* \to E'_*$ be a morphism of simplicial sets and A_* be a simplicial object in a good abelian category; if the morphisms

$$H_m^{\text{sing}}(\omega_*, \mathbf{Z}): H_m^{\text{sing}}(E_*, \mathbf{Z}) \to H_m^{\text{sing}}(E'_*, \mathbf{Z})$$

are isomorphisms, $m = 0, 1, \ldots$ ($\mathbf{Z}$ being the group of rational integers), then the morphisms

$$H_n^{\text{sing}}(\omega_*, A_*): H_n^{\text{sing}}(E_*, A_*) \to H_n^{\text{sing}}(E'_*, A_*)$$

are isomorphisms, $n = 0, 1, \ldots$. The proof uses standard arguments; a summary of them will suffice. First remark: any simplicial object A_* is isomorphic to a quotient of a direct sum of simplicial objects of a special type

$$A_*(F_*, A) = \sum_{F_*} A$$

(F_* being any simplicial set, A being any object in the abelian category). Consequently, it suffices to prove the result for $A_* = A_*(F_*, A)$. Second remark: there is a canonical isomorphism

$$H_n^{\text{sing}}(E_*, A_*(F_*, A)) \cong H_n^{\text{sing}}(E_* \times F_*, A).$$

Consequently, it suffices to verify that the morphisms

$$H_n^{\text{sing}}(E_* \times F_*, A) \to H_n^{\text{sing}}(E'_* \times F_*, A)$$

are all isomorphisms. Third remark: the morphisms

$$H_m^{\text{sing}}(E_* \times F_*, \mathbf{Z}) \to H_m^{\text{sing}}(E'_* \times F_*, \mathbf{Z})$$

are all isomorphisms. Consequently, replacing $E_* \times F_*$ by E_* and $E'_* \times F_*$ by E'_*, it suffices to verify that the morphisms

$$H_n^{\text{sing}}(E_*, A) \to H_n^{\text{sing}}(E'_*, A)$$

are all isomorphisms. Fourth remark: the morphisms above are all isomorphisms if the morphisms

$$H^k_{\text{sing}}(E'_*, \operatorname{Hom}(A, B)) \to H^k_{\text{sing}}(E_*, \operatorname{Hom}(A, B))$$

are all isomorphisms (B being any object in the abelian category). Consequently, it suffices to verify that the morphisms

$$H^k_{\text{sing}}(E'_*, G) \to H^k_{\text{sing}}(E_*, G)$$

are all isomorphisms (G being any abelian group). This conclusion is well known; and now we can use the isomorphisms

$$H_n^{\text{sing}}(E_*, A_*) \to H_n^{\text{sing}}(E'_*, A_*).$$

Now let us come back to the proof of the isomorphism

$$H''_q[\bar{C}_{p,*}(N_*, F)] \to H''_q[\bar{C}_{p,*}(N'_*, F)]$$

which will prove the comparison theorem. Let us use the decomposition

$$(\lambda^0, \ldots, \lambda^p)(\alpha^0)(\alpha^1, \ldots, \alpha^p)$$

of the index and the preceding long remark on singular homology. Rewriting completely the double complexes we are using, we see that it suffices to prove that

$$\sum_{(\lambda^0,\ldots,\lambda^p)} H_q^{\text{sing}}(\langle M_*^{\lambda^0}, \nu_*\rangle_*, A_*^{(\lambda^0,\ldots,\lambda^p)})$$

is an isomorphism; the simplicial object is defined in the following way:

$$A_q^{(\lambda^0,\ldots,\lambda^p)} = \sum_{(\alpha^1,\ldots,\alpha^p)} \bar{F}[\alpha^0, \ldots, \lambda^p],$$

the direct sum of as many copies of $F(M_q^{\lambda p})$ as there are elements in the product set

$$\text{Hom}_{1_q}(M_q^{\lambda^1}, M_q^{\lambda^0}) \times \cdots \times \text{Hom}_{1_q}(M_q^{\lambda^p}, M_q^{\lambda^{p-1}}).$$

We know enough on singular homology to assert that it suffices to prove that

$$H_*^{\text{sing}}(\langle M_*^{\lambda}, \nu_*\rangle_*, \mathbf{Z})$$

is an isomorphism; but that is the hypothesis of the comparison theorem.

5. **Spectral sequence.** In the proof of the comparison theorem we have used one of the two spectral sequences associated to the double complex $\bar{C}_{*,*}(N_*, F)$; and now let us study a little the second one, which relates the homology theory of a simplicial model category to the homology theory of different model categories.

Once more, we consider a simplicial category $(\mathbf{N}, T)$ with models M_*^λ, $\lambda \in \Lambda$. The "*fibre*" $\mathbf{N}_q$ over $[q]$ is a model category, and the set of all those model categories is a good approximation of the simplicial model category. An object in $\mathbf{N}_q$ is an object N in $\mathbf{N}$ with $T(N) = [q]$, a morphism in $\mathbf{N}_q$ is a morphism α in $\mathbf{N}$ with $T(\alpha) = 1_q$, and a model in $\mathbf{N}_q$ is an object M_q^λ, $\lambda \in \Lambda$. Now let us consider a simplicial object N_* in the simplicial category $(\mathbf{N}, T)$ and a functor F from the category $\mathbf{N}$ to a good abelian category. By restriction to the fibre, we get an object N_q in the category $\mathbf{N}_q$ and a functor (still denoted by F) from the category $\mathbf{N}_q$ to the abelian category. The category $\mathbf{N}_q$ has models. Consequently, the homology object $H_p(N_q, F)$ is well defined (see the introduction) for any $p \geq 0$. But there are relations between the different fibres $\mathbf{N}_q$ since they are parts of one category, and there are relations between the different objects N_q since they are parts of one simplicial object. From those relations we can deduce that, for a fixed p, the object $H_p(N_q, F)$ is the q-dimensional part of a simplicial object in the abelian category.

This simplicial object in the abelian category is denoted by $H_p(N_*, F)_*$ (quite different from the object $\bar{H}_p(N_*, F)$ we are studying in this paper). We have the equality

$$H_p(N_*, F)_q = H_p(N_q, F)$$

and there remains to define the morphism

$$\bar{\omega}: H_p(N_q, F) \to H_p(N_{q'}, F)$$

corresponding to the morphism $\omega: [q] \to [q']$. This morphism is defined by a morphism of complexes

$$\bar{\omega}: C_*(N_q, F) \to C_*(N_{q'}, F).$$

In dimension p, the morphism

$$\bar{\omega}: C_p(N_q, F) \to C_p(N_{q'}, F)$$

is defined by the following equality

$$\bar{\omega} \circ [\alpha^0, \lambda^0, \alpha^1, \lambda^1, \ldots, \alpha^p, \lambda^p] = [\bar{\omega}\alpha^0, \lambda^0, \bar{\omega}\alpha^1, \lambda^1, \ldots, \bar{\omega}\alpha^p, \lambda^p] \circ F(\bar{\omega})$$

of the commutative diagram

$$\begin{array}{ccc} F(M_q^{\lambda^p}) & \xrightarrow{F(\bar{\omega})} & F(M_{q'}^{\lambda^p}) \\ \downarrow & & \downarrow \\ C_p(N_q, F) & \xrightarrow{\bar{\omega}} & C_p(N_{q'}, F) \end{array}$$

(as we know, the morphism $\bar{\omega}\alpha^i$ is completely defined by the equality $\bar{\omega}\alpha^i \circ \bar{\omega} = \bar{\omega} \circ \alpha^i$). Thus the simplicial object $H_p(N_*, F)_*$ in the abelian category is defined. We shall see that its homology $H_q[H_p(N_*, F)_*]$ (the homology of the complex

$$\cdots \to H_p(N_*, F)_q \xrightarrow{\Sigma(-1)^i \bar{\epsilon}_q^i} H_p(N_*, F)_{q-1} \to \cdots$$

where $\bar{\epsilon}_q$ is the ith face morphism in dimension q) is an approximation of the object $\bar{H}_n(N_*, F)$.

THEOREM. *Let* $(\mathbf{N}, T)$ *be a simplicial category with models* M_*^λ, $\lambda \in \Lambda$, *let* N_* *be a simplicial object in* $(\mathbf{N}, T)$, *and let* F *be a functor from* $\mathbf{N}$ *to a good abelian category. The homology objects* $\bar{H}_n(N_*, F)$ *are defined for* $n \geq 0$. *Then let* $\mathbf{N}_q$ *be the fibre of* $(\mathbf{N}, T)$ *over* $[q]$ *with models* M_q^λ, $\lambda \in \Lambda$. *The homology objects* $H_p(N_q, F)$ *are defined for* $p, q \geq 0$. *Then there exists a spectral sequence*

$$H_q[H_p(N_*, F)]_* \underset{q}{\Rightarrow} \bar{H}_n(N_*, F)$$

where $H_q[H_p(N_*, F)_*]$ *is the qth homology object of a simplicial object* $H_p(N_*, F)_*$ *in the abelian category, such that*

$$H_p(N_*, F)_q = H_p(N_q, F).$$

PROOF. If C is a double complex with differentials d' and d'', we know that the three different types of homology (H' for the first differential, H'' for the second differential, and H for the total differential) are related by a spectral sequence

$$H''_q[H'_p[C]] \underset{q}{\Rightarrow} H_n[C].$$

The theorem is the special case $C = \bar{C}_{*,*}(N_*, F)$.

The comparison theorem and the spectral sequence theorem give the following corollary.

COROLLARY. *Let* $(\mathbf{N}, T)$ *be a simplicial category with models* M_*^λ, $\lambda \in \Lambda$, *let* $p_*: P_* \to N_*$ *be an equivalence of simplicial objects in the simplicial model category, and let F be a functor from* $\mathbf{N}$ *to a good abelian category. Let us suppose that for each* $m \geq 0$, *there exists an index* $\lambda_m \in \Lambda$ *with* $P_m \simeq M_m^{\lambda_m}$. *Then for each* $n \geq 0$ *the homology object* $\bar{H}_n(N_*, F)$ *is isomorphic to the nth homology object of the following complex*:

$$\cdots \to F(P_n) \xrightarrow{\Sigma(-1)^i F(\bar{\varepsilon}_n^i)} F(P_{n-1}) \to \cdots.$$

PROOF. By the comparison theorem there is an isomorphism

$$\bar{H}_n(N_*, F) \simeq \bar{H}_n(P_*, F).$$

By the spectral sequence theorem there is a spectral sequence

$$H_q[H_p(P_*, F)_*] \underset{q}{\Longrightarrow} \bar{H}_n(P_*, F).$$

But since P_q is isomorphic to one of the models of the fibre $\mathbf{N}_q$, there are the following equalities

$$\begin{aligned} H_p(P_*, F)_q &= H_p(P_q, F) \\ &= F(P_q) \qquad \text{if } p = 0 \\ &= 0 \qquad\quad\;\; \text{if } p \neq 0. \end{aligned}$$

Consequently, the spectral sequence degenerates and there is an isomorphism

$$\bar{H}_n(P_*, F) \simeq H_n[F(P_*)].$$

6. **Generalization.** There are several ways of generalizing the homology objects

$$H_n(N, F) \quad \text{and} \quad \bar{H}_n(N_*, F).$$

Let us describe one of them, which is convenient for the applications: the functor F will be a little more general.

As in the introduction of this paper where the homology object $H_n(N, F)$ is defined, we consider a category $\mathbf{N}$ with models M^λ, $\lambda \in \Lambda$, and an object N in $\mathbf{N}$. Now let us define the category $\mathbf{N}/N$ in the following way: for each morphism

$\pi:P \to N$ in $\mathbf{N}$ there is an object (π, P) in $\mathbf{N}/N$ and for each commutative diagram in $\mathbf{N}$

$$\begin{array}{ccc} \pi:P & \longrightarrow & N \\ \Big\downarrow{\scriptstyle \alpha} & & \Big\downarrow{\scriptstyle 1_N} \\ \pi':P' & \longrightarrow & N \end{array}$$

there is a morphism (π', α) in $\mathbf{N}/N$. If F is a functor from $\mathbf{N}/N$ to a good abelian category (for example a functor from $\mathbf{N}$ to a good abelian category), the homology object $H_n(N, F)$ can still be defined. It suffices to generalize the complex $C_*(N, F)$ by the following equalities:

$$F[\alpha^0, \ldots, \lambda^n] = F(\alpha^0 \cdots \alpha^n, M^{\lambda^n}),$$

$$s_n^n \circ [\alpha^0, \ldots, \lambda^n] = [\alpha^0, \ldots, \lambda^{n-1}] \circ F(\alpha^0 \cdots \alpha^{n-1}, \alpha^n).$$

As in the paragraph of this paper where the homology object $\bar{H}_n(N_*, F)$ is defined, we consider a simplicial category $(\mathbf{N}, T)$ with models M_*^λ, $\lambda \in \Lambda$, and a simplicial object N_* in $(\mathbf{N}, T)$. Now let us define the category $\mathbf{N}/N_*$ in the following way: for each morphism $\pi:P \to N_p$ in $\mathbf{N}$ with $T(P) = [p]$ and $T(\pi) = 1_p$ there is an object (π, P) in $\mathbf{N}/N_*$, and for each commutative diagram in $\mathbf{N}$

$$\begin{array}{ccc} \pi:P & \longrightarrow & N_p \\ \Big\downarrow{\scriptstyle \alpha} & & \Big\downarrow{\scriptstyle \bar{\omega}} \\ \pi':P' & \longrightarrow & N_{p'} \end{array} \qquad \text{with } T(\alpha) = \omega$$

there is a morphism (π', α, π) in $\mathbf{N}/N_*$. If F is a functor from $\mathbf{N}/N_*$ to a good abelian category (for example a functor from $\mathbf{N}$ to a good abelian category), the homology object $\bar{H}_n(N_*, F)$ can still be defined. It suffices *to generalize the double complex* $\bar{C}_{*,*}(N_*, F)$ by the following equalities:

$$\bar{F}[\alpha^0, \ldots, \lambda^p] = F(\alpha^0 \cdots \alpha^p, M_q^{\lambda^p}),$$

$$s'^p_{p,q} \circ [\alpha^0, \ldots, \lambda^p] = [\alpha^0, \ldots, \lambda^{p-1}] \circ F(\alpha^0 \cdots \alpha^{p-1}, \alpha^p, \alpha^0 \cdots \alpha^p),$$

$$s''^j_{p,q} \circ [\alpha^0, \ldots, \lambda^p] = [\tilde{\epsilon}_q^j \alpha^0, \ldots, \lambda^p] \circ F(\tilde{\epsilon}_q^j(\alpha^0 \cdots \alpha^p), \bar{\epsilon}_q^j, \alpha^0 \cdots \alpha^p).$$

Now let us consider the simplicial category $(\mathbf{N}, T)$ with models, a morphism $\nu_*:N_* \to N'_*$ of simplicial objects in the simplicial category, and a functor F from the category $\mathbf{N}/N'_*$ to a good abelian category. From F and ν_* we get a functor, still denoted by F, from the category $\mathbf{N}/N_*$ to the abelian category:

$$F(\pi', \alpha, \pi) = F(\nu_{p'}\pi', \alpha, \nu_p\pi) \quad \text{if} \quad T(\pi) = 1_p \quad \text{and} \quad T(\pi') = 1_{p'}.$$

Consequently, there is a canonical morphism $\bar{H}(N_*, F) \to \bar{H}_n(N'_*, F)$. Our problem is to show that sometimes it is an isomorphism (comparison theorem). By definition the *functor* F is *sweet* if it satisfies the following condition. For each $\lambda \in \Lambda$, there exist a simplicial object A_*^λ in the abelian category, an isomorphism for each object (π, M_q^λ)

$$F(\pi, M_q^\lambda) \longrightarrow A_q^\lambda \qquad \pi : M_q^\lambda \to N'_q$$

and a commutative diagram for each morphism $(\pi', \bar{\omega}, \pi)$

$$\begin{array}{ccccccc} F(\pi, M_q^\lambda) & \xrightarrow{\sim} & A_q^\lambda & \qquad & \pi : M_q^\lambda & \longrightarrow & N'_q \\ {\scriptstyle F(\pi',\bar\omega,\pi)}\downarrow & & \downarrow{\scriptstyle\bar\omega} & & {\scriptstyle\bar\omega}\downarrow & & \downarrow{\scriptstyle\bar\omega} \\ F(\pi', M_{q'}^\lambda) & \xrightarrow{\sim} & A_{q'}^\lambda & & \pi' : M_{q'}^\lambda & \longrightarrow & N'_{q'} \end{array}$$

A functor F given by a functor from **N** to the abelian category is sweet since $F(\pi, M_q^\lambda) = F(M_q^\lambda)$. There is the following comparison theorem.

THEOREM. *Let $\nu_* : N_* \to N'_*$ be an equivalence in a simplicial category* $(\mathbf{N}, T)$ *with models and let F be a sweet functor from the category $\mathbf{N}/N'_*$ to a good abelian category. Then the morphisms*

$$\bar{H}_n(\nu_*, F) : \bar{H}_n(N_*, F) \to \bar{H}_n(N'_*, F) \qquad n \geq 0$$

are isomorphisms.

PROOF. It is a copy of the proof of the comparison theorem we already know. It is obvious as soon as we know that the simplicial object $A_*^{(\lambda^0, \ldots, \lambda^p)}$ is well defined. Here we use the fact that the functor F is sweet and we write

$$A_q^{(\lambda^0, \ldots, \lambda^p)} = \sum_{(\alpha^1, \ldots, \alpha^p)} A_q^{\lambda^p},$$

one copy of $A_q^{\lambda^p}$ for each $(\alpha^1, \alpha^2, \ldots, \alpha^p)$. Then for each α^0, there is an isomorphism

$$A_q^{(\lambda^0, \ldots, \lambda^p)} \cong \sum_{(\alpha^1, \ldots, \alpha^p)} \bar{F}[\alpha^0, \ldots, \lambda^p]$$

and we can reproduce the proof step by step.

There is no problem for generalizing the spectral sequence theorem. We consider a simplicial category $(\mathbf{N}, T)$ with models, a simplicial object N_* in the simplicial category, and a functor F from $\mathbf{N}/N_*$ to a good abelian category. The fibre $\mathbf{N}_q$ is a category with models and by restriction the functor F gives a functor from $\mathbf{N}_q/N_q$ to the abelian category. Thus the homology object $H_p(N_q, F)$ is well defined. It belongs to a simplicial object $H_p(N_*, F)_*$ in the abelian category:

$$H_p(N_*, F)_q = H_p(N_q, F).$$

The morphism

$$\bar{\omega} : H_p(N_q, F) \to H_p(N_{q'}, F)$$

is deduced from the morphism

$$-1 : C_p(N_q, F) \to C_p(N_{q'}, F)$$

defined by the following equality:

$$\bar{\omega} \circ [\alpha^0, \lambda^0, \ldots, \alpha^p, \lambda^p] = [\bar{\omega}\alpha^0, \lambda^0, \ldots, \bar{\omega}\alpha^p, \lambda^p] \circ F(\bar{\omega}(\alpha^0 \cdots \alpha^p), , -1\alpha^0 \cdots \alpha^p).$$

There is the following spectral sequence theorem.

THEOREM. *Let* $(\mathbf{N}, T)$ *be a simplicial category with models* M_*^λ, $\lambda \in \Lambda$, *let* N_* *be a simplicial object in* $(\mathbf{N}, T)$, *and let* F *be a functor from* $\mathbf{N}/N_*$ *to a good abelian category. The homology objects* $\bar{H}_n(N_*, F)$ *are defined for* $n \geq 0$. *Then let* N_q *be the fibre of* $(\mathbf{N}, T)$ *over* $[q]$ *with models* M_q^λ, $\lambda \in \Lambda$. *The homology objects* $H_p(N_q, F)$ *are defined for* $p, q \geq 0$. *Then there exists a spectral sequence*

$$H_q[H_p(N_*, F)_*] \underset{q}{\Longrightarrow} \bar{H}_n(N_*, F)$$

where $H_q[H_p(N_*, F)_*]$ *is the qth homology object of a simplicial object* $H_p(N_*, F)_*$ *in the abelian category such that* $H_p(N_*, F)_q = H_p(N_q, F)$.

The comparison theorem and the spectral sequence theorem give the following corollary.

COROLLARY. *Let* $(\mathbf{N}, T)$ *be a simplicial category with models* M_*^λ, $\lambda \in \Lambda$, *let* $p_* : P_* \to N_*$ *be an equivalence of simplicial objects in the simplicial model category, and let* F *be a sweet functor from* $\mathbf{N}/N_*$ *to a good abelian category. Let us suppose that for each* $m \geq 0$, *there exists an index* $\lambda_m \in \Lambda$ *with* $P_m \cong M_m^{\lambda_m}$. *Then for each* $n \geq 0$ *the homology object* $\bar{H}_n(N_*, F)$ *is isomorphic to the nth homology object of the following complex:*

$$\cdots \to F(p_n, P_n) \xrightarrow{\Sigma(-1)^i F(p_{n-1}, \bar{\epsilon}_n^i, p_n)} F(p_{n-1}, P_{n-1}) \to \cdots.$$

7. **Simplicial resolutions.** We have seen in the preceding corollary that it is easier to compute the homology objects $\bar{H}_n(N_*, F)$ if there is a simplicial object P_* with some good properties: P_* is "almost" a model and P_* is "almost" N_*. Thus it seems natural to introduce the following definition. In a simplicial category $(\mathbf{N}, T)$ with models M_*^λ, $\lambda \in \Lambda$, a *simplicial resolution* of a simplicial object N_* is an equivalence $p_* : P_* \to N_*$, the domain of which has the following property: for each $m \geq 0$ there is an isomorphism $P_m \cong M_m^{\lambda_m}$ for an index $\lambda_m \in \Lambda$. As we know, a morphism $p_* : P_* \to N_*$ of simplicial objects is an equivalence if the following condition is satisfied for any model M_*^λ: the homomorphism

$$H_*^{\text{sing}}(\langle M_*^\lambda, P_* \rangle_*, \mathbf{Z}) \to H_*^{\text{sing}}(\langle M_*^\lambda, N_* \rangle_*, \mathbf{Z})$$

is an isomorphism. The simplicial set $\langle M_*, Q_* \rangle_*$ is defined by the equality

$$\langle M_*, Q_* \rangle_n = \text{Hom}_{1_n}(M_n, Q_n).$$

Our problem is to find a method for the construction of simplicial resolutions. To solve it, it seems more adequate to use homotopy groups instead of homology groups in the definition of an equivalence. Let us now summarize what we have to know of homotopy theory.

Let us consider a simplicial abelian group G_*. The mth *homotopy group* $\pi_m(G_*)$ of G_* is by definition the mth homology group of the complex

$$\cdots \to \operatorname{Ker} \bar{\epsilon}^1_m \cap \cdots \cap \operatorname{Ker} \bar{\epsilon}^m_m \xrightarrow{\bar{\epsilon}^0_m} \operatorname{Ker} \bar{\epsilon}^1_{m-1} \cap \cdots \cap \operatorname{Ker} \bar{\epsilon}^{m-1}_{m-1} \to \cdots.$$

In other words, an element of $\pi_m(G_*)$ is represented by an m-simplex $g \in G_m$ with $\bar{\epsilon}^0_m(g) = 0, \ldots, \bar{\epsilon}^m_m(g) = 0$ and two m-simplices g' and g'' represent the same element of $\pi_m(G_*)$ if there is an $m+1$-simplex $g \in G_{m+1}$ with

$$\bar{\epsilon}^0_{m+1}(g) = g' - g'', \bar{\epsilon}^1_{m+1}(g) = 0, \ldots, \bar{\epsilon}^{m+1}_{m+1}(g) = 0.$$

The mth homotopy group $\pi_m(G_*)$ is also isomorphic to the mth homology group of the complex

$$\cdots \to G_m \xrightarrow{\Sigma(-1)^i \bar{\epsilon}^i_m} G_{m-1} \to \cdots.$$

Now let us consider a morphism $\omega_* : G_* \to K_*$ of simplicial abelian groups and the corresponding homomorphisms:

$$H^{\text{sing}}_n(\omega_*, \mathbf{Z}) : H^{\text{sing}}_n(G_*, \mathbf{Z}) \to H^{\text{sing}}_n(K_*, \mathbf{Z})$$
$$\pi_m(\omega_*) : \pi_m(G_*) \to \pi_m(K_*).$$

They are related by the following assertion: $H^{\text{sing}}_n(\omega_*, \mathbf{Z})$ is an isomorphism for any $n \geq 0$ if (and only if) $\pi_m(\omega_*)$ is an isomorphism for any $m \geq 0$.

8. **Construction of simplicial resolutions (Part I).** The definition of a simplicial resolution of a simplicial object N_* involves only the following objects: N_q and M^λ_q, $q \geq 0$ and $\lambda \in \Lambda$. Consequently, to solve the problem of the existence of a simplicial resolution, we can suppose that the category $\mathbf{N}$ is small; let the morphisms form a set of cardinality $\mathscr{C}$ (infinite and greater than the cardinality of Λ).

In a simplicial category $(\mathbf{N}, T)$, a direct sum is defined in the following way. For a set of objects N^k, $k \in K$, with $T(N^k) = [n]$, the direct sum consists of an object $\bigvee_{i \in K} N^i$ and of morphisms $\beta^k : N^k \to \bigvee N^i$, $k \in K$, with $T(\bigvee N^i) = [n]$ and $T(\beta^k) = 1_n$. The following universal property is satisfied: all the canonical morphisms

$$\operatorname{Hom}_\omega (\bigvee N^k, N) \to \Pi \operatorname{Hom}_\omega (N^k, N)$$

are isomorphisms. If the category $\mathbf{N}$ is small of cardinality $\mathscr{C}$, we say that the simplicial category $(\mathbf{N}, T)$ has direct sums, if the direct sums, in the sense defined above, exist as soon as the cardinality of K is smaller than $\mathscr{C}$. For a set of simplicial objects N^k_*, $k \in K$ the direct sum is a simplicial object $\bigvee N^i_*$ defined by the following equality: $(\bigvee N^i_*)_n = \bigvee N^i_n$.

A cogroup structure for a simplicial object M_* in a simplicial category $(\mathbf{N}, T)$ is a morphism of simplicial objects $\mu_* : M_* \to M_* \vee M_*$ satisfying the usual cogroup equalities. In other words, it is a morphism μ_* of simplicial objects such that μ_q gives a group structure to the set $\operatorname{Hom}_{1_q} (M_q, N)$ for each $q \geq 0$ and for each N with $T(N) = [q]$. If furthermore all the maps

$$\operatorname{Hom}_{1_q} (M_q, N) \to \operatorname{Hom}_\omega (M_p, N)$$

are bijections, it is easy to verify that μ_* gives a simplicial group structure to the simplicial set $\langle M_*, N_* \rangle_*$.

Now here is a theorem proving the existence of simplicial resolutions in simplicial model categories of a certain type. This type is not very general but sufficient for several applications and convenient for an exposition.

THEOREM. *Let* $(\mathbf{N}, T)$ *be a simplicial category with models* M_*^λ, $\lambda \in \Lambda$, *satisfying the following conditions:*

(1) *the category* $\mathbf{N}$ *is small,*

(2) *the simplicial category* $(\mathbf{N}, T)$ *has direct sums,*

(3) *the direct sum of a set of models is a model,*

(4) *each model has at least one abelian cogroup structure.*

Then any simplicial object N_* *has a simplicial resolution* $p_*: P_* \to N_*$.

PROOF. Before beginning the proof, we choose an abelian cogroup structure μ_*^λ for each model M_*^λ. In the examples coming from algebra, the models have much to do with free algebras, free modules and so on, and the choice of one abelian cogroup structure is equivalent to the choice of one set of free generators.

Now any simplicial morphism $p_*: P_* \to N_*$ gives morphism of simplicial abelian groups

$$\langle M_*^\lambda, p_* \rangle_* : \langle M_*^\lambda, P_* \rangle_* \to \langle M_*^\lambda, N_* \rangle_* \quad , \quad \lambda \in \Lambda$$

and consequently morphisms of abelian groups

$$\pi_n(\langle M_*^\lambda, p_* \rangle_*): \pi_n(\langle M_*^\lambda, P_* \rangle_*) \to \pi_n(\langle M_*^\lambda, N_* \rangle_*), \qquad \lambda \in \Lambda, n \geq 0.$$

The morphism p_* is a simplical resolution of N_* if and only if on the one hand all the morphisms $\pi_n(\langle M_*^\lambda, p_* \rangle_*)$ are isomorphisms and on the other hand for each $m \geq 0$, there is an isomorphism $P_m \cong M_m^{\lambda_m}$ for an index $\lambda_m \in \Lambda$. We are using homotopy groups; therefore it is quite natural to introduce an algebraic analogue of the topological procedure of attaching cells. An object M_n^λ will be an n-cell, and we shall have the possibility of modifying homotopy groups in such a way that the theorem can be proved.

9. **Attaching cells.** We consider a simplicial category $(\mathbf{N}, T)$, a simplicial object X_*, and a simplicial object M_* satisfying the condition we already know: all the maps

$$\mathrm{Hom}_{1_q}(M_q, N) \to \mathrm{Hom}_\omega(M_p, N)$$

are isomorphisms. Then the simplicial set $\langle M_*, X_* \rangle_*$ is well defined. An *n-crown* k is a set of morphisms

$$k^i: M_{n-1} \to X_{n-1}, \qquad T(k^i) = 1_{n-1}, \qquad 0 \leq i \leq n$$

satisfying the following equalities:

$$\epsilon_{n-1}^i(k^j) = \epsilon_{n-1}^{j-1}(k^i), \qquad 0 \leq i < j \leq n$$

(no morphism if $n = 0$ and no equality if $n = 1$). Let us denote by $S_{p,q,}$ the set of all morphisms $[p] \to [q]$ corresponding to the order preserving surjective maps from $(0, 1, \ldots, q)$ to $(0, 1, \ldots, p)$. Then we can rewrite the definition of an n-crown. For each morphism $\alpha: [n] \to [q]$, $q \geq 0$, $\alpha \notin S_{n,q}$ there is a morphism $k[\alpha]: M_q \to X_q$, $T(k[\alpha]) = 1_q$, and for each pair of morphisms α as above and $\omega: [q] \to [p]$ there is an equality $\bar{\omega}(k[\alpha]) = k[\omega\alpha]$. Then k^i is equal to $k[\epsilon_n^i]$.

Now we can modify the simplicial object X_* by means of the n-crown k and get a new simplicial object kX_*. Let us suppose that the simplicial category has sufficiently many direct sums. Then by definition

$$kX_m = X_m \vee \left(\bigvee_{\sigma \in S_{n,m}} M_m^{(\sigma)} \right) \quad \text{with} \quad M_*^{(\sigma)} \cong M_*.$$

For a morphism $\omega: [q] \to [p]$, the morphism $\bar{\omega}: kX_q \to kX_p$ is defined componentwise in the following way. On the component X_q of kX_q, this morphism $\bar{\omega}$ is given by the following commutative diagram:

$$\begin{array}{ccc} X_q & \xrightarrow{\bar{\omega}} & X_p \\ \downarrow & & \downarrow \\ kX_q & \xrightarrow{\bar{\omega}} & kX_p \end{array}$$

On the component $M_q^{(\sigma)}$ of kX_q, this morphism $\bar{\omega}$ is given by one of the following commutative diagrams:

$$\begin{array}{ccc} M_q^{(\sigma)} \cong M_q & \xrightarrow{\bar{\omega}} & M_p \cong M_p^{(\omega\sigma)} \\ \downarrow & & \downarrow \\ kX_q & \xrightarrow{\bar{\omega}} & kX_p \end{array}$$

if $\omega\sigma$ belongs to $S_{n,p}$, and

$$\begin{array}{ccccc} M_q^{(\sigma)} \cong M_q & \xrightarrow{\bar{\omega}} & M_p & \xrightarrow{k[\omega\sigma]} & X_p \\ \downarrow & & & & \downarrow \\ kX_q & & \xrightarrow{\bar{\omega}} & & kX_p \end{array}$$

if $\omega\sigma$ does not belong to $S_{n,p}$. Thus the simplicial object kX_* is well defined. There is a natural morphism $k_*: X_* \to kX_*$.

Now let us consider a morphism of simplicial objects $x_*: X_* \to N_*$. A *complete n-crown* $\hat{k}$ (with respect to x_*) consists of an n-crown k (with respect to X_*) and of a morphism $k^+: M_n \to N_n$, $T(k^+) = l_n$, such that the following equalities hold:

$$x_{n-1} \circ k^i = \bar{\epsilon}_n^i(k^+) \qquad 0 \leq i \leq n.$$

Then there is a morphism of simplicial objects $\hat{k}x_*: kX_* \to N_*$ defined componentwise in the following way. In dimension m, on the component X_m of kX_m, the morphism $\hat{k}x_m$ is given by the following commutative diagram

$$\begin{array}{ccc} X_m & \xrightarrow{x_m} & N_m \\ \downarrow & & \downarrow {\scriptstyle 1_{N_m}} \\ kX_m & \xrightarrow{\hat{k}x_m} & N_m \end{array}$$

In dimension m, on the component $M_m^{(\sigma)}$ of kX_m, the morphism $\hat{k}x_m$ is given by the following commutative diagram

$$\begin{array}{ccc} M_m^{(\sigma)} \cong M_m & \xrightarrow{\bar{\sigma}(k^+)} & N_m \\ \downarrow & & \downarrow {\scriptstyle 1_{N_m}} \\ kX_m & \xrightarrow{\hat{k}x_m} & N_m \end{array}$$

Thus the simplicial morphism $\hat{k}x_*$ is well defined. We can summarize this construction in a lemma.

LEMMA. *Let* $(\mathbf{N}, T)$ *be a simplicial category with one model* M_* *and sufficiently many direct sums. Let* $x_*: X_* \to N_*$ *be a morphism of simplicial objects and let* $\hat{k}$ *be a complete n-crown with respect to* x_*. *Then there is a decomposition of the morphism* x_*

$$X_* \xrightarrow{k_*} kX_* \xrightarrow{\hat{k}x_*} N_*$$

having the following properties:

(1) *for any* $m \geq 0$, *the object* kX_m *is the direct sum of* X_m *and of a finite number of copies of* M_m;

(2) *for* $m < n$, *the object* kX_m *is equal to the object* X_m;

(3) *the object* kX_n *is equal to the direct sum* $X_n \vee M_n$;

(4) *on the component* M_n *the morphism* $\bar{\epsilon}_n^i: kX_n \to kX_{n-1}$ *is equal to the morphism* $k^i \circ \bar{\epsilon}_n^i: M_n \to X_{n-1}$;

(5) *on the component* M_n *the morphism* $\hat{k}x_n: kX_n \to N_n$ *is equal to the morphism* $k^+: M_n \to N_n$.

PROOF. The set $S_{n,m}$ has a finite number of elements, only one if $n = m$ and zero if $m < n$. The proof of the last two properties is immediate from the definitions.

What we have done with one model M_* and one complete n-crown $\hat{k}$ can be reproduced with any set of models M_*^λ, $\lambda \in \Lambda$, and any set of complete n-crowns $\hat{k}_\gamma$, $\gamma \in \Gamma$ (the complete n-crown $\hat{k}_\gamma$ is defined with respect to the model $M_*^{\lambda_\gamma}$). It is enough to consider the model

$$M_* = \bigvee_{\gamma \in \Gamma} M_*^{\lambda_\gamma}$$

and the complete n-crown $\hat{k}$ defined in the following way: on the component $M_{n-1}^{\lambda\gamma}$ of M_{n-1} the k^i is equal to the morphism k_γ^i and on the component $M_n^{\lambda\gamma}$ of M_n the morphism k^+ is equal to the morphism k_γ^+. Then we can rewrite the lemma and we get the following properties of the decomposition $x_* = \hat{k}x_* \circ k_*$:

(a) $kX_m = X_m$ if $m < n$ and $kX_n = X_n \vee (\bigvee_{\gamma\in\Gamma} M_n^{\lambda\gamma})$;

(b) on the component $M_n^{\lambda\gamma}$, the morphism $\bar{\epsilon}_n^i$ is equal to $k_\gamma^i \circ \bar{\epsilon}_n^i$;

(c) on the component $M_n^{\lambda\gamma}$, the morphism $\hat{k}x_n$ is equal to k_γ^+.

10. **Construction of simplicial resolutions (Part II).** Now we have to establish the existence of simplicial resolutions. All the hypotheses of the theorem are satisfied and we shall use the following remark about the decomposition

$$X_* \xrightarrow{k_*} kX_* \xrightarrow{\hat{k}x_*} N_*, \qquad \hat{k}x_* \circ k_* = x_*,$$

or more precisely about the homomorphisms

$$\pi_m^\lambda = \pi_m(\langle M_*^\lambda, x_*\rangle_*) : \pi_m(\langle M_*^\lambda, X_*\rangle_*) \to \pi_m(\langle M_*^\lambda, N_*\rangle_*)$$

$$\hat{k}\pi_m^\lambda = \pi_m(\langle M_*^\lambda, \hat{k}x_*\rangle_*) : \pi_m(\langle M_*^\lambda, kX_*\rangle_*) \to \pi_m(\langle M_*^\lambda, N_*\rangle_*)$$

with $\lambda \in \Lambda$ and $m \geq 0$. The following four properties are satisfied:

(1) if π_m^λ is an epimorphism ($m \geq 0$), then $\hat{k}\pi_m^\lambda$ is an epimorphism;

(2) if π_m^λ is a monomorphism ($m < n - 1$), then $\hat{k}\pi_m^\lambda$ is a monomorphism;

(3) it is possible to choose $\hat{k}$ such that $\hat{k}\pi_n^\lambda$ is an epimorphism for all λ's;

(4) it is possible to choose $\hat{k}$ such that $\hat{k}\pi_{n-1}^\lambda$ is a monomorphism for all λ's.

Here $\hat{k}$ denotes any set of complete n-crowns with respect to the models M_*^λ, $\lambda \in \Lambda$, and the simplicial morphism x_*. The first property is formal. The second property is a consequence of the equality $kX_i = X_i$ for $i < n$. The third property is satisfied if, for example, the set of complete n-crowns we are considering contains all complete n-crowns $\hat{k}_\gamma$ such that $k_\gamma^i = 0$, $0 \leq i \leq n$. The fourth property is satisfied if, for example, the set of complete n-crowns we are considering contains all complete n-crowns $\hat{k}_\gamma$ such that $k_\gamma^i = 0$, $0 < i \leq n$. Actually the third property involves commutative diagrams of the following type:

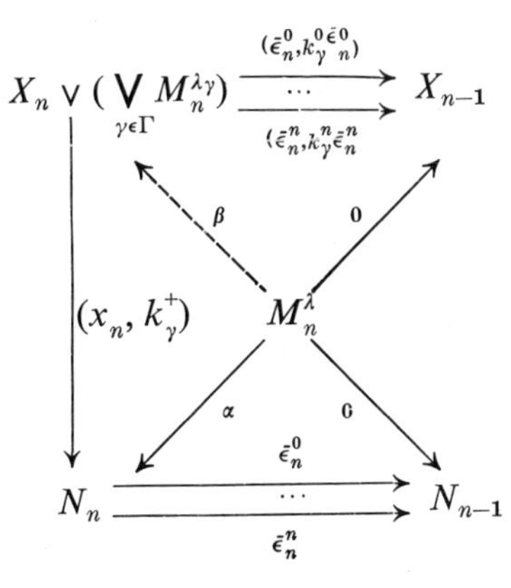

The third property is satisfied if for any λ and any α with $T(\alpha) = 1_n$, there is a β with $T(\beta) = 1_n$. In the simplest way of establishing the third property, the morphism β is one of the canonical morphisms of the direct sum. Actually the fourth property involves pairs of commutative diagrams of the following type:

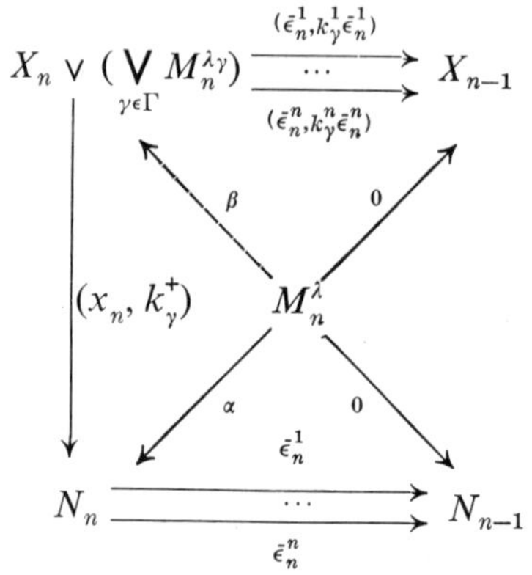

and

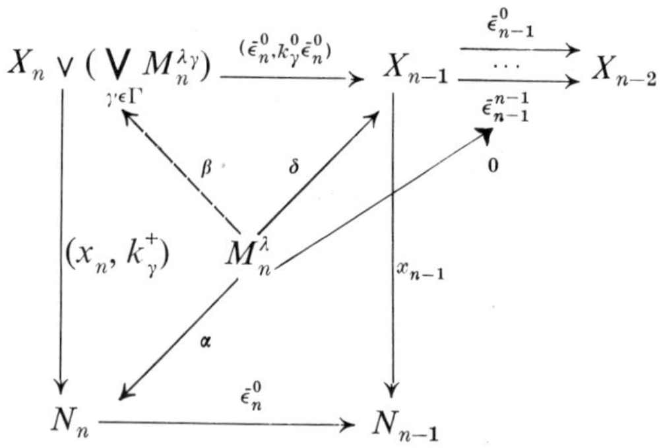

The fourth property is satisfied if for any λ, any α with $T(\alpha) = 1_n$ and any δ with $T(\delta) = \epsilon_n^0$, there is a β with $T(\beta) = 1_n$. In the simplest way of establishing the fourth property, the morphism β is one of the canonical morphism of the direct sum. Notice the ambiguous use of 0 in the diagrams above.

The four properties of the procedure of *attaching cells* give a short proof of the theorem (existence of simplicial resolutions). The simplicial object N_* is given and a simplicial resolution $p_* : P_* \to N_*$ has to be constructed. The construction is made step by step. Let us describe the nth step. We already have a morphism of simplicial objects $p_*^{n-1} : P_*^{n-1} \to N_*$ with the following property:

$$\pi_m(\langle M_*^\lambda, P_*^{n-1}\rangle_*) \to \pi_m(\langle M_*^\lambda, N_*\rangle_*)$$

is an isomorphism if $m < n - 1$ and an epimorphism if $m = n - 1$. Then we choose a set of complete n-crowns with respect to p_*^{n-1} such that the properties described at the beginning of this section are satisfied. We use this set of complete n-crowns for attaching cells and we get a new morphism of simplicial objects

$p_*^n : P_*^n \to N_*$ with the following property:

$$\pi_m(\langle M_*^\lambda, P_*^n\rangle_*) \to \pi_m(\langle M_*^\lambda, N_*\rangle_*)$$

is an isomorphism if $m < n$ and an epimorphism if $m = n$. The 0*th* step begins with $P_*^{-1} = 0$. Putting all steps together, we get the following picture:

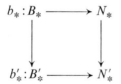

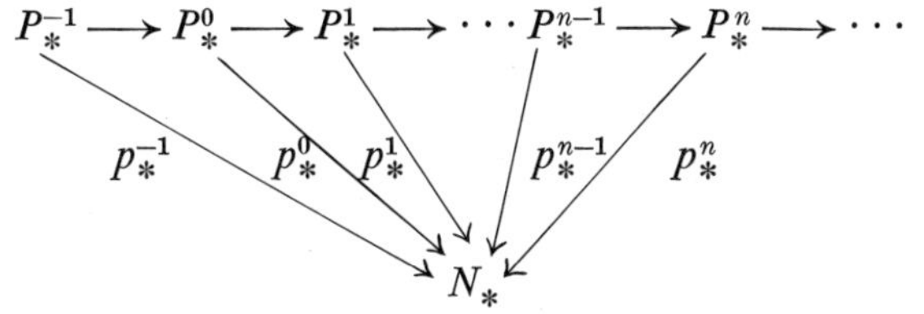

and we define $P_* = \lim_\rightarrow P_*^n$ and $p_* = \lim_\rightarrow p_*^n$. This is well defined since the morphism $P_m^{n-1} \to P_m^n$ is an isomorphism if $n > m$. Furthermore, this remark shows that $p_* : P_* \to N_*$ is a simplicial resolution and the theorem is proved.

Actually we can prove a little more than the preceding theorem, without any further work.

THEOREM. *Let* $(\mathbf{N}, T)$ *be a simplicial category with models* M_*^λ, $\lambda \in \Lambda$, *satisfying the following conditions*:

(1) *the category* $\mathbf{N}$ *is small*;

(2) *the simplicial category* $(\mathbf{N}, T)$ *has direct sums*;

(3) *the direct sum of a set of models is a model*;

(4) *each model has at least one abelian cogroup structure.*

Let $b_* : B_* \to N_*$ *be a simplicial morphism such that for each* $m \geq 0$ *there exists an isomorphism* $B_m \simeq M_m^{\lambda_m}$ *for an index* $\lambda_m \in \Lambda$. *Then there is a decomposition of the simplicial morphism*

$$B_* \xrightarrow{i_*} P_* \xrightarrow{p_*} N_* \qquad p_* i_* = b_*$$

such that

(1) *the simplicial morphism* p_* *is a simplicial resolution of* N_*;

(2) *for each* $n \geq 0$ *there exists an isomorphism* $P_n \simeq B_n \vee M_n^{\bar{\lambda}_n}$ *for an index* $\bar{\lambda}_n \in \Lambda$ *such that the morphism* $i_n : B_n \to P_n$ *corresponds to the canonical morphism* $B_n \to B_n \vee M_n^{\bar{\lambda}_n}$.

Furthermore, the decomposition (i_*, p_*) *can be obtained in a natural way out of the morphism* b_*.

PROOF. We get a decomposition (i_*, p_*) of b_* if we repeat the preceding proof in beginning with $P_*^{-1} = B_*$ and $p_*^{-1} = b_*$. Now the naturality is understood in the following way. For each simplicial morphism b_*, a decomposition (i_*, p_*) is given, and for each commutative square of simplicial morphisms

$$\begin{array}{ccc} b_* : B_* & \longrightarrow & N_* \\ \downarrow & & \downarrow \\ b'_* : B'_* & \longrightarrow & N'_* \end{array}$$

a morphism $P_* \to P'_*$ is given such that the following diagram is commutative:

$$\begin{array}{ccccc} B_* & \longrightarrow & P_* & \longrightarrow & N_* \\ \downarrow & & \downarrow & & \downarrow \\ B'_* & \longrightarrow & P'_* & \longrightarrow & N'_* \end{array} .$$

It suffices to prove that each step of the proof of the existence of a decomposition can be done in a natural way. One of the possible natural ways can be described as follows. Let us use the diagrams appearing at the beginning of this section, for the dimension n. The set of complete n-crowns we use consists of one complete n-crown for each α appearing in the general diagram corresponding to the third property and of one complete n-crown for each pair α/δ appearing in the pair of general diagrams corresponding to the fourth property, and that in such a way that β in both cases is nothing but the corresponding canonical morphism of the direct sum.

References

1. M. André, *Méthode simpliciale en algèbre homologique et algèbre commutative*, Lecture Notes in Mathematics, no. 32, Springer-Verlag, Berlin, 1967.

2. H. Cartan and S. Eilenberg, *Homological algebra*, Princeton University Press, Princeton, N.J., 1956.

3. A. Dold and D. Puppe, *Homologie nicht-additiver Funktoren: Anwendungen*, Ann. Inst. Fourier (Grenoble) **11** (1961), 201–312.

4. D. Kan, *A combinatorial definition of homotopy groups*, Ann. of Math. **67** (1958), 282–312.

5. D. Quillen, *Homotopical algebra*, Lecture Notes in Mathematics, no. 43, Springer-Verlag, Berlin, 1967.

The University of Chicago

Battelle Institute, Geneva, Switzerland

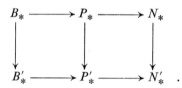

On completing bicartesian squares

P. J. Hilton and I. S. Pressman

The completion of a filtration in an abelian category $\mathscr{A}$ was studied in **[2, 4]**. Thus given a filtration of an object X in $\mathscr{A}$, together with its annihilating co-filtration,

$$\cdots \subseteq X^{p-1} \subseteq X^p \subseteq \cdots \subseteq X \twoheadrightarrow \cdots \twoheadrightarrow X_p \twoheadrightarrow X_{p+1} \twoheadrightarrow \cdots, \quad -\infty < p < +\infty,$$

so that $X/X^p = X_p$, we may obtain the diagrams

$$\begin{array}{ccccc} X^p & \subseteq X & \twoheadrightarrow & X_p \\ \| & \uparrow{\scriptstyle\lambda} & & \uparrow \\ X^p & \subseteq X^\infty & \twoheadrightarrow & X_p^\infty \\ \downarrow & \downarrow{\scriptstyle\rho^\infty} & & \| \\ X^p_{-\infty} & \subseteq X^\infty_{-\infty} & \twoheadrightarrow & X^\infty_p \end{array} \qquad \begin{array}{ccccc} X^p & \subseteq X & \twoheadrightarrow & X_p \\ \downarrow & \downarrow{\scriptstyle\rho} & & \| \\ X^p_{-\infty} & \subseteq X_{-\infty} & \twoheadrightarrow & X_p \\ \| & \uparrow{\scriptstyle\lambda_{-\infty}} & & \uparrow \\ X^p_{-\infty} & \subseteq X^\infty_{-\infty} & \twoheadrightarrow & X^\infty_p \end{array}$$

where the (common) third line represents the completion of the original filtration,

$$X^\infty = \varinjlim X^p, \qquad X_{-\infty} = \varprojlim X_p,$$

and

$$X^\infty_{-\infty} = \varprojlim X^\infty_p = \varinjlim X^p_{-\infty}.$$

Moreover it was proved in **[2]** (and stated in Theorem 2.3(iii) of **[4]**) that the square

$$\text{(0)} \qquad \begin{array}{ccc} X^\infty & \xrightarrow{\lambda} & X \\ \downarrow{\scriptstyle\rho^\infty} & & \downarrow{\scriptstyle\rho} \\ X^\infty_{-\infty} & \xrightarrow{\lambda_{-\infty}} & X_{-\infty} \end{array}$$

is bicartesian (that is, a pull-back and a push-out; see [**3**]). If we look at a particular aspect of this result we see that it asserts that, starting with any filtration and constructing the limits X^{∞}, $X_{-\infty}$, the composable pair of morphisms (λ, ρ) can be embedded in a bicartesian square. This therefore suggests the question: given any composable pair of morphisms (f, g) in $\mathscr{A}$,

$$\begin{array}{ccc} A & \xrightarrow{\ f\ } & B \\ & & \Big\downarrow{\scriptstyle g} \\ & & E, \end{array}$$

to find necessary and sufficient conditions under which we can find a bicartesian square

$$\begin{array}{ccc} A & \xrightarrow{\ f\ } & B \\ {\scriptstyle d}\Big\downarrow & & \Big\downarrow{\scriptstyle g} \\ D & \xrightarrow{\ e\ } & E. \end{array} \tag{1}$$

The theorem on filtrations also raises the cognate question of how to characterize the 'completions' (1) of the pair (f, g), and, in particular, how to pick out, in a canonical way, the completion (0). We do not discuss this question in this note beyond remarking that it is a natural generalization of the construction of the Ext group. For there is a very obvious and natural way to collect the completions (1) of (f, g) into equivalence classes, and if $B = 0$ then a completion (1) is just a short exact sequence (s.e.s.) $A \overset{d}{\rightarrowtail} D \overset{e}{\twoheadrightarrow} E$, and the equivalence classes are precisely the elements of $\mathrm{Ext}^1(E, A)$. This observation suggests the plausibility of looking for some natural algebraic structure in the set of equivalence classes of completions (1), and we hope to return to this question later.

Another obvious special case of our problem is that in which $E = 0$. Then we see (and, indeed, it follows immediately from Theorem 2 of this note) that a completion (1) exists if and only if f is a projection of A onto a direct summand B, and that $d: A \to D$ is then an arbitrary projection onto a complementary summand D. Thus the set of completions is equivalent to the set of complements of B; the dual statement holds, of course, if $A = 0$.

In this note we obtain necessary and sufficient conditions for the existence of (1). These conditions imply, of course, properties of the terms involved in the completion process of a filtration in $\mathscr{A}$; we hope also to study these properties in a subsequent note.

We also remark that the question of the existence of a completion (1), for special cases, has been considered in [**5**, III. 3.8 and XII. 5] and [**6**].

In the second part of this note we establish necessary and sufficient conditions for the uniqueness of the completion (1) of (f, g), given that such a completion exists.

We now establish notation. All objects and morphisms mentioned will belong to a fixed abelian category $\mathscr{A}$. A short exact sequence (s.e.s.) of $\mathscr{A}$

$$0 \to A \xrightarrow{a} B \xrightarrow{b} C \to 0$$

will be written as $b \,[]\, a$ or $A \rightarrowtail^{a} B \xrightarrow{b}\!\!\!\!\rightarrow C$; its class in Ext (C, A) is $\{b \,[]\, a\}$. $b\ a$ means that the morphism a is the kernel of b; $b\ a$ means that b is the cokernel of a. Every map $f: A \to B$ gives rise, canonically, to the sequence

$$K_f \overset{\kappa^f}{\rightarrowtail} A \overset{\lambda^f}{\twoheadrightarrow} I_f \overset{\mu^f}{\rightarrowtail} B \overset{\nu^f}{\twoheadrightarrow} C_f$$

where $\lambda^f \,[]\, \kappa^f$,$\nu^f \,[]\, \mu^f$, $\mu^f\lambda^f = f$, $f\,[\kappa^f,\ \nu^f]f$, $K_f = \ker f$, $I_f = \operatorname{im} f$, and $C_f = \operatorname{cok} f$.

In [7] it was shown that there was an E-H functor $(\tau, \partial): \mathscr{A}^2 \to \mathscr{A}$ which to each pair of composable maps (f, g) assigned the exact sequence

$$(2) \qquad 0 \to K_f \xrightarrow{k'} K_{gf} \xrightarrow{k''} K_g \xrightarrow{\partial_0} C_f \xrightarrow{c'} C_{gf} \xrightarrow{c''} C_g \to 0$$

where $\tau_0(f) = K_f$, $\tau_{-1}(f) = C_f$, and $\tau_j(f) = 0$ for $j \neq 0, -1$. k' was described as $\tau_0\begin{pmatrix}1\\ g\end{pmatrix}$, etc., and $\partial_0(f, g) = \partial_0 = \nu^f\kappa^g$. When $(f, g) = (\lambda^f, \mu^f)$, k' and c'' are taken to be the identity maps of K_f and C_f respectively.

The composable pair (f, g) gives rise to the commutative diagram (3),

$$(3)\qquad \begin{array}{ccccccccc}
K_f & \overset{\kappa^f}{\rightarrowtail} & A & \overset{\lambda^f}{\twoheadrightarrow} & I_f & \overset{\mu^f}{\rightarrowtail} & B & \overset{\nu^f}{\twoheadrightarrow} & C_f \\
\downarrow{\scriptstyle k'} & & \| & & \downarrow{\scriptstyle i'} & & \downarrow{\scriptstyle g} & & \downarrow{\scriptstyle c'} \\
K_{gf} & \overset{\kappa^{gf}}{\rightarrowtail} & A & \overset{\lambda^{gf}}{\twoheadrightarrow} & I_{gf} & \overset{\mu^{gf}}{\rightarrowtail} & E & \overset{\nu^{gf}}{\twoheadrightarrow} & C_{gf} \\
\downarrow{\scriptstyle k''} & & \downarrow{\scriptstyle f} & & \downarrow{\scriptstyle j''} & & \| & & \downarrow{\scriptstyle c''} \\
K_g & \overset{\kappa^g}{\rightarrowtail} & B & \overset{\lambda^g}{\twoheadrightarrow} & I_g & \overset{\mu^g}{\rightarrowtail} & E & \overset{\nu^g}{\twoheadrightarrow} & C_g
\end{array}$$

If d, e, f and g in (1) are each factored canonically into an epic λ and monic μ, then (1) gives rise to a commutative diagram (4) with four commutative squares. The center of the diagram can clearly be taken to be I_{gf} because $gf: A \to E$ factors through it, is epic onto it, monic from it, and $\lambda^g\mu^f = j''i'$. This last follows because in (3) $\lambda^g\mu^f\lambda^f = \lambda^g f = j''\lambda^{gf} = j''i'\lambda^f$, and λ^f is epic.

$$(4)\qquad \begin{array}{ccccc}
A & \overset{\lambda^f}{\twoheadrightarrow} & I_f & \overset{\mu^f}{\rightarrowtail} & B \\
\downarrow{\scriptstyle \lambda^d} & \Gamma^1 & \downarrow{\scriptstyle i'} & \Gamma^2 & \downarrow{\scriptstyle \lambda^g} \\
I_d & \overset{j'}{\twoheadrightarrow} & I_{gf} & \overset{j''}{\rightarrowtail} & I_g \\
\downarrow{\scriptstyle \mu^d} & \Gamma^3 & \downarrow{\scriptstyle i''} & \Gamma^4 & \downarrow{\scriptstyle \mu^g} \\
D & \overset{\lambda^e}{\twoheadrightarrow} & I_e & \overset{\mu^e}{\rightarrowtail} & E
\end{array}$$

By [**3**, Theorem 3.7] if each Γ^i is bicartesian, then so is (1). Conversely, by [**3**, Corollary 3.11] if (1) is exact, then Γ^1 is cocartesian, Γ^2 and Γ^3 are bicartesian, and Γ^4 is cartesian. If (1) is bicartesian, so is each Γ^i.

LEMMA 1. *$K_f \xrightarrow{k'} K_{gf} \xrightarrow{k''} K_g$ is a split s.e.s. if* (1) *is cartesian.*

PROOF. Theorem 3.5 [**3**] gives an isomorphism $\ker d \cong \ker g$. That is,

$$\tau_0\begin{pmatrix} f \\ e \end{pmatrix} : \tau_0(d) \xrightarrow{\cong} \tau_0(g).$$

But

$$\tau_0\begin{pmatrix} f \\ e \end{pmatrix} = \tau_0\begin{pmatrix} f \\ 1 \end{pmatrix}\tau_0\begin{pmatrix} 1 \\ e \end{pmatrix} : \tau_0(d) \to \tau_0(gf) \to \tau_0(g), \qquad \text{and } k'' = \tau_0\begin{pmatrix} f \\ 1 \end{pmatrix}.$$

Therefore, k'' has a right inverse

$$s = \tau_0\begin{pmatrix} 1 \\ \ \end{pmatrix}\tau_0\begin{pmatrix} f \\ e \end{pmatrix}^{-1}.$$

LEMMA 2. *If* (1) *is bicartesian, then*

(i) $\partial_0(f, g) = \nu^f \kappa^g = 0$,
(ii) *$c'' \,[\,]\, c'$ is a split s.e.s.*
(iii) *$k'' \,[\,]\, k'$ is a split s.e.s., and*
(iv) *the exact sequence* (5) *represents the zero class of* $\operatorname{Ext}^2(C_g, K_f)$.

$$0 \to K_f \xrightarrow{\lambda^d \kappa^f} I_d \xrightarrow{i'' j'} I_e \xrightarrow{\nu^g \mu^e} C_g \to 0. \tag{5}$$

PROOF. Lemma 1 and its dual establish the first three conditions. Consider, the Ext-sequence (6) given by the s.e.s. $j' \,[\,]\, \lambda^d \kappa^f$, that is, $\alpha^1 = (\lambda^d \kappa^f)_*$, $\beta^1 = j'_*$

$$\cdots \to \operatorname{Hom}(C_g, I_{gf}) \xrightarrow{\gamma^0} \operatorname{Ext}^1(C_g, K_f) \xrightarrow{\alpha^1} \operatorname{Ext}^1(C_g, I_d) \xrightarrow{\beta^1} \operatorname{Ext}^1(C_g, I_{gf}) \xrightarrow{\gamma^1} \operatorname{Ext}^2(C_g, K_f) \to \cdots. \tag{6}$$

We have $\{\nu^g e \,[\,]\, \mu^d\} \in \operatorname{Ext}^1(C_g, I_d)$ and $\nu^g e \,[\,]\, \mu^d$ pushes out under j' to $\nu^g \mu^e \,[\,]\, i''$. This passes over to (5) by the action of the boundary operator. By the exactness of (6), (5) must represent the zero class.

Lemma 2 gives the clue to our main theorem. We say that the pair (f, g) is *completable* if a bicartesian square (1) exists, and we then call (d, e) a *completion* of (f, g). Now we may form the diagram

$$\begin{array}{ccccc} A & \xrightarrow{\lambda^f} & I_f & \xrightarrow{\mu^f} & B \\ & & \downarrow{\scriptstyle i'} & \Gamma^2 & \downarrow{\scriptstyle \lambda^g} \\ & & I_{gf} & \xrightarrow{j''} & I_g \\ & & & & \downarrow{\scriptstyle \mu^g} \\ & & & & E. \end{array} \tag{7}$$

Then in view of the relation of (1) and (4) it is clear that (f, g) is completable if and only if the following four conditions are satisfied:

($\mathrm{i_c}$) Γ^2 is bicartesian;

($\mathrm{ii_c}$) (j'', μ^g) is completable;

($\mathrm{iii_c}$) (λ^f, i') is completable;

($\mathrm{iv_c}$) for some completion[1] (λ^d, j') of (λ^f, i'), and some completion (i'', μ^e) of (j'', μ^g), (j', i'') is completable.

Thus the following theorem establishes necessary and sufficient conditions that (f, g) be completable.

THEOREM 1. *Conditions* (α_c) *hold if and only if conditions* (α) *of Lemma* 2 *hold,* $\alpha = \mathrm{i, ii, iii, iv}$.

PROOF. Lemma 2 shows that conditions (α_c) imply conditions (α). Conversely suppose conditions (α) hold. It is easy to see that condition (i) implies condition ($\mathrm{i_c}$). We then have $\lambda : I_g \twoheadrightarrow C_f$ such that $\lambda\lambda^g = \nu^f$, $\lambda \square j''$, and $\mu : K_g \rightarrowtail I_f$ such that $\mu^f \mu = \kappa^g$, $i' \square \mu$.

Consider the diagram

$$\begin{array}{ccccc}
I_{gf} & \xrightarrow{j''} & I_g & \xrightarrow{\lambda} & C_f \\
 & & \downarrow{\scriptstyle \mu^g} & & \downarrow{\scriptstyle c'} \\
 & & E & \xrightarrow{\nu^{gf}} & C_{gf} \\
 & & \downarrow{\scriptstyle \nu^g} & & \downarrow{\scriptstyle c''} \\
 & & C_g & = & C_g .
\end{array}$$

This diagram shows that in the exact sequence

$$\mathrm{Ext}^1(C_g, I_{gf}) \xrightarrow{j''_*} \mathrm{Ext}^1(C_g, I_g) \xrightarrow{\lambda_*} \mathrm{Ext}^1(C_g, C_f)$$

we have $\lambda_*\{\nu^g \square \mu^g\} = \{c'' \square c'\}$. Thus if (ii) holds, $\lambda_*\{\nu^g \square \mu^g\} = 0$, so that $\{\nu^g \square \mu^g\} = j''_*\{\nu^g\mu^e \square i''\}$ for some push-out, and hence bicartesian square,

$$\begin{array}{ccc}
I_{gf} & \xrightarrow{j''} & I_g \\
\downarrow{\scriptstyle i''} & & \downarrow{\scriptstyle \mu^g} \\
I_e & \xrightarrow{\mu^e} & E .
\end{array}$$

This establishes condition ($\mathrm{ii_c}$). Similarly condition (iii) establishes condition ($\mathrm{iii_c}$).

---

[1] This notation is legitimate since the completion of (λ^f, i') consists of epics; that of (j'', μ^g) of monics.

Now suppose further that condition (iv) holds. Reversing the steps of the argument in Lemma 2, we see that $\{\nu^g\mu^e \square i''\}$ passes to zero under γ^1, so that $\{\nu^g\mu^e \square i''\} \in j'_* \operatorname{Ext}^1(C_g, I_d)$, say,

$$\{\nu^g\mu^e \square i''\} = j'_*\{\nu^g\mu^e\lambda^e \square \mu^d\}$$

for some push-out, and hence bicartesian, square

$$\begin{array}{ccc} I_d & \xrightarrow{j'} & I_{gf} \\ \Big\downarrow{\scriptstyle \mu^d} & & \Big\downarrow \\ D & \xrightarrow{\lambda^e} & I_e. \end{array}$$

This establishes condition (iv$_c$) and hence the theorem.

REMARK. The logical relation between the conditions (α) and (α_c) is as follows. Conditions (i) and (i$_c$) are equivalent. Condition (ii) includes condition (i) and is equivalent to conditions (i$_c$) and (ii$_c$); similarly condition (iii) is equivalent to (i$_c$) and (iii$_c$). Finally if we assume (i), (ii), (iii) (or (i$_c$), (ii$_c$), (iii$_c$)) hold, then conditions (iv) and (iv$_c$) are equivalent.

The proof of Theorem 1 itself suggests the following reformulation of condition (iv). Condition (ii) is equivalent to the assertion that $\{\nu^g \square \mu^g\} \in j''_* \operatorname{Ext}^1(C_g, I_{gf})$ and condition (iii) is equivalent to the assertion that $\{\lambda^f \square \kappa^f\} \in i'^* \operatorname{Ext}^1(I_{gf}, K_f)$. By taking the set of counterimages in $\operatorname{Ext}^1(C_g, I_{gf})$ and $\operatorname{Ext}^1(I_{gf}, K_f)$ and forming the Yoneda product we produce an *obstruction set* in $\operatorname{Ext}^2(C_g, K_f)$. Then

condition (iv$_c$) *asserts that* $0 \in$ *obstruction set* $\subseteq \operatorname{Ext}^2(C_g, K_f)$.

Summing up, we have proved

THEOREM 2. *The composable pair* (f, g),

$$\begin{array}{ccc} A & \xrightarrow{f} & B \\ & & \Big\downarrow{\scriptstyle g} \\ & & E \end{array}$$

is completable if and only if

(a) $c'' \square c'$ *is a split s.e.s.*, $C_f \xrightarrow{c'} C_{gf} \xrightarrow{c''} C_g$,

(b) $k'' \square k'$ *is a split s.e.s.*, $K_f \xrightarrow{k'} K_{gf} \xrightarrow{k''} K_g$, *and*

(c) $0 \in$ *obstruction set* $\subseteq \operatorname{Ext}^2(C_g, K_f)$.

COROLLARY 1. (i) *If* K_f *is injective then condition* (a) *is necessary and sufficient for the completability of* (f, g).

(ii) *If* C_g *is projective then condition* (b) *is necessary and sufficient for the completability of* (f, g).

(iii) *If* $\mathscr{A}$ *is a category of modules over a ring of dimension* 1 *then conditions* (a) *and* (b) *are necessary and sufficient for the completability of* (f, g).

COROLLARY 2. *If f is epic and g is monic then we may identify I_f, I_{gf}, I_g with B,*

$$\{\nu^g \square g\} \in \mathrm{Ext}^1(C_g, B), \quad \{f \square \kappa^f\} \in \mathrm{Ext}^1(B, K_f)$$

and (f, g) is completable if and only if the Yoneda product $\{\nu^g \square g\}\{f \square \kappa^f\}$ vanishes.

REMARKS. (a) Case (i) of Corollary 1 covers the case when f is monic, case (ii) the case when g is epic.

(b) Case (iii) of Corollary 1 covers the case when $\mathscr{A}$ is the category of abelian groups. The 'second obstruction' then automatically vanishes. That the second obstruction does not always vanish is easily seen; as an obvious example, let π be free abelian of rank 2, $\Lambda = Z[\pi]$, $\mathscr{A}$ the category of (right) Λ-modules. Since $\mathrm{Ext}^1_\Lambda(Z, Z) = Z \oplus Z$ and the two generators may be chosen so that their Yoneda product generates $\mathrm{Ext}^2_\Lambda(Z, Z) = Z$, we simply choose f epic, g monic so that $\{\nu^g \square g\}$, $\{f \square \kappa^f\}$ are these chosen generators. Corollary 2 then shows that (f, g) is not completable in this case.

Of course no claim is made that the completion is unique. Since our problem generalizes that of obtaining $\mathrm{Ext}^1(E, A)$ (the Case $B = 0$), it follows that, in general, no uniqueness is to be expected. We now proceed to study the problem of when, in fact, we do have uniqueness.

It will be assumed henceforth that (f, g) *has a completion* and that the bicartesian square (BCS) (1) is a fixed completion of the pair of morphisms (f, g). We now explain what we should understand by the uniqueness of (1).

Let

$$\begin{array}{ccc} A & \xrightarrow{f} & B \\ \Big\downarrow{\scriptstyle d_r} & & \Big\downarrow{\scriptstyle g} \\ D_r & \xrightarrow{e_r} & E, \end{array} \qquad r = 1, 2$$

be BCS. We say that they are *congruent* (or that the completions (d_r, e_r) are *congruent*) if there exist automorphisms $\alpha: A \to A$, $\eta: E \to E$, and a morphism $h: D_1 \to D_2$ such that $f\alpha = f$, $\eta g = g$, and the diagram

$$\begin{array}{ccccc} A & \xrightarrow{d_1} & D_1 & \xrightarrow{e_1} & E \\ \Big\downarrow{\scriptstyle \alpha} & & \Big\downarrow{\scriptstyle h} & & \Big\downarrow{\scriptstyle \eta} \\ A & \xrightarrow{d_2} & D_2 & \xrightarrow{e_2} & E \end{array}$$

is commutative. It is easy to see that h is then an isomorphism, so that congruence is an equivalence relation among the completions of (f, g). We say that the completion of (f, g) is unique if all completions belong to the same congruence class. We now prove some results about the congruence relation.

PROPOSITION 1. *The following three statements are equivalent:* (i) *the completions (d_1, e_1), (d_2, e_2) are congruent;* (ii) *there exists an automorphism $\alpha: A \to A$ and an isomorphism $h: D_1 \to D_2$ such that $f\alpha = f$, $hd_1 = d_2\alpha$;* (iii) *there* *exists an automorphism* $\eta : E \rightarrow E$ *and an isomorphism* $h : D_1 \rightarrow D_2$ *such that* $\eta g = g$, $e_2 h = \eta e_1$.

PROOF. It is evidently sufficient to show that (ii) $\Rightarrow$ (i). Now since (d_2, e_2) completes (f, g), it follows that $(d_2\alpha, e_2)$ completes $(f\alpha, g)$, that is, (hd_1, e_2) completes (f, g). It then follows that (d_1, e_2h) completes (f, g). But (d_1, e_1) completes (f, g), so we have two push-outs of (f, d_1). Thus there exists an automorphism $\eta : E \rightarrow E$ such that $\eta g = g$, $\eta e_1 = e_2 h$.

Now we may regard (7) as the *fixed part* of (1). Writing λ^i, μ^j for i', j'', we may draw the following diagram representing two completions of (f, g).

$$\begin{array}{ccccc}
A & \xrightarrow{\lambda^f} & I_f & \xrightarrow{\mu^f} & B \\
\downarrow{\scriptstyle \lambda_d{}^r} & \Gamma^1_r & \downarrow{\scriptstyle \lambda^i} & \Gamma^2 & \downarrow{\scriptstyle \lambda^g} \\
I_{dr} & \xrightarrow{\lambda_j{}^r} & I_{gf} & \xrightarrow{\mu^j} & I_g \\
\downarrow{\scriptstyle \mu_d{}^r} & \Gamma^3_r & \downarrow{\scriptstyle \mu_i{}^r} & \Gamma^4_r & \downarrow{\scriptstyle \mu^g} \\
D_r & \xrightarrow{\lambda_e{}^r} & I_{er} & \xrightarrow{\mu_e{}^r} & E
\end{array} \qquad r = 1, 2. \tag{8}$$

PROPOSITION 2. *If the completions* (d_r, e_r), $r = 1, 2$, *are congruent, then* Γ^1_1 *is congruent to* Γ^1_2 *and* Γ^4_1 *is congruent to* Γ^4_2.

PROOF. By splitting down the middle we deduce from the given congruence a commutative diagram

$$\begin{array}{ccccccccc}
A & \xrightarrow{\lambda_d{}^1} & I_{d1} & \xrightarrow{\mu_d{}^1} & D_1 & \xrightarrow{\lambda_e{}^1} & I_{e1} & \xrightarrow{\mu_e{}^1} & E \\
\downarrow{\scriptstyle \alpha} & & \downarrow{\scriptstyle u} & & \downarrow{\scriptstyle h} & & \downarrow{\scriptstyle v} & & \downarrow{\scriptstyle \eta} \\
A & \xrightarrow[\lambda_d{}^2]{} & I_{d2} & \xrightarrow[\mu_d{}^2]{} & D_2 & \xrightarrow[\lambda_e{}^2]{} & I_{e2} & \xrightarrow[\mu_e{}^2]{} & E
\end{array} \tag{9}$$

with all vertical arrows isomorphisms. Also $\lambda^f\alpha = \lambda^f$ since $f\alpha = f$, and $\eta\mu^g = \mu^g$ since $\eta g = g$. We now apply Proposition 1 to the 'outside' squares of (9). This completes the proof and establishes the commutative diagram

$$\begin{array}{ccccc}
I_{d1} & \xrightarrow{\lambda_j{}^1} & I_{gf} & \xrightarrow{\mu_i{}^1} & I_{e1} \\
\downarrow{\scriptstyle u} & & \Vert & & \downarrow{\scriptstyle v} \\
I_{d2} & \xrightarrow{\lambda_j{}^2} & I_{gf} & \xrightarrow{\mu_i{}^2} & I_{e2};
\end{array} \tag{10}$$

for the only automorphism θ of I_{gf} such that $\theta\lambda^i = \lambda^i$ is the identity and, likewise, $\mu^j\theta = \mu^j$ forces $\theta = 1$. Diagrams (9) and (10) also serve to describe the relation between the squares Γ^3_1 and Γ^3_2 when the completions (d_r, e_r) are congruent. It is, moreover, clear that the function $(\mu, \lambda) \rightarrow (\mu u^{-1}, v\lambda)$ then sets up a one-one correspondence between congruence classes of completions of (λ^j_1, μ^i_1) and congruence classes of completions of (λ^j_2, μ^i_2).

We now prove a converse of these results. Suppose in (8) that Γ_1^1 is congruent to Γ_2^1 and that Γ_1^4 is congruent to Γ_2^4. Thus we have a diagram

$$(11)\qquad \begin{array}{ccccccccc}
A & \xrightarrow{\lambda_d^1} & I_{d1} & \xrightarrow{\lambda_j^1} & I_{ef} & \xrightarrow{\mu_i^1} & I_{e1} & \xrightarrow{\mu_e^1} & E \\
\downarrow{\scriptstyle \alpha} & & \downarrow{\scriptstyle u} & & \| & & \downarrow{\scriptstyle v} & & \downarrow{\scriptstyle \eta} \\
A & \xrightarrow[\lambda_d^2]{} & I_{d2} & \xrightarrow[\lambda_j^2]{} & I_{gf} & \xrightarrow[\mu_i^2]{} & I_{e2} & \xrightarrow[\mu_e^2]{} & E,
\end{array}$$

$$\lambda^f\alpha = \lambda^f,\ \eta\mu^g = \mu^g.$$

Let us further suppose that Γ_1^3 and Γ_2^3 are (u, v)-*congruent*, meaning that the completions (μ_2^d, λ_2^e) and $(\mu_1^d u^{-1}, v\lambda_1^e)$ of (λ_2^j, μ_2^i) are congruent. We then prove

PROPOSITION 3. *Under these hypotheses the completions* (d_r, e_r), $r = 1, 2$, *of* (f, g) *are congruent.*

PROOF. We have $f\alpha = f$, $\eta g = g$. Further we have a diagram

$$\begin{array}{ccccc}
I_{d1} & \xrightarrow{\mu_d^1} & D_1 & \xrightarrow{\lambda_d^1} & I_{e1} \\
\downarrow{\scriptstyle su} & & \downarrow{\scriptstyle h} & & \downarrow{\scriptstyle tv} \\
I_{d2} & \xrightarrow[\mu_d^2]{} & D_2 & \xrightarrow[\lambda_d^2]{} & I_{e2},
\end{array}$$

where h is an isomorphism and s, t are automorphisms of I_{d2}, I_{e2} such that $\lambda_2^j s = \lambda_2^j$, $t\mu_2^i = \mu_2^i$. Now $(\lambda_2^d, \lambda_2^j)$ is a completion of (λ^f, λ^i). Since $\lambda_2^j s = \lambda_2^j$ it follows that $(s\lambda_2^d, \lambda_2^j)$ is also a completion of (λ^f, λ^i). Thus (λ^f, λ_2^d) and $(\lambda^f, s\lambda_2^d)$ are both pull-backs of (λ^i, λ_2^j), so that there exists an automorphism ξ of A such that $\lambda^f\xi = \lambda^f$ and $\lambda_2^d\xi = s\lambda_2^d$. Consider the square

$$\begin{array}{ccc}
A & \xrightarrow{d_1} & D_1 \\
\downarrow{\scriptstyle \xi\alpha} & & \downarrow{\scriptstyle h} \\
A & \xrightarrow[d_2]{} & D_2.
\end{array}$$

Then $f\xi\alpha = f$ and $hd_1 = h\mu_1^d\lambda_1^d = \mu_2^d su\lambda_1^d = \mu_2^d s\lambda_2^d\alpha = \mu_2^d\lambda_2^d\xi\alpha = d_2\xi\alpha$. We apply Proposition 1 to complete the proof.

We now approach the congruence problem from a different point of view. Let us suppose that g is monic and consider the diagram

$$\begin{array}{ccc}
A & \xrightarrow{f} & B \\
 & & \downarrow{\scriptstyle g} \\
 & & E \\
 & & \downarrow{\scriptstyle \nu^g} \\
 & & C_g.
\end{array}$$

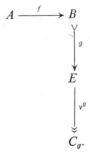

Then f induces $f_*:\text{Ext}^1(C_g, A) \to \text{Ext}^1(C_g, B)$, and if $\sigma = \{\nu^g \,[]\, g\} \in \text{Ext}^1(C_g, B)$ then the completability of (f, g) simply states that σ is in the image of f_*. Let $\text{Aut}_f A$ be the group consisting of those automorphisms α of A for which $f\alpha = f$ and let $\text{Aut}_g E$ be similarly defined. Plainly $\text{Aut}_g E$ acts as a group of automorphisms of C_g, so that $\text{Aut}_f A$ and $\text{Aut}_g E$ both act as a group of automorphisms of $\text{Ext}^1(C_g, A)$. Indeed, the actions commute, so that $\text{Aut}_f A \times \text{Aut}_g E$ acts as a group of automorphisms of $\text{Ext}^1(C_g, A)$. We will write ${}^\alpha\theta^\eta$ for the effect of (α, η) on $\theta \in \text{Ext}^1(C_g, A)$.

PROPOSITION 4. *If* $f_*(\theta) = \sigma$ *then* $f_*({}^\alpha\theta^\eta) = \sigma$.

PROOF. Since $f\alpha = f$ it is immediate that $f_*({}^\alpha\theta) = f_*(\theta)$. Now $\text{Aut}_g E$ acts as the identity on σ, so that $f_*(\theta^\eta) = f_*(\theta)^\eta = \sigma^\eta = \sigma$.

PROPOSITION 5. *Let* $\theta_1, \theta_2 \in f_*^{-1}(\sigma)$. *Then we have diagrams*

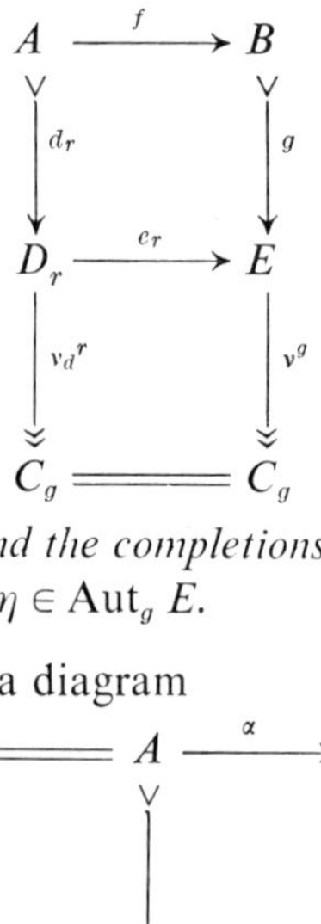

where $\theta_r = \{\nu_r^d \,[]\, d_r\}, r = 1, 2$; *and the completions* (d_r, e_r) *are congruent if and only if* $\theta_2 = {}^\alpha\theta_1^\eta$ *for some* $\alpha \in \text{Aut}_f A$, $\eta \in \text{Aut}_g E$.

PROOF. If $\theta_2 = {}^\alpha\theta_1^\eta$ we have a diagram

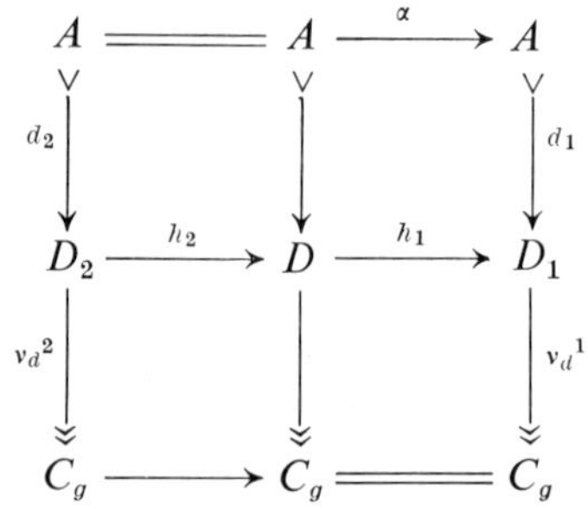

where all horizontal morphisms are isomorphisms and $f\alpha = f$. Apply Proposition 1. Conversely, if the completions (d_r, e_r) are congruent, $r = 1, 2$, then we have a diagram

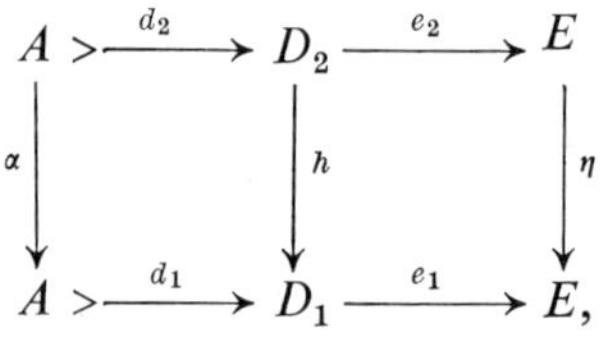

$f\alpha = f, \quad \eta g = g,$

inducing

$$\begin{array}{ccccc} A & \overset{d_2}{\rightarrowtail} & D_2 & \overset{\nu_d^{\,2}}{\twoheadrightarrow} & C_g \\ \downarrow{\scriptstyle \alpha} & & \downarrow{\scriptstyle h} & & \downarrow{\scriptstyle \theta} \\ A & \overset{d_1}{\rightarrowtail} & D_1 & \overset{\nu_d^{\,1}}{\twoheadrightarrow} & C_g, \end{array}$$

and it remains to show that the square

$$\begin{array}{ccc} E & \overset{\eta}{\longrightarrow} & E \\ \downarrow{\scriptstyle \nu^g} & & \downarrow{\scriptstyle \nu^g} \\ C_g & \overset{\theta}{\longrightarrow} & C_g \end{array} \tag{12}$$

commutes. Now $\theta\nu^g g = 0$ and $\nu^g\eta g = \nu^g g = 0$; also $\theta\nu^g e_2 = \theta\nu_2^d = \nu_1^d h = \nu^g e_1 h = \nu^g \eta e_2$. Thus the commutativity of (12) follows from the push-out property of a BCS.

Plainly Propositions 4, 5 have duals (when f is epic) which we will not enunciate explicitly. Together they provide a complete answer to the congruence classification and hence to the uniqueness problem when g is monic or f is epic.

We now revert to the general case. It follows from Proposition 2 that if (Γ_1^1, Γ_1^4) and (Γ_2^1, Γ_2^4) both permit the construction of a Γ^3 completing a BCS, then the completed square cannot be unique unless Γ_1^1, Γ_2^1 are congruent and Γ_1^4, Γ_2^4 are congruent. It also follows from the remark following the proof of Proposition 2 that the uniqueness of Γ^3 depends only on the congruence classes of Γ^1 and Γ^4. We now prove

PROPOSITION 6. *Given Γ^1, Γ^2, Γ^4, the constructability of Γ^3 depends only on the congruence classes of Γ^1 and Γ^4.*

PROOF. Suppose we consider (8) without Γ_r^3, $r = 1, 2$, and let Γ_1^1, Γ_2^1 be congruent, Γ_1^4, Γ_2^4 congruent. Let $\theta_{1r} = \{\nu_r^i \,[]\, \mu_r^i\}$, $\xi_1 = \{\nu^g \,[]\, \mu^g\}$, $\theta_r^1 = \{\lambda_r^j \,[]\, \kappa_r^j\}$, $\xi^1 = \{\lambda^f \,[]\, \kappa^f\}$, $r = 1, 2$, so that $\mu_*^j(\theta_{1r}) = \xi_1$, $\lambda^{i*}(\theta_r^1) = \xi^1$. By Theorem 2 we may construct Γ_r^3 if and only if $\theta_r^1\theta_{1r} = 0$. Now, by Proposition 5, $\theta_{12} = \theta_{11}^\eta$, $\eta \in \mathrm{Aut}_g E$, and, by the dual of Proposition 5, $\theta_2^1 = {}^\alpha\theta_1^1$, $\alpha \in \mathrm{Aut}_f A$. Thus $\theta_2^1\theta_{12} = {}^\alpha(\theta_1^1\theta_{11})^\eta$ and the proposition is proved.

Now for any morphism $k : M \to N$ in $\mathscr{A}$ we may construct the exact sequence

$$K_k \overset{\kappa^k}{\rightarrowtail} M \overset{k}{\longrightarrow} N \overset{\nu^k}{\twoheadrightarrow} C_k$$

determining $\{k\} \in \mathrm{Ext}^2(C_k, K_k)$. Moreover if $x \in \mathrm{Hom}(X, C_k)$ there is an induced element $\{k\}x$ in $\mathrm{Ext}^2(X, K_k)$ and if $y \in \mathrm{Hom}(K_k, Y)$ there is an induced element $y\{k\} \in \mathrm{Ext}^2(C_k, Y)$.

We extract from (3) and (4) the s.e.s's

$$I_{gf} \overset{\mu^j}{\rightarrowtail} I_g \overset{\nu^j}{\twoheadrightarrow} C_f, \tag{13}$$

$$K_g \overset{\kappa^i}{\rightarrowtail} I_f \overset{\lambda^i}{\twoheadrightarrow} I_{gf}, \tag{14}$$

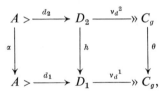

which depend only on f and g, and the induced sequences

$$(15)\quad \mathrm{Hom}\,(C_g, I_g) \xrightarrow{\nu_j{}^*} \mathrm{Hom}\,(C_g, C_f) \xrightarrow{\partial_*} \mathrm{Ext}^1\,(C_g, I_{gf}) \xrightarrow{\mu_j{}^*} \mathrm{Ext}^1\,(C_g, I_g),$$

$$(16)\quad \mathrm{Hom}\,(I_f, K_f) \xrightarrow{\kappa^i{}^*} \mathrm{Hom}\,(K_g, K_f) \xrightarrow{\partial_*} \mathrm{Ext}^1\,(I_{gf}, K_f) \xrightarrow{\lambda^i{}^*} \mathrm{Ext}^1\,(I_f, K_f).$$

We suppose (1) constructed and set $\theta_1 = \{\nu^i \square \mu^i\}$, $\xi_1 = \{\nu^g \square \mu^g\}$, $\theta^1 = \{\lambda^j \square \kappa^j\}$, $\xi^1 = \{\lambda^f \square \kappa^f\}$ as in the proof of Proposition 6, so that

$$\mu^j_*(\theta_1) = \xi_1, \qquad \lambda^{i*}(\theta^1) = \xi^1, \quad \text{and} \quad \theta^1\theta_1 = 0$$

(since (1) exists).

Let $\phi \in \mathrm{Hom}\,(C_g, C_f)$, $\psi \in \mathrm{Hom}\,(K_g, K_f)$. We make some computations of Yoneda products.

PROPOSITION 7. $(\partial^*\psi)(\partial_*\phi) = 0$.

PROOF. We obtain $(\partial^*\psi)(\partial_*\phi)$ from the diagram

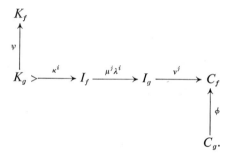

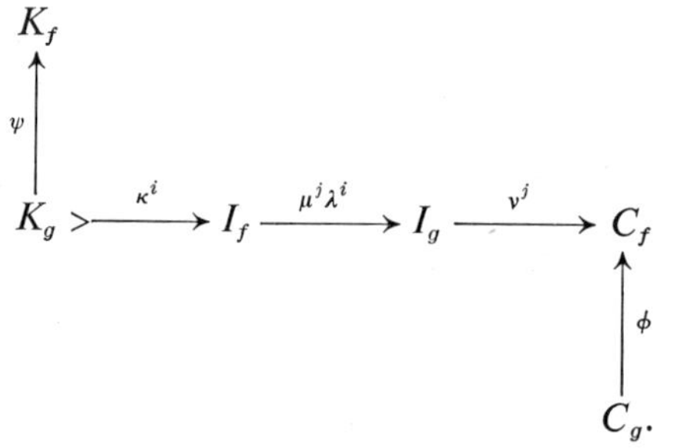

But since (λ^i, μ^j) is completable, $\{\mu^j\lambda^i\} = 0$ by Theorem 2, so $(\partial^*\psi)(\partial_*\phi) = 0$.

PROPOSITION 8. $\theta^1(\partial_*\phi) = \{f\}\phi$, $(\partial^*\psi)\theta_1 = \psi\{g\}$.

PROOF. It is plain from the definition of $\partial_*\phi$ and of the Yoneda product that $\theta^1(\partial_*\phi) = \{j\}\phi$. But plainly $\{j\} = \{f\}$. Similarly, $(\partial^*\psi)\theta_1 = \psi\{i\}$ and $\{i\} = \{g\}$.

From these last propositions we infer

PROPOSITION 9. $(\theta^1 + \partial^*\psi)(\theta_1 + \partial_*\phi) = \{f\}\phi + \psi\{g\}$.

We saw in Proposition 6 how $\mathrm{Aut}_f\, A$ operates on $\mathrm{Ext}^1\,(I_{gf}, K_f)$ by acting on K_f and likewise how $\mathrm{Aut}_g\, E$ operates on $\mathrm{Ext}^1\,(C_g, I_{gf})$. From Propositions 6 and 9 we infer

PROPOSITION 10. $\{f\}\phi + \psi\{g\} = 0$ *if* $\partial_*\phi \in (\mathrm{Aut}_g\, E - 1)\, 1\theta$ *and* $\partial^*\psi \in (\mathrm{Aut}_f\, A - 1)\, 1\theta$.

We may now state the main theorem on uniqueness.

THEOREM 3. *There is a unique congruence class of completions* (1) *of* (f, g) if *and only if the following two conditions are satisfied:*

(a) $$\{f\}\phi + \psi\{g\} = 0 \Rightarrow \partial_*\phi \in (\mathrm{Aut}_g E - 1)\, 1\theta,$$
$$\partial^*\psi \in (\mathrm{Aut}_f A - 1)\, 1\theta$$

(b) (λ^j, μ^i) *has a unique completion.*

PROOF. It follows from (10) that noncongruent completions of (λ^j, μ^i) lead to noncongruent completions of (f, g). Thus if the completion of (f, g) is unique (b) must hold. But also, by Propositions 2 and 5, it must be impossible to find ϕ, ψ such that $\theta_1 + \partial_*\phi \notin (\mathrm{Aut}_g E)(\theta_1)$ or $\theta^1 + \partial^*\psi \notin (\mathrm{Aut}_f A)(\theta^1)$ and $(\theta^1 + \partial^*\psi)(\theta_1 + \partial_*\phi) = 0$. This is guaranteed by (a).

Conversely if (a) holds we are forced into the choices made, up to congruence, or Γ^1 and Γ^4, in order that Γ^3 can be constructed. In the light of Proposition 3, (b) then guarantees the uniqueness of the congruence class of (1).

Naturally, the validity of (b) in any particular case may be tested by applying Proposition 5 or its dual.

REMARK. Of course, Proposition 5, or its dual, is much easier to apply than Theorem 3 if g is monic or f epic. However, it should be noted that the congruence class of (1) may be unique without (λ^f, λ^i) or (μ^j, μ^g) admitting *unique* completions, so that Proposition 5 cannot be applied directly to the general case. However, and rather paradoxically, it *can* be applied when f is monic or g epic! For since Γ^2 is always unique, the uniqueness of (1) hinges only on the uniqueness of Γ^4 if f is monic (and we may then apply Proposition 5 to μ^g and $j: I_{gf} \to I_g$), or on the uniqueness of Γ^1 if g is epic (and we may then apply the dual of Proposition 5 to λ^f and $i: I_f \to I_{gf}$).

BIBLIOGRAPHY

1. D. A. Buchsbaum, *A note on homology in categories*, Ann. of Math. **69** (1959), 66–74.
2. B. Eckmann and P. J. Hilton, *Filtrations, associated graded objects and completions*, Math. Z. **98** (1967), 319–354.
3. P. J. Hilton, *Correspondences and exact squares*. Proc. Conf. Categorical Algebra, La Jolla 1965, 254–271.
4. ———, *Filtrations*, Cahiers de Topologie et Géometrie Différentielle, Paris, 1967, 243–253.
5. S. MacLane, *Homology*, Berlin-Göttingen-Heidelberg, Springer, 1963.
6. B. Mitchell, *Theory of categories*, Academic Press, New York, 1965.
7. I. S. Pressman, *Functors whose domain is a category of morphisms*, Acta Math. **118** (1967), 223–249.

E. T. H. ZÜRICH, AND
COURANT INSTITUTE, N.Y.U., NEW YORK
E. T. H. ZÜRICH AND
OHIO STATE UNIVERSITY

A categorical setting for the Baer extension theory

Murray Gerstenhaber[1]

The purpose of this paper is to describe a class of categories **C** within which the Baer extension theory is meaningful. That is, within **C** we may speak of a "singular extension of an object A by an A-module M", and the equivalence classes of these will form an additive group, denoted $\mathscr{E}^2_{\mathbf{C}}(A, M)$. The present theory is actually adequate to discuss the next higher cohomology group, $\mathscr{E}^3_{\mathbf{C}}(A, M)$, which contains the obstructions to extension problems, but this will be deferred to a later paper.

The major problem in the Baer Theory is to choose a set of axioms weak enough to hold simultaneously for all the known cases where singular extensions can be made into a group. In particular, the axioms must admit the categories of sheaves of groups, sheaves of rings (associative, Lie, and commutative associative) and topological groups as models. In doing this we have been aided by unpublished work of John Moore and have elected to call the categories defined by the present axioms "Moore categories". The axioms are probably not the most efficient possible, but are at least workable.

Since most of the principles are very well known, many of the minor proofs are omitted.

1. **The axioms.** A *Moore category* **C** is one which is pointed, has kernels and cokernels, and which satisfies four additional sets of axiom grouped as follows:

1. Two self-dual axioms: The 3×3 lemma, and a criterion for a morphism to be normal.
2. Three axioms on fibered products (the first of which assures their existence).
3. An axiom on split extensions.
4. A foundational axiom insuring that certain classes are sets.

A *zero-sequence* $A \to B \to C$ is one in which the composite morphism is zero. The sequence is an *extension* (of C by A) if $A \to B$ is the kernel of $B \to C$ and

[1] The author gratefully acknowledges the support of the NSF through grant GP-8648 to the University of Pennsylvania.

$B \to C$ is the cokernel of $A \to B$. Following the terminology of MacLane, a monomorphism of C will be called briefly a "monic" and an epimorphism an "epic". A monic $A \to B$ which is the kernel of some morphism $B \to C$ is *normal.* An epic $B \to C$ which is the cokernel of some morphism $A \to B$ is *conormal*, but for simplicity, we call this, too, "*normal*".

It is a simple consequence of the fact that $\mathbf{C}$ is pointed and has kernels and cokernels, that every morphism $A \to B$ has an essentially unique *analysis*,

$$K \to A \to I \to I' \to B \to J \tag{1.1}$$

in which the composite $A \to I \to I' \to B$ is the original morphism, $K \to A$ is the kernel of $A \to B$ and $B \to J$ its cokernel, and $K \to A \to I$, $I \to B \to J$ are extensions.

Axiom 1. (The 3×3 lemma). Suppose that we have a commutative diagram:

$$\begin{array}{ccccc}
A' & \longrightarrow & A & \longrightarrow & A'' \\
\downarrow & & \downarrow & & \downarrow \\
B' & \longrightarrow & B & \longrightarrow & B'' \\
\downarrow & & \downarrow & & \vdots \\
C' & \longrightarrow & C & \longrightarrow & C''
\end{array}$$

If all three rows and any two columns are extensions, and if the third column is a zero-sequence, then it, too, is an extension.

The 3×3 lemma has in particular the following consequences, whose proofs are easy exercises in diagrams.

(i) Let the solid part of the following diagram be given and commutative,

$$\begin{array}{ccccc}
A' & \longrightarrow & A & \longrightarrow & A'' \\
\downarrow & & \downarrow & & \downarrow \\
B' & \longrightarrow & B & \longrightarrow & B'' \\
\downarrow & & \downarrow & & \downarrow \\
C' & \longrightarrow & C & \dashrightarrow & C'',
\end{array}$$

and suppose that the first two rows and columns are extensions and that $A'' \to B''$ and $C' \to C$ are normal epic and normal monic, respectively. Then the dotted part can be filled in (in an essentially unique way) to make the diagram commutative with all rows and columns extensions. Since the 3×3 lemma is self-dual one has an analogous result when the upper left corner of the diagram is missing.

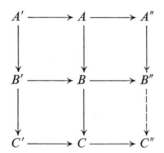

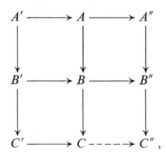

(ii) Suppose that we have a commutative diagram

$$\begin{array}{ccccc} A' & \xrightarrow{a'} & A & \xrightarrow{a} & A'' \\ \downarrow{\scriptstyle f} & & \downarrow{\scriptstyle g} & & \downarrow{\scriptstyle h} \\ B' & \xrightarrow{b'} & B & \xrightarrow{b} & B'' \end{array} \tag{1.2}$$

in which the rows are extensions and f is an isomorphism. If g is a normal epic and $k:K \to A$ a kernel of g, then h is a normal epic and ak is a kernel for h. If h is a normal monic and $j:B'' \to J$ a cokernel for h, then g is a normal monic and jb is a cokernel for g.

There is an obvious dual statement in which the diagram is the same but h is assumed to be an isomorphism. Concerning diagram (1.2) we now assume the following:

AXIOM 1.2. In the diagram (1.2) with exact rows, if h is an isomorphism and f a normal monic, then $b'f$ is normal monic. Dually, if f is an isomorphism and h a normal epic then ha is normal epic.

The axioms on fibered products will imply that a composite of normal epics is normal, so the second part of Axiom 1.2 need not be assumed independently. In general, of course, a composite of normal monics is not normal.

Axiom 2 on fibered products (pull backs) has three parts.

AXIOM 2.1. There exist fibered products.

The fibered product of two morphisms $u:U \to A$ and $v:V \to A$ will be denoted $U \times_A V$, the associated map $U \times_A V \to U$ being denoted p_U, and similarly for p_V. When $U = V$ we write p_1, p_2, respectively. If we have morphisms $f:B \to U$ and $g:B \to V$ such that if $uf = vg$, then the associated morphism $B \to U \times_A V$ is denoted $f \vee g$. When A is zero, the fibered product $U \times_A V$ is just the direct product $U \times V$. It is a familiar fact that the existence of fibered products implies that every finite diagram has an inverse limit.

We do *not* assume that the fibered sum of morphisms $A \to U$ and $A \to V$ exists, but if it does, then the object in it is denoted $U *_A V$.

One can show without difficulty that if we have a commutative diagram

$$\begin{array}{ccccc} A' & \xrightarrow{a'} & A & \xrightarrow{a} & A'' \\ \downarrow{\scriptstyle f} & & \downarrow{\scriptstyle g} & & \downarrow{\scriptstyle h} \\ B' & \xrightarrow{b'} & B & \xrightarrow{b} & B'' \end{array}$$

in which the rows are extensions, h is an isomorphism, and f is monic, then the natural morphism $f \vee a' : A' \to B' \times_B A$ is an isomorphism. Dually, if f is an isomorphism and h a normal epic, then the right square of the diagram is a push out, so B'' will serve for $B *_A A''$.

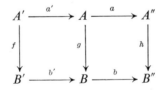

AXIOM 2.2. Pull-backs of normal morphisms are normal. That is, in the diagram

$$\begin{array}{ccc} B \times_A C & \longrightarrow & C \\ \downarrow & & \downarrow \\ B & \longrightarrow & A \end{array}$$

if $B \to A$ is normal monic or normal epic, then $B \times_A C \to C$ is normal monic or normal epic, respectively.

Axiom 2.2 implies, in particular, that we can pull back extensions, i.e. that if in the following commutative diagram

$$\begin{array}{ccccc} K & \xrightarrow{i \vee 0} & B \times_A C & \xrightarrow{p_C} & C \\ \| & & \downarrow & & \downarrow \\ K & \xrightarrow{i} & B & \longrightarrow & A \end{array}$$

the bottom row is an extension, then so is the top. It is clear that $i \vee 0 = \ker p_C$.

Axiom 2.2 also implies that if we have an extension $B' \to B \to B''$ and a morphism $A \to B$ then there is an essentially unique commutative diagram

$$(1.3)\qquad \begin{array}{ccccc} B' \times_B A & \xrightarrow{p_A} & A & \xrightarrow{a'} & A'' \\ \downarrow{\scriptstyle p_{B'}} & & \downarrow{\scriptstyle v} & & \downarrow{\scriptstyle w} \\ B' & \xrightarrow{b} & B & \xrightarrow{b'} & B'' \end{array}$$

whose top row is an extension. It is easy to verify that if $A \to B$ is a normal epic, and $k: K \to B' \times_B A$ a kernel of $p_{B'}$ then $p_A k$ is a kernel of v, and therefore w is an isomorphism. From this one concludes that a composite of normal epics is normal. For if we have normal epics $v: A \to B$ and $b': B \to B''$, let $b: B' \to B$ be a kernel of b', arrange the morphisms as in the diagram (1.3), and take a' to be a cokernel of p_A. Then w exists and by the preceding remark is an isomorphism, so $wa': A \to B''$ is also a cokernel of p_A. But $wa' = b'v$.

Concerning diagram (1.3), we must assume

AXIOM 2.3. Intersection of normal subobjects is normal. If in (1.3), $v: A \to B$ is normal monic, then so are $bp_{B'}$, and w.

A sequence $A \xrightarrow{a} B \xrightarrow{b} C$ is *split* if $a = \ker b$ and if there is a *splitting morphism* $s: C \to B$ such that $sb = 1_C$. For brevity, we denote $\mathrm{Hom}_C(X, Y)$ simply by (X, Y). If $A \xrightarrow{a} B \underset{s}{\overset{b}{\rightleftarrows}} C$ is a split sequence and U arbitrary, then there is a natural set map

$$(B, U) \to (A, U) \times (C, U)$$

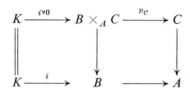

defined by $u \mapsto (ua, us)$.

AXIOM 3. If $A \to B \to C$ is split then the natural set map

$$(B, U) \to (A, U) \times (C, U) \text{ is 1-1.}$$

It follows that the split sequence $A \xrightarrow{a} B \underset{s}{\overset{b}{\rightleftarrows}} C$ is an extension and that $b =$ coker a. For if we have $B \xrightarrow{u} U$ with $ua = 0$ then the images of u and usb in $(A, U) \times (C, U)$ are identical, namely $(0, us)$. Therefore $u = usb$.

The morphism $1_A \vee 0 : A \to A \times C$ will be denoted simply i_A; similarly we set $0 \vee 1_C = i_C$. The sequence $A \xrightarrow{i_A} A \times C \xrightarrow{p_C} C$ trivially has kernel i_A and is split by i_C, hence is an extension. (Note that if A and C have a coproduct $A * C$ then Axiom 3 implies that the natural morphism $A * C \to A \times C$ is an epic.)

A *morphism* of extensions (of B by A) is a commutative diagram:

$$\begin{array}{ccccc} A & \longrightarrow & X & \longrightarrow & B \\ \| & & \downarrow & & \| \\ A & \longrightarrow & Y & \longrightarrow & B \end{array} \tag{1.4}$$

Applying the 3×3 lemma to the diagram obtained by prefixing a row of zeros to (1.4), one concludes that $0 \to X \to Y$ is an extension and hence, since $0 \to X$ already has one cokernel, namely 1_X, that $X \to Y$ is an isomorphism. The relation of having a morphism between two extensions is therefore an equivalence. As part of the foundational axioms, we assume:

AXIOM 4.1. There is a representative *set* under this equivalence relation for the extensions of B by A.

Loosely, the equivalence classes of extensions of B by A form a set. It is usual to define two monics $X \to A$ and $Y \to A$ to be equivalent if there is an isomorphism $X \to Y$ making the diagram

$$\begin{array}{ccc} X & \longrightarrow & Y \\ & \searrow \quad \swarrow & \\ & A & \end{array}$$

commute. We require:

AXIOM 4.2. The equivalence classes of monics $X \to A$ form a set, and dually, the equivalence classes of epics $A \to X$ form a set. (These are the "subobjects" and "quotient objects" of A, respectively.)

Here are some elementary consequences of the axioms: Suppose we have morphisms $u : U \to A$ and $v : V \to A$. Then we also have $u \times v : U \times V \to A \times A$ and $\Delta : A \to A \times A$ where Δ denotes the diagonal morphism $i_1 \vee i_2$. It is then trivial from the definition of the fibered product that there is a canonical isomorphism $U \times_A V \to (U \times V) \times_{A \times A} A$. It is even more trivial that if we have a commuting diagram

$$\begin{array}{ccc} A & \longrightarrow & V \\ \downarrow & & \downarrow \\ U & \longrightarrow & B \end{array}$$

where $A \to U$ is monic, then $A \to U \times_B V$ is also monic.

LEMMA 1.1. *If $A' \to A \to A''$ and $B' \to B \to B''$ are extensions, then so is $A' \times B' \to A \times B \to A'' \times B''$.*

PROOF. Apply the 3×3 lemma to the diagram whose first and last rows are the two given extensions and whose middle row is $A' \times B' \to A \times B \to A'' \times B''$. ∎

The following is fundamental.

LEMMA 1.2. *Suppose that we have a morphism of the form $K \times L \to A$ such that the composite morphisms*

$$K \xrightarrow{1_K \vee 0} K \times L \to A$$

and

$$L \xrightarrow{0 \vee 1_L} K \times L \to A$$

are both normal monics. Suppose further that the fibered product of $K \to A$ and $L \to A$ is zero. Then there is a commutative diagram

$$\begin{array}{ccccc} 0 & \longrightarrow & K & = & K \\ \downarrow & & \downarrow{\scriptstyle k} & & \downarrow \\ L & \xrightarrow{l} & A & \longrightarrow & B \\ \| & & \downarrow & & \downarrow \\ L & \longrightarrow & C & \longrightarrow & D \end{array} \tag{1.5}$$

in which the rows and columns are all extensions; $K \times L \to A$ is then the kernel of the composite morphism $A \to B \to D$, and in particular, $K \times L \to A$ is a normal monic.

PROOF. Let $A \to B$ be a cokernel of $L \to A$ and $A \to C$ be a cokernel of $K \to A$. Then the composite morphisms $K \to A \to B$ and $L \to A \to C$ are normal by Axiom 2.3. The first two rows and columns of diagram (1.5) can then be filled in to be exact, after which the lower right corner can be added, as remarked after Axiom 1.1.

Now pull back the extension $L \to A \to B$ by the morphism $K \to B$, getting a commutative diagram

$$\begin{array}{ccccc} L & \longrightarrow & A \times_B K & \longrightarrow & K \\ \| & & \downarrow & & \downarrow \\ L & \longrightarrow & A & \longrightarrow & B \end{array}$$

in which the rows are extensions and the vertical morphisms are normal monics. The upper row is split by the morphism $k \vee 1_K : K \to A \times_B K$. Letting $A \to D$ be the composite morphism $A \to B \to D$, we have a commutative diagram

$$\begin{array}{ccccc} L & \longrightarrow & A \times_B K & \longrightarrow & K \\ \| & & \downarrow & & \downarrow \\ L & \longrightarrow & A & \longrightarrow & B \\ \downarrow & & \downarrow & & \downarrow \\ 0 & \longrightarrow & D & = & D \end{array}$$

in which all rows and the two outer columns are extensions and the center column is a zero sequence. It, too, is therefore an extension. It is sufficient, therefore, to show that there is an isomorphism $j : K \times L \to A \times_B K$ such that the composite $K \times L \xrightarrow{j} A \times_B K \xrightarrow{p_A} A$ is our original morphism $i : K \times L \to A$. Define $j : K \times L \to A \times_B K$ to be $(k \vee 1_K) \times (l \vee 0)$. That $p_A\, j$ is identical with i follows immediately from Axiom 3. To see that j is an isomorphism, observe that we have a morphism of extensions:

$$\begin{array}{ccccc} L & \longrightarrow & K \times L & \longrightarrow & K \\ \| & & \downarrow{\scriptstyle j} & & \| \\ L & \longrightarrow & A \times_B K & \longrightarrow & K \end{array}$$

As remarked earlier, the 3×3 lemma then implies that j is an isomorphism. ∎

The class of Moore categories includes the categories of sheaves of (1) associative algebras (2) Lie algebras (3) commutative associative algebras, and (4) groups over a fixed topological space X, which may, of course, be a point. It includes also the category of topological groups, where a morphism $f : G \to H$ is a continuous group morphism such that the relative topology of fG in H is the same as the topology induced from G.

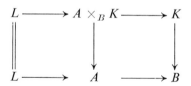

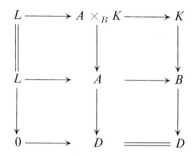

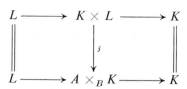

If **C** is a Moore category and **D** a subcategory which contains kernels, cokernels, and fibered products (in **C**) of **D**-morphisms $A \to B$, $C \to B$, then **D** is trivially again a Moore category. This yields many more examples, e.g. sheaves of finite groups, solvable groups, and nilpotent associative algebras, compact topological groups, and many others.

2. **Abelian objects.** An *abelian object* of **C** is a group in the category, i.e. an object M such that (T, M) is an abelian group for all objects T of **C** and $T \rightsquigarrow (T, M)$ is a contravariant functor from **C** to the category of abelian groups. We shall see that the full subcategory of abelian objects is an additive category with certain additional properties but is not necessarily abelian. Trivially the zero element of (T, M) is the zero morphism $T \to M$.

LEMMA 2.1. *If M is abelian and U, V arbitrary, then the canonical set map $(U \times V, M) \to (U, M) \times (V, M)$ is a bijection.*

PROOF. It is 1-1 by Axiom 3. If we have $u: U \to M$ and $v: V \to M$, then (u, v) is the image of $u \cdot p_U + v \cdot p_V$. ∎

Observe that there exist natural morphisms $-1: M \to M$ (the negative of $1_M \in (M, M)$), add: $M \times M \to M$ defined by add $= p_1 + p_2$ (the sum of the projections on the two factors), and $t: M \times M \to M \times M$, the interchange of factors. One has add $\cdot\, t =$ add. All the axioms for an ordinary abelian group can be expressed in terms of commutative diagrams involving the underlying set of the group and its products with itself two and three times. Precisely the same axioms hold here, except that $M \times M$ is now the product in the category **C**. Conversely, an object M of **C** endowed with morphisms 0, $-1 \in (M, M)$ and add: $M \times M \to M$ which formally satisfy the group axioms and the equation add $\cdot\, t =$ add is easily seen to be an abelian object. (In particular, if we have $f, g \in (U, M)$, then $f + g$ is the composite $U \xrightarrow{f \vee g} M \times M \xrightarrow{\text{add}} M$.) It is further easy to see that if M, N are abelian, then so is $M \times N$, and $(U, M \times N) = (U, M) \times (U, N)$.

When necessary to distinguish, we write add_M for the "addition map" $M \times M \to M$.

LEMMA 2.2. *If $g: M \to N$ is a **C**-morphism with M, N abelian, then*

$$\text{add}_N \cdot (g \times g) = g \cdot \text{add}_M : M \times M \to M. \tag{2.1}$$

PROOF. By Axiom 3 it is sufficient to show that if both sides of (2.1) are preceded by $i_1: M \to M \times M$ then the composites are the same, and similarly for $i_2: M \to M \times M$. All these composites are just g. ∎

The lemma says that any **C**-morphism $g: M \to N$ preserves addition, hence is a "categorical group morphism".

COROLLARY. *If M and N are abelian and U arbitrary, then the canonical composition map*

$$(M, N) \times (U, M) \to (U, N)$$

given by $(g, f) \mapsto gf$ is biadditive.

PROOF. Additivity as a function of g is part of the "functoriality" in the definition of a group object, additivity as a function of f follows from the preceding. ∎

LEMMA 2.3. *If M is abelian then*

$$M \xrightarrow{1 \vee -1} M \times M \xrightarrow{\text{add}} M$$

is an extension.

PROOF. Define $m: M \times M \to M \times M$ by $m = p_1 \times \text{add}$; this is an automorphism, for $m^{-1} = p_1 \times (1 \vee -1)$. The following diagram commutes:

$$\begin{array}{ccccc} M & \xrightarrow{1\vee-1} & M \times M & \xrightarrow{\text{add}} & M \\ \Vert & & \downarrow{\scriptstyle m} & & \Vert \\ M & \xrightarrow{i_1} & M \times M & \xrightarrow{p_2} & M \end{array}$$

The bottom row is an extension, hence so is the top. ∎

An object M is a *monoid* in the category $\mathbf{C}$ if (U, M) is an additive monoid (i.e. has a commutative, associative addition with zero element) which is contravariant as a functor of U. It is trivial to verify that all the results of this section which are meaningful for monoids are in fact true for them.

LEMMA 2.4. *Let $U \xrightarrow{u} M \xrightarrow{v} V$ be an extension with M a monoid. Then U and V are monoids.*

PROOF. The composite morphism

$$U \times U \xrightarrow{u \times u} M \times M \xrightarrow{\text{add}} M \xrightarrow{V} V$$

is zero, using Axiom 3. Therefore, we can fill in the dotted arrows uniquely to make the following diagram commute:

$$\begin{array}{ccccc} U \times U & \xrightarrow{u \times u} & M \times M & \xrightarrow{v \times v} & V \times V \\ {\scriptstyle \text{add}_U}\downarrow & & \downarrow{\scriptstyle \text{add}_M} & & \downarrow{\scriptstyle \text{add}_V} \\ U & \xrightarrow{u} & M & \xrightarrow{v} & V \end{array}$$

The commutativity and associativity of add_U and add_V are easy exercises. ∎

THEOREM 2.1. *Suppose that we have $K \xrightarrow{k} M \xrightarrow{m} N \xrightarrow{j} J$ where M and N are abelian and $k = \ker m$, $j = \operatorname{coker} m$. Then K and J are abelian.*

PROOF. From the preceding it remains only to show that there exists a morphism $-1_K: K \to K$ such that $1_K + (-1_K) = 0$, and similarly for J. Since -1_N is an automorphism of N, k also serves as a kernel for $-1_N \cdot m = m \cdot (-1_M)$. In the following diagram, k is therefore the kernel of both rows, so the dotted arrow can be filled in uniquely:

$$K \xrightarrow{k} M \xrightarrow{m \cdot -1_M} N$$
$$\downarrow -1_K \quad \downarrow -1_M \quad \downarrow$$
$$K \xrightarrow{k} M \xrightarrow{m} N$$

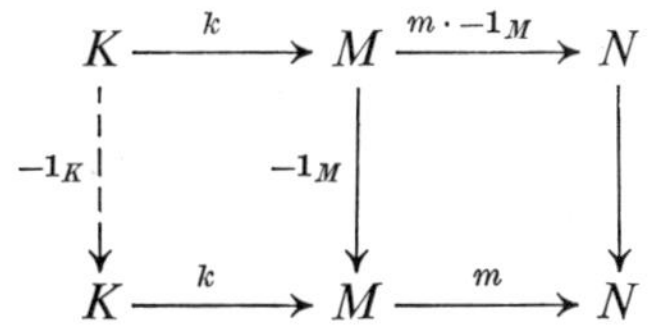

The existence of -1_J follows similarly. Finally,

$$k(1_K + (-1_K)) = k + k \cdot (-1_K) = k + (-1_M) \cdot k = 0;$$

as k is a monic, it follows that $1_K + (-1_K) = 0$, and similarly for -1_J. ∎

It follows that the class **Ab C** of abelian objects of **C**, together with all **C**-morphisms between them, is an additive category which in addition satisfies Abel-1 of MacLane [1, p. 254]. Further, in the analysis (cf. (1.1)) of a morphism $M \to N$ with M, N abelian, namely

(2.2) $$K \to M \to I \to I' \to N \to J$$

all objects are abelian.

THEOREM 2.2. *In the analysis* (2.2), $I \to I'$ *is a monic.*

PROOF. It is sufficient, since **Ab C** is additive, to prove that if a composite morphism $T \to I \to I'$ is 0 then $T \to I$ is 0. In the diagram

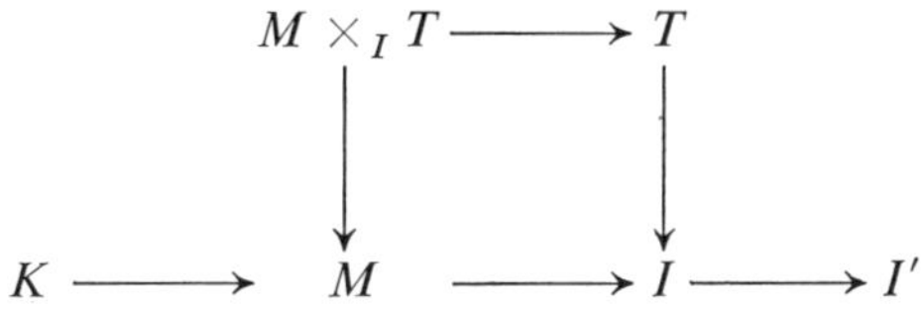

the morphism $M \to I$ is a normal epic, hence, by Axiom 2.2, so is $M \times_I T \to T$. It is therefore sufficient to show that the composite $M \times_I T \to T \to I = M \times_I T \to M \to I$ is zero. But composing the former with $I \to I'$ gives 0 and $K \to M$ is the kernel of both $M \to N$ and $M \to I$, hence of $M \to I \to I'$, so the morphism $M \times_I T \to M$ can be factored through K. Therefore $M \times_I T \to M \to I$ is indeed zero. ∎

It follows that every morphism $M \to N$ of abelian objects has a factorization (normal epic) · (monic), and therefore **Ab C** satisfies also Abel-2, of MacLane. Of the axioms for an abelian category, only Abel-2 (all monics and epics are normal) fails. As an example, let **C** be the category whose objects are topological groups in which, as in §1, $f: G \to H$ is a morphism if it is continuous, a group morphism, and if the canonical map $G/\ker f \to fG$ is an isomorphism of topological groups. This category satisfies the axioms and **Ab C** is the full subcategory of abelian groups. The inclusion of an abelian group into its completion is both monic and epic, but not normal.

3. **Structures and Modules.** An *A-structure* on an object C is a split extension $C \to B \to A$, together with a definite choice of splitting morphism $s: A \to C$. We identify the A structures defined by this split sequence and another $C \to B' \rightleftarrows A$ if there is a commutative diagram:

$$\begin{array}{ccccc} C & \longrightarrow & B & \rightleftarrows & A \\ \| & & \downarrow & & \| \\ C & \longrightarrow & B' & \rightleftarrows & A \end{array}$$

By abuse of language we may speak of "the A-structure C" when all else is understood. To reduce the number of symbols, a split sequence may be denoted by some notation as $C \to CA \rightleftarrows A$ to suggest that CA is a semidirect product of C and A. If C and D are A-structures, then an "A-structure morphism $c: C \to D$" is a commutative diagram:

$$\begin{array}{ccccc} C & \longrightarrow & CA & \longrightarrow & A \\ {\scriptstyle c}\downarrow & & \downarrow & & \| \\ D & \longrightarrow & DA & \longrightarrow & A \end{array}$$

With these definitions, A-structures form a pointed category whose zero element is $0 \to A = A$.

THEOREM 3.1. *Kernels and cokernels exist in the category of A-structures. Specifically, if $C \to D$ is a morphism of A-structures, then there is a commutative diagram*

$$\begin{array}{ccccc} K & \longrightarrow & KA & \rightleftarrows & A \\ \downarrow & & \downarrow & & \| \\ C & \longrightarrow & CA & \rightleftarrows & A \\ \downarrow & & \downarrow & & \| \\ D & \longrightarrow & DA & \rightleftarrows & A \\ \downarrow & & \downarrow & & \| \\ J & \longrightarrow & JA & \rightleftarrows & A \end{array}$$

in which all rows are split extensions and $K \to C$, $D \to J$ are kernel and cokernel respectively, of $C \to D$.

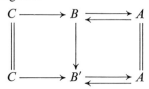

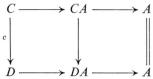

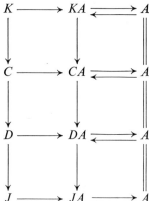

PROOF. There are given, in particular, morphisms $CA \to DA$ and $A \to DA$, the latter being the splitting morphism. Set $KA = CA \times_{DA} A$. In $K \to CA \times_{DA} A \to A$ the fact that the first morphism is the kernel of the second follows by diagram chasing. The sequence is split by $s \vee 1_A$, where s is the splitting morphism $A \to CA$, hence is an extension. It is easy to verify that this is the kernel of the given A-structure morphism. Now factor the morphism $CA \to DA$ into $CA \to I \to I' \to DA$ where $CA \to I$ is a normal epic and $I' \to DA$ a normal monic. Then we have a diagram:

$$\begin{array}{ccccc}
C & \longrightarrow & CA & \rightleftarrows & A \\
\downarrow & & \downarrow & & \| \\
D \times_{DA} I & \longrightarrow & I & \longrightarrow & A \\
\downarrow & & \downarrow & & \| \\
D \times_{DA} I' & \longrightarrow & I' & \longrightarrow & A \\
\downarrow & & \downarrow & & \| \\
D & \longrightarrow & DA & \rightleftarrows & A
\end{array}$$

It is easy to check that in each of the middle rows the first morphism is a kernel of the second. The sequences split, the splitting morphism of the second now being, for example, the composite morphism $A \to CA \to I$. Since $CA \to I$ and $I' \to DA$ are normal, it follows that the left column of the diagram is actually the canonical factorization of $C \to D$.

Finally, by virtue of the preceding, to exhibit cokernels it is sufficient to show that the bottom A-structure morphism in the diagram has a cokernel. But by Axiom 2.3, the composite morphism $D \times_{DA} I' \to D \to DA$ has a cokernel; denote it by $DA \to JA$. Then $J \to JA \to A$ is the desired sequence. ∎

Note that if we have A-structures C and D and a morphism $C \to D$ which can be extended to an A-structure morphism (i.e. to a commutative diagram) then by Axiom 3 that extension is unique. Therefore the set $(C, D)_A$ of A-structure morphisms from C to D is a subset of (C, D). We may therefore, by abuse of language, speak of an "A-structure morphism $C \to D$" and the analysis of this morphism may also, without ambiguity, be viewed as belonging to A. If we have A-structure morphisms $C \to B$ and $D \to B$ then it is easy to check that their fibered product is $C \times_B D \to CA \times_{BA} DA \to A$, which is split by $s \vee s'$ where $s: A \to CA$ and $s': A \to DA$ are the respective splitting maps. Verification that the category of A-structures and morphisms satisfies the remainder of the Moore category axioms is tedious but essentially trivial.

An *A-module* is an A-structure $M \to MA \rightleftarrows A$ in which M is abelian. This is the same thing as an abelian object in the Moore category of A-structures, so all the results of §2 apply. An A-structure morphism $M \to N$ between abelian objects will be called briefly an A-morphism.

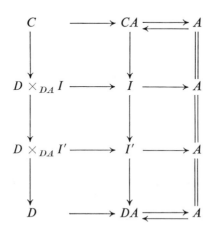

4. **Singular extensions; the Baer theory.** An extension $M \xrightarrow{i} B \to A$ is *singular* if M is abelian. If we form the extension $M \times M \to B \times B \to A \times A$ and pull back by the diagonal morphism $\Delta : A \to A \times A$, then since $(B \times B) \times_A A = B \times_A B$, we obtain an extension of the form $M \times M \to B \times_A B \to A$. Together with the original extension this gives a commutative diagram

$$
\begin{array}{ccccc}
M & \longrightarrow & B & \longrightarrow & A \\
\Big\downarrow{\scriptstyle \Delta_M} & & \Big\downarrow{\scriptstyle \Delta_B} & & \Big\| \\
M \times M & \longrightarrow & B \times_A B & \longrightarrow & A
\end{array}
$$

where $\Delta_B = 1_B \vee 1_B$. Now Δ_M is a normal monic so it follows from Axiom 1.2 that the composite monic $M \to M \times M \to B \times_A B$ is normal; we denote this also by Δ. We can therefore form a commutative diagram

$$
\begin{array}{ccccc}
M & = & M & \longrightarrow & 0 \\
\Big\downarrow{\scriptstyle 1 \vee -1} & & \Big\downarrow{\scriptstyle \Delta} & & \Big\downarrow \\
M \times M & \xrightarrow{ix-i} & B \times_A B & \longrightarrow & A \\
\Big\downarrow{\scriptstyle \text{add}} & & \Big\downarrow{\scriptstyle u} & & \Big\| \\
M & \longrightarrow & U & \longrightarrow & A
\end{array}
$$

where $-i = i \cdot (-1_M)$, U is the cokernel of Δ, and where all three columns and the first two rows are extensions. Since the last row is a zero sequence, it follows that it, too, is an extension. This extension splits. For the composite morphism

$$M \longrightarrow B \xrightarrow{\Delta_B} B \times_A B \xrightarrow{U} U$$

is zero, which permits the composite morphism $B \to U$ in the foregoing to be factored through A. Writing $U = MA$, to the singular extension $M \to B \to A$ we have thus associated, as usual, an A-module structure on M. If an A-module structure on M is understood in advance, then we say that $M \to B \to A$ is an *extension of A by the A-module M* if the module structure just obtained is the same as that given in advance.

The Baer Theory asserts that if an A-module structure $M \to MA \rightleftarrows A$ is given, then the equivalence classes of extensions of A by the module M form a group, the zero element of which is the class of the given split extension $M \to MA \rightleftarrows A$. Addition of classes having representatives $M \xrightarrow{i} B \longrightarrow A$ and $M \xrightarrow{j} C \longrightarrow A$, respectively, is defined by filling in the dotted parts of the diagram

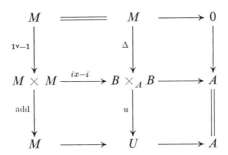

$$\begin{array}{ccccc}
M & = & M & \longrightarrow & 0 \\
\downarrow{\scriptstyle 1\vee -1} & & \downarrow{\scriptstyle i\vee -j} & & \downarrow \\
M \times M & \longrightarrow & B \times_A C & \longrightarrow & A \\
\downarrow{\scriptstyle \mathrm{add}} & & \downarrow{\scriptstyle u} & & \| \\
M & \dashrightarrow & U & \dashrightarrow & A
\end{array}$$

and taking the class of the bottom row, assuming, of course, that it is an extension. To be able to do this, it is necessary and sufficient to know that $i \vee -j$ is a normal monic; one then takes u to be a cokernel of $i \vee -j$. The negative of the class of $M \xrightarrow{i} B \longrightarrow A$ will then be the class of $M \xrightarrow{-i} B \longrightarrow A$. There are a number of tedious but essentially trivial things which must be verified, the proofs of all of which depend on the same basic principles. Most important of these is the fact that if we have three extensions $E_B{:}M \xrightarrow{i} B \longrightarrow C$, $E_C{:}M \xrightarrow{j} C \longrightarrow A$ and $E_D{:}M \xrightarrow{k} D \longrightarrow A$ and if $E_B + E_C$, $E_C + E_D$ are defined, then

(4.1) $$(E_B + E_C) + E_D = E_B + (E_C + E_D)$$

in the sense that both sides are defined and are equal, at least up to equivalence class. We confine ourselves to a sketch of the proof of this one fact.

Denote $(B \times C \times D) \times_A (A \times A \times A)$ by $B \times_A C \times_A D$; this is the inverse limit of the diagram consisting of the three morphisms $B \to A$, $C \to A$ and $D \to A$. Then we have an extension

$$M \times M \times M \to B \times_A C \times_A D \to A.$$

The kernel, f, of the addition map

$$(p_1 + p_2 + p_3){:}M \times M \times M \to M$$

is isomorphic to $M \times M$ in such a way that the composite

$$M \xrightarrow{i_1} M \times M \xrightarrow{f} M \times M \times M \to B \times_A C \times_A D$$

is $i \vee -j \vee 0$, and such that replacing i_1 by i_2, one has $0 \vee j \vee -k$. Now, denote an extension in the class of $E_B + E_C$ by $M \to (BC) \to A$, and similarly for (CD). By the hypothesis that $E_B + E_C$ exists we know that $i \vee -j$ and $j \vee -k$ are normal. Therefore, by Lemma 1.2, we have a commutative diagram of extensions:

$$\begin{array}{ccccc}
0 & \longrightarrow & M & = & M \\
\downarrow & & \downarrow{\scriptstyle 0\vee j\vee -k} & & \downarrow \\
M & \xrightarrow{i\vee -j\vee 0} & B \times_A C \times_A D & \longrightarrow & (BC) \times_A D \\
\| & & \downarrow & & \downarrow \\
M & \longrightarrow & B \times_A (CD) & \longrightarrow & (BCD)
\end{array}$$

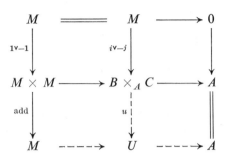

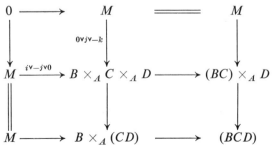

Here (BCD) is for the moment an object determined by the diagram but it is easy to see that (4.1) is equivalent to the commutativity of the lower right corner of the diagram and that (BCD) appears in $E_B + E_C + E_D: M \to (BCD) \to A$.

At this point the remaining verifications necessary to show that we actually have a group of extension classes follow standard lines.

It is trivial to observe that we can define the derivations of A into an A-module M. These are the automorphisms of the split extension $M \to MA \to A$, i.e. the morphisms $MA \to MA$ such that the following diagram commutes:

$$\begin{array}{ccccc} M & \longrightarrow & MA & \longrightarrow & A \\ \| & & \downarrow & & \| \\ M & \longrightarrow & MA & \longrightarrow & A \end{array}$$

The description of the next higher cohomology group after the Baer group is deferred to another paper.

Reference

1. Saunders MacLane, *Homology*, Academic Press, New York, 1963.

University of Pennsylvania

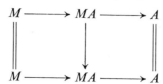

On the (co-) homology of commutative rings

Daniel Quillen[1]

This paper is devoted to a summary of the main results of a (co-) homology theory of commutative rings due independently to André [**1**] and the author [**19**]. The theory has reached a degree of completeness that there is little doubt that it is the correct cohomology theory for commutative rings. It is perhaps worthwhile to relate this theory to definitions made by other authors.

For any morphism $A \to B$ of commutative rings and B-module M, let $\mathrm{Der}_A(B, M)$ be the B-module of A-algebra derivations of B with values in M, and let $\mathrm{Exalcomm}_A(B, M)$ be the B-module of infinitesimal A-algebra extensions of B by M ([**11**, Chapter IV]). It is known that the functors Der and Exalcomm possess the following two properties:

0.1. (*Transitivity*). Given morphisms $A \to B \to C$ of commutative rings and a C-module M there is a six-term exact sequence

$$0 \to \mathrm{Der}_B(C, M) \to \mathrm{Der}_A(C, M) \to \mathrm{Der}_A(B, M)$$
$$\to \mathrm{Exalcomm}_B(C, M) \to \mathrm{Exalcomm}_A(C, M) \to \mathrm{Exalcomm}_A(B, M).$$

0.2. (*Flat base change*). Given morphisms $A \to B$, $A \to A'$ of commutative rings and an $A' \otimes_A B$ module M, there are isomorphisms

$$\mathrm{Der}_{A'}(A' \otimes_A B, M) \simeq \mathrm{Der}_A(B, M)$$
$$\mathrm{Exalcomm}_{A'}(A' \otimes_A B, M) \simeq \mathrm{Exalcomm}_A(B, M) \quad \text{if } \mathrm{Tor}_1^A(A', B) = 0.$$

The cohomology theory to be presented here associates to each morphism $A \to B$ of commutative rings and B-module M cohomology B-modules $D^q(B/A, M)$ for each integer $q \geq 0$, coinciding with $\mathrm{Der}_A(B, M)$ and $\mathrm{Exalcom}_A(B, M)$ for $q = 0$ and 1 respectively, and extending in the obvious way the properties 0.1–0.2.

In [**13**] Harrison gave a definition of commutative ring cohomology using a

[1] Supported by NSF GP-6959 and the Alfred P. Sloan Foundation.

subcomplex of the Hochschild complex for computing associative algebra cohomology. Harrison restricted himself to the case where the ground ring A is a field; his definition does not extend to a general morphism $A \to B$ since one needs at least that B be projective as an A-module in order that Harrison's H^2 coincide with $\mathrm{Exalcomm}_A(B, M)$. However, it can be shown (§9 below) that Harrison's complex may be used to compute the correct cohomology $D^*(B/A, M)$ when B is projective as an A-module and when B is of characteristic 0, that is, an algebra over the rational numbers.

In **[16]**, Lichtenbaum and Schlessinger define a satisfactory cohomology theory in dimensions $q \leq 2$ in the sense that their cohomology extends the exact sequence of 0.1 to nine terms. Their method uses a free differential graded anticommutative A-algebra resolution of B. It may be shown that their cohomology group $T^q(B/A, M)$ coincides with our $D^q(B/A, M)$ for $q \leq 2$ and that free differential graded anticommutative A-algebra resolutions of B may be used to compute $D^*(B/A, M)$ when B is of characteristic zero (§9).

In **[21]** it is shown how the transitivity and flat base change properties of the Lichtenbaum-Schlessinger $T^q(B/A, M)$ for $q \leq 2$ (hence also of our $D^q(B/A, M)$) yield a satisfactory deformation theory for commutative algebras. Moreover the Lichtenbaum-Schlessinger cohomology is known to be the same as the theory of Gerstenhaber **[8]** specialized to commutative rings.

The cohomology groups $D^*(B/A, M)$ are defined as suitable nonabelian derived functors of the functor $X \mapsto \mathrm{Der}_A(X, M)$ on the category of A-algebras over B, where the derived functors are defined by applying the functor dimension-wise to a free (semi-) simplicial A-algebra resolution of B and taking the cohomology of the associated cosimplicial B-module. A particular example of such a resolution is furnished by the standard construction associated to the (sets)—(A-algebras) cotriple. Thus the cohomology $D^*(B/A, M)$ is a kind of cotriple cohomology **[5]**, however, from the cotriple point of view one loses sight of the possibility of choosing a resolution for computing the derived functors with special properties, a technique vital for proving the transitivity and flat base change properties of the commutative ring cohomology.

1. **Derived functors.** In a category closed under finite limits and having enough projective objects, it is possible to define left derived functors of any functor from the category to an abelian category generalizing those of Dold-Puppe [7]. We give the construction of these derived functors in this section.

An object P of a category $\mathscr{C}$ is said to be *projective* if for any effective epimorphism $p: X \to Y$, $\mathrm{Hom}(P, p): \mathrm{Hom}(P, X) \to \mathrm{Hom}(P, Y)$ is surjective. $\mathscr{C}$ is said to have *enough projective objects* if for every object X there is an effective epimorphism $P \to X$ with P projective. When $\mathscr{C}$ is closed under finite projective limits and has enough projective objects, then the effective epimorphisms are the maps p for which $\mathrm{Hom}(P, p)$ is surjective for all projective objects P [**18,** II, §4, Proposition 2]; in particular the effective epimorphisms are universally effective.

Let $s\mathscr{C}$ be the category of (semi-) simplicial objects over $\mathscr{C}$. We always identify an object X of $\mathscr{C}$ with the constant simplicial object whose simplicial operators are all equal to the identity morphism of X. For $X, Y \in \text{Ob } s\mathscr{C}$, let $\text{Hom}(X, Y)$ be the "function complex" simplicial set of morphisms from X to Y; a homotopy from a morphism $f: X \to Y$ to g is thus an element h of $\text{Hom}(X, Y)_1$ with $d_1 h = f$ and $d_0 h = g$. Assuming $\mathscr{C}$ is closed under finite limits, it is possible to define [18, II, §1, Proposition 2] for each object X of $s\mathscr{C}$ and simplicial set K having only finitely many nondegenerate simplices generalized "cylinder" and "function-space" objects X_K and X^K satisfying the formulas

(1.1) $\qquad \text{Hom}(X_K, Y) = \text{Hom}_{s(\text{sets})}(K, \text{Hom}(X, Y)) = \text{Hom}(X, Y^K).$

In particular if I is the standard 1-simplex, there are canonical maps $d_0^X, d_1^X: X^I \to X$ such that to give a simplicial homotopy joining $f: Y \to X$ to g is the same as giving a map $h: Y \to X^I$ such that $d_1 h = f$ and $d_0 h = g$. Also there is a canonical map $s_0: X \to X^I$ representing the constant homotopy from id_X to itself.

We say that a morphism $i: A \to B$ in $\mathscr{C}$ has the *left lifting property* (LLP) with respect to a morphism $p: X \to Y$, or that p has the *right lifting property* (RLP) with respect to i, if the dotted arrow exists in any commutative square of solid arrows

(1.2)
$$\begin{array}{ccc} A & \xrightarrow{\alpha} & X \\ {\scriptstyle i}\downarrow & \nearrow & \downarrow{\scriptstyle p} \\ B & \xrightarrow{\beta} & Y \end{array}$$

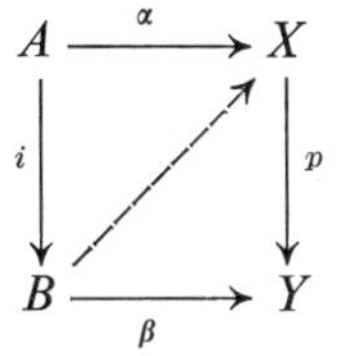

A morphism of simplicial sets will be called an *acyclic fibration* ("acyclic" is much preferable to the term "trivial" used in [18]) if it has the RLP with respect to any monomorphism of simplicial sets, or equivalently if it is a Kan fibration which is surjective and whose fibers are contractible. A morphism of simplicial objects over a category $\mathscr{C}$ will be called an *acyclic fibration* if for any projective object P, the induced morphism $\text{Hom}(P, X) \to \text{Hom}(P, Y)$ is an acyclic fibration of simplicial sets. Finally we call a morphism in $s\mathscr{C}$ a *cofibration* if it has the LLP with respect to all acyclic fibrations.

PROPOSITION 1.3. *Assume $\mathscr{C}$ is closed under finite projective limits and let $i: A \to B$ be a cofibration and $p: X \to Y$ an acyclic fibration in $s\mathscr{C}$. Then given a commutative square 1.2 any two choices for the dotted arrow are homotopic under A and over B. More generally given homotopies $\bar{\alpha}: A \to X^I, \bar{\beta}: B \to Y^I$ compatible with i and p (i.e. $p^I \bar{\alpha} = \bar{\beta} i$), and given dotted arrows $\theta, \theta': B \to X$ with $\theta i = d_1 \bar{\alpha}, \theta' i = d_0 \bar{\alpha}, p\theta = d_1 \bar{\beta}, p\theta' = d_0 \bar{\beta}$, there is a homotopy $\bar{\theta}: B \to X^I$ joining θ to θ' compatible with $\bar{\alpha}$ and $\bar{\beta}$ (i.e. $\bar{\theta} i = \bar{\alpha}$ and $p^I \bar{\theta} = \bar{\beta}$).*

$\bar{\theta}$ is obtained as the dotted arrow in the square

$$\begin{array}{ccc} A & \xrightarrow{\alpha} & X^I \\ {\scriptstyle i}\downarrow & & \downarrow{\scriptstyle (d_1^X, d_0^X, p^I)} \\ B & \xrightarrow{(\theta,\theta',\bar{\beta})} & X^I \times_{(Y^I)} (Y^I) \end{array}$$

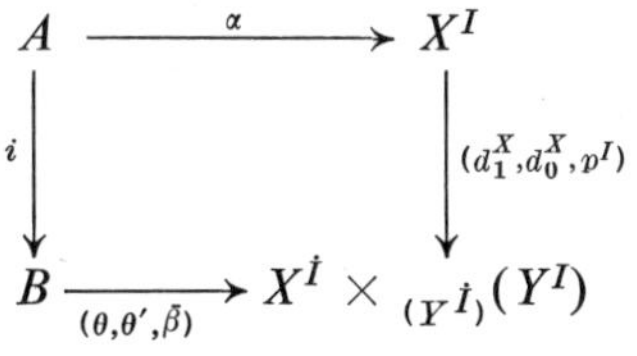

where $X^I = X \times X$. To see that (d_1^X, d_0^X, p^I) is an acyclic fibration one applies the functor Hom $(P, ?)$, P a projective object of $\mathscr{C}$; using 1.1 one is reduced to the case of simplicial sets which may then be handled by means of the formula $L_K = L \times K$.

PROPOSITION 1.4. *Let $\mathscr{C}$ be a category closed under finite limits and having sufficiently many projective objects. Then any morphism f of $s\mathscr{C}$ may be factored $f = pi$ where i is cofibrant and p is an acyclic fibration.*

This factorization is constructed by a simplicial analogue of the familiar process in algebraic topology of killing homotopy groups by attaching cells. For details see [**18,** II, §4, Proposition 3]. By standard arguments using 1.3, the factorization is unique up to homotopy equivalence, and up to homotopy depends functorially on the morphism f.

Suppose now that $\mathscr{C}$ satisfies the hypotheses of 1.4. If X is an object of $\mathscr{C}$, we may define a simplicial *resolution* of X to be an acyclic fibration $Q \to X$. A simplicial object P will be called *cofibrant* if the map $\phi \to P$ is a cofibration where ϕ is the initial object of $\mathscr{C}$. (The term "projective" instead of cofibrant is perhaps more in keeping with tradition except that a cofibrant P need not be projective in the category $s\mathscr{C}$.) By 1.3 and 1.4 cofibrant resolutions of X exist, are unique up to homotopy equivalence, and up to homotopy depend functorially on X. Therefore if $F: \mathscr{C} \to \mathscr{A}$ is a functor, where $\mathscr{A}$ is an abelian category, we may define left derived functors by $(L_n F)(X) = H_n F(P)$, where as usual the homology of a simplicial object in an abelian category is the homology of the associated chain complex with $d = \Sigma(-1)^i d_i$, or of the normalized subcomplex.

If $\mathscr{C}$ is an abelian category with enough projectives, then by [7] the normalization $N: s\mathscr{C} \to \text{Ch}(\mathscr{C})$ is an equivalence of $s\mathscr{C}$ with the category of chain complexes of $\mathscr{C}$. One may show (see [**18,** II, §3, Proposition 3]) that a map $f: X \to Y$ in $s\mathscr{C}$ is an acyclic fibration iff Nf is surjective inducing isomorphisms on homology, and that f is a cofibration iff Nf is injective with cokernel having projective objects of $\mathscr{C}$ in each dimenson. Therefore a cofibrant resolution of X in the sense described above is the same as a projective resolution of X in the usual sense, so the derived functors just defined coincide with those of Dold-Puppe [7].

2. **Homology and cohomology for universal algebras.** Let X be an object in a category $\mathscr{C}$ closed under finite projective limits and having sufficiently many projective objects. Let M be an abelian group object of the category $\mathscr{C}/X$ of objects over X and let P be a cofibrant simplicial resolution of X as in §1. We define the *cohomology groups* of X *with values in* M by the formula

$$D^q(X, M) = H^q\{\text{Hom}_{\mathscr{C}/X}(P, M)\}, \tag{2.1}$$

where H^q denotes the homology of a cosimplicial abelian group with respect to the differential $\delta = \Sigma(-1)^i\delta_i$. By 1.3 the cohomology is independent of the choice of P and depends functorially on the pair (X, M), where a morphism $(X, M) \to (X', M')$ is a pair consisting of a morphism $X' \to X$ in $\mathscr{C}$ and a morphism $X' \times_X M \to M'$ of abelian group objects over X'.

Suppose now that $\mathscr{C}$ is an *algebraic category* by which we mean a category closed under inductive limits and having a set of small projective generators. For example if $\mathscr{C}$ has a single small projective generator, $\mathscr{C}$ is equivalent to the category of universal algebras defined by a set of operations and relations [**15**]. Then it may be shown that for any object X of $\mathscr{C}$, the category $(\mathscr{C}/X)_{ab}$ of abelian group objects over X is an abelian algebraic category and that moreover the abelianization functor $\mathrm{Ab}:\mathscr{C}/X \to (\mathscr{C}/X)_{ab}$, left adjoint to the forgetful functor, exists. Choosing a simplicial cofibrant resolution P of X, we define a simplicial object of $(\mathscr{C}/X)_{ab}$ by the formula

$$\mathbf{L}\mathrm{Ab}(X) = \mathrm{Ab}(P). \tag{2.2}$$

$\mathbf{L}\mathrm{Ab}\,(X)$ depends up to homotopy equivalence only on X and therefore gives rise via the normalization functor to an object of the derived category of $(\mathscr{C}/X)_{ab}$. $\mathbf{L}\mathrm{Ab}\,(X)$ should be thought of as being analogous to the complex of chains of a space X since one may calculate the cohomology of X with values in M by the formula

$$D^q(X, M) = H^q\{\mathrm{Hom}_{(\mathscr{C}/X)_{ab}}(\mathbf{L}\mathrm{Ab}\,(X), M)\}. \tag{2.3}$$

We define the qth *homology* object of X to be $D_q(X) = H_q\{\mathbf{L}\mathrm{Ab}\,(X)\}$.

The following proposition summarizes the properties of homology and cohomology in this general setting.

PROPOSITION 2.4. (i) *$D^*(X, M)$ is a cohomological functor of the object M of $(\mathscr{C}/X)_{ab}$.*

(ii) *There is a universal coefficient spectral sequence*

$$E_2^{pq} = \mathrm{Ext}^p_{(\mathscr{C}/X)_{ab}}(D_q(X), M) \Rightarrow D^{p+q}(X, M).$$

(iii) *If X is a projective object of $\mathscr{C}$, then $D_q(X) = D^q(X, M) = 0$ for all $q > 0$ and abelian group objects M over X.*

(iv) $D^0(X, M) = \mathrm{Hom}_{\mathscr{C}/X}(X, M)$, $D^1(X, M) =$ *isomorphism classes of objects Y over X which are torsors for M, i.e. endowed with a right action $Y \times_X M \to Y$ of M such that there exists an effective epimorphism $Z \to X$ such that $Z \times_X Y$ is isomorphic to $Z \times_X M$ with its natural right $Z \times_X M$ action.*

(v) *If X_i, $i \in I$ is a filtered inductive system of objects of $\mathscr{C}$, then* $\varinjlim D_q(X_i) \simeq D_q(\varinjlim X_i)$.

If P_i, $i \in I$ is a set of small projective generators for $\mathscr{C}$, then one obtains a simplicial cofibrant resolution of any object of $\mathscr{C}$ by taking the standard construction relative to the functor $X \to \prod_i \mathrm{Hom}_{\mathscr{C}}(P_i, X)$ from $\mathscr{C}$ to Sets/I and its left adjoint. Thus the cohomology groups (2.1) are a kind of cotriple cohomology [**5**] and they coincide with the ones defined by André [**1**] using the set of P_i for models.

The cohomology groups $D^q(X, M)$ are also a special case of a very general definition of cohomology due to Grothendieck [**26**] as follows. The class of effective epimorphisms of $\mathscr{C}$ is stable under composition and base change and thus we obtain a pretopology in $\mathscr{C}$ by defining a covering of X to be a family consisting of a single effective epimorphism $Y \to X$. Representable functors are sheaves for the associated topology, so if M is an abelian group object over X, the functor $h_M = \mathrm{Hom}_{\mathscr{C}/X}(?, M)$ is a sheaf on $\mathscr{C}/X$ with the induced topology and sheaf cohomology groups $H^q(\mathscr{C}/X, h_M)$ are defined. A proof that $D^q(X, M) \simeq H^q(\mathscr{C}/X, h_M)$ is given in [**18**, II, §5]; alternatively it follows from a general result of Verdier on the calculation of sheaf cohomology by a modified Cech procedure [**26**, Appendix]. The advantage of the Grothendieck definition is that it applies even when there are not enough projectives in $\mathscr{C}$, e.g. categories of sheaves of algebras, provided the effective epimorphisms are closed under base change. When this last condition fails one may still consider other topologies e.g. (see [**12**, Exposé IV, 3.4]).

3. **Associative algebra cohomology.** Let A be a commutative ring with identity and let $\mathscr{C} = A$-Ass be the category of associative not necessarily commutative A-algebras with identity. In this section we will compute the cohomology of an object B of $\mathscr{C}$ in terms of Ext functors.

Let M be a B-bimodule, i.e. a left $B \otimes_A B^0$-module where B^0 is the opposed algebra to B, and let $B \oplus M$ denote the semidirect product A-algebra with multiplication $(b_1 \oplus m_1)(b_2 \oplus m_2) = b_1 b_2 \oplus (b_1 m_2 + m_1 b_2)$. Then if C is an A-algebra over B we have canonical isomorphisms

$$\begin{aligned} \mathrm{Hom}_{\mathscr{C}/B}(C, B \oplus M) &\simeq \mathrm{Der}_A(C, M) \\ &\cong \mathrm{Hom}_{C \otimes_A C^0}(D_{C/A}, M) \\ &\cong \mathrm{Hom}_{B \otimes_A B^0}((B \otimes_A B^0) \otimes_{(C \otimes_A C^0)} D_{C/A}, M) \end{aligned} \tag{3.1}$$

where $\mathrm{Der}_A(C, M)$ is the abelian group of derivations of the A-algebra C with values in M regarded as a C-bimodule via the structural map $C \to B$ of C, and where $D_{C/A}$ is the C-bimodule of differentials of the A-algebra C. There is a canonical A-algebra derivation $d: C \to D_{C/A}$ which is universal for all derivations; one may prove that there is an exact sequence of C-bimodules

$$0 \longrightarrow D_{C/A} \xrightarrow{i} C \otimes_A C^0 \xrightarrow{\mu} C \longrightarrow 0 \tag{3.2}$$

where $\mu(x \otimes y) = xy$ and id $(x) = 1 \otimes x - x \otimes 1$.

As $C \mapsto \mathrm{Der}_A(C, M)$ is a functor from $\mathscr{C}/B$ to abelian groups, it follows from 3.1 that $B \oplus M$ is naturally an abelian group object of $\mathscr{C}/B$. Conversely one shows easily that any abelian group object (more generally any monoid object) X of $\mathscr{C}/B$ is isomorphic to $B \oplus M$ with $M =$ kernel of structural map $X \to B$. Moreover in this way one obtains an equivalence of $(\mathscr{C}/B)_{ab}$ with the category of B-bimodules. Identifying these categories, 3.1 shows that the abelianization functor $\mathrm{Ab}: \mathscr{C}/B \to (\mathscr{C}/B)_{ab}$ is given by $C \mapsto (B \otimes_A B^0) \otimes_{(C \otimes_A C^0)} D_{C/A}$, hence

$$D^q_{A\text{-Ass}}(B, M) \cong H^q\{\mathrm{Hom}_{P \otimes_A P^0}(D_{P/A}, M)\} \tag{3.3}$$

$$\mathrm{LAb}(B) \cong (B \otimes_A B^0) \otimes_{(P \otimes_A P^0)} D_{P/A}, \tag{3.4}$$

where P is a simplicial projective A-algebra resolution of B.

In order to put 3.3 in a more agreeable form, one shows that the right-hand side (since $D_{P/A}$ is a free P-bimodule) is isomorphic to the group $\mathrm{Ext}_{P\otimes_A P^0}{}^q(D_{P/A}, M)$ of morphisms from $D_{P/A}$ to the qth suspension $\Sigma^q M$ of M in the homotopy category of simplicial $P \otimes_A P^0$-modules [**18**, II, §6]. From the exact sequence 3.2, with C replaced by P, one obtains a long exact sequence yielding isomorphisms

$$D^q_{A\text{-Ass}}(B, M) \cong \mathrm{Ext}_{P\otimes_A P^0}{}^{q+1}(B, M) \quad \text{if } q > 0. \tag{3.5}$$

Since $P \to B$ induces isomorphisms on homology the right-hand side is isomorphic to $\mathrm{Ext}_{P\otimes_A P^0}{}^{q+1}(B, M)$. Now the homotopy category $\mathrm{Ho}(\mathcal{M}_R)$ of simplicial modules over a simplicial ring R depends only on the weak homotopy type of R in the sense that if $R \to R'$ is a map of simplicial rings inducing isomorphisms on homology, then the adjoint functors of extension and restriction of scalars induce an equivalence of $\mathrm{Ho}(\mathcal{M}_R)$ and $\mathrm{Ho}(\mathcal{M}_{R'})$. Consequently $\mathrm{Ext}_{P\otimes_A P^0}{}^*(B, M) \cong \mathrm{Ext}_{B\overset{\mathbf{L}}{\otimes}_A B^0}{}^*(B, M)$, where $B \overset{\mathbf{L}}{\otimes}_A B^0$ denotes any simplicial ring of the same weak homotopy type over $B \otimes_A B^0$ as $P \otimes_A P^0$. In particular if $\mathrm{Tor}_q^A(B, B) = 0$ for $q > 0$, then we may take $B \overset{\mathbf{L}}{\otimes}_A B^0$ to be $B \otimes_A B$, whence the category $\mathrm{Ho}(\mathcal{M}_{B\otimes_A B^0})$ is by Dold-Puppe just the full subcategory of the derived category of B-bimodules consisting of chain complexes, and hence $\mathrm{Ext}_{P\otimes_A P^0}{}^*(B, M)$ is isomorphic to the usual $\mathrm{Ext}_{B\otimes_A B^0}{}^*(B, M)$ of homological algebra. Thus combining 3.5 and 2.4 (iv), we have

PROPOSITION 3.6. *If B is an associative A-algebra and M is a B-bimodule, then*

$$\begin{aligned} D^q_{A\text{-Ass}}(B, M) &\cong \mathrm{Der}_A(B, M) && \text{if } q = 0, \\ &\cong \mathrm{Ext}_{B\overset{\mathbf{L}}{\otimes}_A B^0}{}^{q+1}(B, M) && \text{if } q > 0, \end{aligned}$$

where we may take $B \overset{\mathbf{L}}{\otimes}_A B \cong B \otimes_A B$ if $\mathrm{Tor}_q^A(B, B) = 0$ for $q > 0$.

Therefore if A is a field the associative algebra cohomology $D^*_{A\text{-Ass}}$ is with minor alterations the same as the Hochschild cohomology. This result should be compared with Barr [**3**], which indicates that the groups $\mathrm{Ext}_{B\overset{\mathbf{L}}{\otimes}_A B^0}{}^*(B, M)$ may be calculated by the differential graded algebra constructed by Shukla [**23**].

By essentially the same method used to prove 3·6 one may prove the following, where A-Lie is the category of Lie algebras over A, and where one identifies $(A\text{-Lie}/\mathfrak{I})_{ab}$ with the category of $\mathfrak{I}$-modules.

PROPOSITION 3.7. *If $\mathfrak{I}$ is a Lie algebra over A and M is a $\mathfrak{I}$-module, then*

$$\begin{aligned} D^q_{A\text{-Lie}}(\mathfrak{I}, M) &= \mathrm{Der}_A(\mathfrak{I}, M) && \text{if } q = 0, \\ &= \mathrm{Ext}^{q+1}_{U(P)}(A, M) && \text{if } q > 0, \end{aligned}$$

where P is a simplicial cofibrant A-Lie algebra resolution of $\mathfrak{I}$ and where U is the universal enveloping algebra functor. Furthermore if $\mathfrak{I}$ is flat as an A-module, then $U(P)$ may be replaced by $U(\mathfrak{I})$.

4. **Cohomology of commutative rings, the relative cotangent complex.** For the rest of this paper all rings shall be understood to be commutative with unit. Let

A be a ring, let $\mathscr{C}$ be the category of A-algebras and let B be an A-algebra. If M is a B-module let $B \oplus M$ be the trivial A-algebra extension of B by M given by the formulas $(b_1 \oplus m_1)(b_2 \oplus m_2) = b_1b_2 \oplus (b_1m_2 + b_2m_1)$, $a(b \oplus m) = ab \oplus am$. Then there are canonical isomorphisms

$$\text{(4.1)} \qquad \operatorname{Hom}_{\mathscr{C}/B}(C, B \oplus M) \simeq \operatorname{Der}_A(C, M) \simeq \operatorname{Hom}_B(B \otimes_C \Omega_{C/A}, M)$$

where $\operatorname{Der}_A(C, M)$ is the B-module of derivations of the A-algebra C with values in M, and where $\Omega_{C/A} = J/J^2$, $J = \operatorname{Ker}\{C \otimes_A C \to C\}$, is the C-module of Kähler differentials of the A-algebra C. The first formula shows that $B \oplus M$ is an abelian group object of $\mathscr{C}/B$, in fact a B-module object. One shows easily that the functor $M \mapsto B \oplus M$ is an equivalence of the category of B-modules with $(\mathscr{C}/B)_{ab}$. Identifying these two categories, the second part of 4.1 shows that the abelianization functor $\operatorname{Ab}: \mathscr{C}/B \to (\mathscr{C}/B)_{ab}$ is given by $\mathscr{C} \mapsto B \otimes_A \Omega_{C/A}$.

In accordance with §2 the *cohomology of the A-algebra B with values in the B-module M* is given by

$$\text{(4.2)} \qquad D^q(B/A, M) = H^q\{\operatorname{Der}_A(P, M)\}$$

where P is a simplicial cofibrant A-algebra resolution of B and where the expression on the right is the cohomology of the cosimplicial B-module $k \mapsto \operatorname{Der}_A(P_k, M)$. $D^*(B/A, M)$ is a cohomological functor of M and in low dimensions is, after interpreting 2.4(iv) given with the notation of the introduction by

$$\text{(4.3)} \qquad \begin{aligned} D^0(B/A, M) &= \operatorname{Der}_A(B, M) \\ D^1(B/A, M) &= \operatorname{Exalcomm}_A(B, M). \end{aligned}$$

The homology B-modules $D_q(B/A)$ of §2 are defined to be the homology of the simplicial B-module

$$\text{(4.4)} \qquad \operatorname{LAb}(B) = B \otimes_P \Omega_{P/A}.$$

We shall denote this simplical B-module by $\mathbf{L}_{B/A}$ in the following and call it the *cotangent complex* of the A-algebra B or of B relative to A. It is a cofibrant simplicial B-module, whose normalization is a chain complex of projective B-modules independent up to homotopy equivalence of the choice of P, and which therefore represents an object unique up to isomorphism of the derived category of the category of B-modules. Combining 4.1 and 4.2 we have

$$\text{(4.5)} \qquad D^q(B/A, M) = H^q\{\operatorname{Hom}_B(\mathbf{L}_{B/A}, M)\}.$$

Using the tensor product operation on B-modules, it is possible to introduce the *homology of the A-algebra B with values in M*

$$\text{(4.6)} \qquad D_q(B/A, M) = H_q(\mathbf{L}_{B/A} \otimes_B M)$$

related to the homology $D_q(B/A) = D_q(B/A, B)$ by a universal coefficient spectral sequence

$$\text{(4.7)} \qquad E^2_{pq} = \operatorname{Tor}^B_p(D_q(B/A), M) \Rightarrow D_{p+q}(B/A, M).$$

However, one sees from 4.5 and 4.6 that the cotangent complex is a finer invariant than either the cohomology or homology functors, and is therefore the basic object of study. This point of view is due to Grothendieck **[9]**.

Suppose now that

$$\begin{array}{ccc} A & \longrightarrow & B \\ \downarrow & & \downarrow \\ A' & \longrightarrow & B' \end{array} \tag{4.8}$$

is a commutative square of rings and that P (resp. P') is a cofibrant simplicial A- (resp. A'-) algebra resolution of B (resp. B'). By 1.3 there is a morphism $P \to P'$ of simplicial A-algebras over B' unique up to homotopy, hence a morphism of B' modules $B' \otimes_B (B \otimes_P \Omega_{P/A}) \to B' \otimes_{P'} \Omega_{P'/A'}$. Thus there is a morphism of simplicial B'-modules

$$B' \otimes_B \mathbf{L}_{B/A} \to \mathbf{L}_{B'/A'} \tag{4.9}$$

unique up to homotopy associated to 4.8 and consequently for any B'-module M' restriction homomorphisms

$$\begin{gathered} D^q(B'/A', M') \to D^q(B/A, M') \\ D_q(B/A, M') \to D_q(B'/A', M'). \end{gathered} \tag{4.10}$$

If $\{A_i \to B_i, i \in I\}$ is a filtered inductive system of morphisms of rings, then for each i let P_i be the cofibrant A_i-algebra resolution of B_i obtained using the standard construction associated to the pair of adjoint functors: "underlying set of an A_i-algebra" and "free A_i-algebra generated by the set." Using these resolutions to calculate the cotangent complex, one obtains the formulas

$$\begin{gathered} \mathbf{L}_{\lim B_i/\lim A_i} = \lim \mathbf{L}_{B_i/A_i} \\ D_q(\lim B_i/\lim A_i, \lim M_i) = \lim D_q(B_i/A_i, M_i) \end{gathered} \tag{4.11}$$

where $i \to M_i$ is a module for the functor $i \mapsto B_i$.

By analyzing the construction of a cofibrant A-algebra resolution P of B, one sees that if A is noetherian and B is an A-algebra of finite type, then P may be chosen so that P_k is a polynomial ring with finitely many variables for each $k \geq 0$. Using this P to calculate the cotangent complex one obtains

PROPOSITION 4.12. *If A is noetherian and B is an A-algebra of finite type, then $\mathbf{L}_{B/A}$ is isomorphic in the derived category of B-modules to a complex which is a free finite type B-module in each dimension. Consequently if M is a B-module of finite type, then for each q, $D^q(B/A, M)$ and $D_q(B/A, M)$ are B-modules of finite type.*

5. **The transitivity, flat base change, and vanishing theorems.** In this section we give the basic properties of the cohomology of commutative rings signaled in the introduction. These will be stated also as properties of the relative cotangent complex $\mathbf{L}_{B/A}$, which is identified via the normalization functor with an object of the

derived category of B-modules. We recall that the derived category **[25]** is endowed with a "suspension" or "translation" functor, here denoted by Σ, and that any exact sequence of complexes gives, via an analogue of the Puppe sequence construction in algebraic topology, a "distinguished triangle" in the derived category.

THEOREM 5.1. (*Transitivity*) *If $A \to B \to C$ are morphisms of rings, then there is a canonical distinguished triangle in the derived category of C-modules*

$$C \otimes_B \mathbf{L}_{B/A} \longrightarrow \mathbf{L}_{C/A} \longrightarrow \mathbf{L}_{C/B} \xrightarrow{\delta} \Sigma\{C \otimes_B \mathbf{L}_{B/A}\}.$$

Consequently if M is a C-module, there are exact sequences

$$0 \longrightarrow D^0(C/B, M) \longrightarrow D^0(C/A, M) \longrightarrow D^0(B/A, M) \xrightarrow{\delta} D^1(C/B, M) \longrightarrow \cdots.$$

The proof uses in an essential way the possibility of calculating the cotangent complex using any cofibrant resolution. Thus let P be a cofibrant simplicial A-algebra resolution of B and let Q be a cofibrant simplicial P-algebra resolution of C, i.e. $P \to Q \to C$ is a factorization into a cofibration followed by an acyclic fibration of the composition $P \to B \to C$. Then we have a diagram of simplicial rings

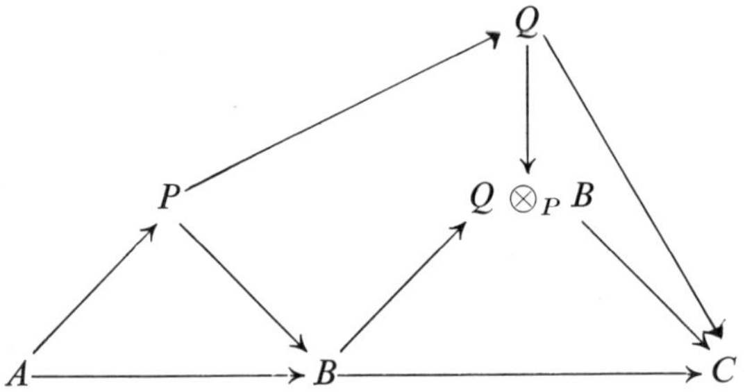

and an exact sequence of C-modules

$$0 \to C \otimes_B (B \otimes_P \Omega_{P/A}) \to C \otimes_Q \Omega_{Q/A} \to C \otimes_{Q \otimes_P B} \Omega_{Q \otimes_P B/B} \to 0.$$

The first and second term may be identified with $C \otimes_B \mathbf{L}_{B/A}$ and $\mathbf{L}_{C/A}$ respectively, the distinguished triangle associated to this exact sequence is the desired one in 5.1 once we identify the last term with $\mathbf{L}_{C/B}$. For this it suffices to show that $Q \otimes_P B$ is a simplicial cofibrant B-algebra resolution of C. $B \to Q \otimes_P B$ is a cofibration since it is the base change of the cofibration $P \to Q$. To see that $Q \otimes_P B \to C$ induces isomorphisms on homology, it suffices to do the same for the map $Q \to Q \otimes_P B$, which results by applying the functor $Q \otimes_P ?$ to the acyclic fibration $P \to B$. Now the homological properties of the tensor product may be analyzed in terms of Künneth spectral sequences **[18,** II, 6]

$$\text{(5.2)} \qquad \begin{aligned} E^2_{pq} &= H_p\{\mathrm{Tor}^R_q(X, Y)\} \Rightarrow H_{p+q}(X \overset{\mathbf{L}}{\otimes}_R Y) \\ E^2_{pq} &= \mathrm{Tor}^{H(R)}_p(H(X), H(Y))_q \Rightarrow H_{p+q}(X \overset{\mathbf{L}}{\otimes}_R Y) \end{aligned}$$

where X (resp. Y) is a left (resp. right) simplicial module over a not necessarily commutative simplicial ring R. Applying these we get a diagram

$$\begin{array}{ccccc} H_p\{\mathrm{Tor}_q^P(Q, P) & \Rightarrow & H_{p+q}(Q \overset{\mathbf{L}}{\otimes}_P P) & \Leftarrow & \mathrm{Tor}_p^{H(P)}(H(Q), H(P))_q \\ \downarrow & & \downarrow & & \downarrow \cong \\ H_p\{\mathrm{Tor}_q^P(Q, B)\} & \Rightarrow & H_{p+q}(Q \overset{\mathbf{L}}{\otimes}_P B) & \Leftarrow & \mathrm{Tor}_p^{H(P)}(H(Q), H(B))_q. \end{array}$$

As $P \to Q$ is a cofibration, Q in each dimension is a retract of a free P-algebra, hence Q is projective as P-module and the Tor's on the left vanish for $q > 0$. Thus the spectral sequences on the left degenerate showing that $H_*(Q) \xrightarrow{\sim} H_*(Q \otimes_P B)$ and concluding the proof of the theorem.

THEOREM 5.3. (*Flat base change.*) *If B and C are A-algebras such that* $\mathrm{Tor}_q^A(B, C) = 0$ *for $q > 0$, then there are isomorphisms in the derived category of $B \otimes_A C$-modules*

$$\mathbf{L}_{B \otimes_A C/C} \cong B \otimes_A \mathbf{L}_{C/A}$$

$$\mathbf{L}_{B \otimes_A C/A} \cong (\mathbf{L}_{B/A} \otimes_A C) \oplus (B \otimes_A \mathbf{L}_{C/A}).$$

Consequently if M is a $B \otimes_A C$-module, there are isomorphisms

$$D^q(B \otimes_A C/C, M) \cong D^q(B/A, M)$$

$$D^q(B \otimes_A C/A, M) \cong D^q(B/A, M) \oplus D^q(C/A, M).$$

If P is a cofibrant A-algebra resolution of B and Q is a cofibrant A-algebra resolution of C, then using the Künneth spectral sequences 5.2 and the fact that P and Q are projective as A-modules, one sees that $B \otimes_A Q$ (resp. $P \otimes_A Q$) is a cofibrant B-(resp. A-) algebra resolution of $B \otimes_A C$. Using these resolutions one obtains the formulas of the theorem.

The following vanishing properties of the cotangent complex may be deduced with the aid of the spectral sequence of §6. However, André has shown by very beautiful arguments that they follow directly from the transitivity and flat base change theorems, so we state them here. Recall that a chain complex K of B-modules is said to have projective (resp. Tor) dimension $\leq n$ if $H^q\{\mathrm{Hom}_B(K, M)\} = 0$ (resp. $H_q(K \otimes_B M) = 0$) for $q > n$ and all B-modules M.

THEOREM 5.4. (i) *If S is a multiplicative system in A, then* $\mathbf{L}_{S^{-1}A/A} = 0$.

(ii) *If* Spec $B \to$ Spec A *is étale* [**11**, IV], *then* $\mathbf{L}_{B/A} = 0$

(iii) *If* Spec $B \to$ Spec A *is smooth* [**11**, IV], *then* $\mathbf{L}_{B/A}$ *is isomorphic in the derived category to* $\Omega_{B/A}$ *and has projective dimension* 0.

(iv) *If* Spec $B \to$ Spec A *is a local complete intersection morphism* [**10**], *then* $\mathbf{L}_{B/A}$ *has projective dimension* ≤ 1.

The assertions (ii)-(iv) admit the following partial converses.

THEOREM 5.5. *If A is noetherian and B is an A-algebra of finite type, then*

(i) Spec $B \to$ Spec A *is smooth* (*resp. étale*) $\Leftrightarrow D^q(B/A, M) = 0$ *for all B-modules M of finite type and* $q = 1$ (*resp.* $q = 0, 1$).

(ii) Spec $B \to$ Spec A *is a local complete intersection morphism* $\Leftrightarrow D^2(B/A, M) = 0$ *for all B-modules M of finite type.*

5.5 (i) is Grothendieck's infinitesimal criterion for smoothness [**11**, IV, §17], (ii) is due to Lichtenbaum-Schlessinger [**16**].

Unsolved Problem. Characterize the morphisms $A \to B$, where A is noetherian and B is an A-algebra of finite type, for which the cotangent complex is of finite projective dimension. What computations we have been able to make show that this is rare and support the following conjectures:

Conjecture 5.6. If $\mathbf{L}_{B/A}$ is of finite projective dimension then it is of projective dimension ≤ 2.

Conjecture 5.7. If $\mathbf{L}_{B/A}$ is of finite projective dimension and if B is of finite Tor dimension as an A-module, then Spec $B \to$ Spec A is a local complete intersection morphism.

6. **The fundamental spectral sequence.** In this section we retain the notations of the preceding two sections except that certain algebras of homology such as $\mathrm{Tor}^A_*(B, B)$ are not commutative but rather anticommutative with respect to the grading. With this exception which will always be clear from the context, rings are understood to be commutative.

If $A \to B$ is a morphism of rings and M is a B-module, and if we express B as a quotient of a polynomial ring P over A, then the transitivity exact sequence 5.1 applied to $A \to P \to B$ combined with the fact that cohomology vanishes on projectives (2.4, *iii*) yields isomorphisms

$$D^q(B/P, M) \xrightarrow{\sim} D^q(B/A, M) \qquad q \geq 2. \tag{6.1}$$

This shows that the calculation of the cohomology in dimensions ≥ 2 reduces to the case in which $B = A/I$ where I is an ideal in A. In this case there is a spectral sequence relating the cotangent complex $\mathbf{L}_{B/A}$ to the algebra $\mathrm{Tor}^A_*(B, B)$ which may be used to obtain information about the former.

If R is a simplicial ring and X is a simplicial R-module, and if F is a functor on the category of pairs (B, M) consisting of a ring and a module with dihomomorphisms for morphisms in the category, then by $F(R, X)$ we denote the simplicial object obtained by applying F dimension-wise. Examples of such F are the symmetric and exterior algebra functors

$$\begin{aligned} S^B M &= \bigoplus_{n=0}^{\infty} S^B_n M \\ \Lambda^B M &= \bigoplus_{n=0}^{\infty} \Lambda^B_n M \end{aligned} \tag{6.2}$$

where the latter is anticommutative.

THEOREM 6.3. *If $A \to B$ is a morphism of rings such that $B \otimes_A B \xrightarrow{\sim} B$ (for example if $B = S^{-1}(A/I)$, I an ideal in A, S a multiplicative system in A/I), then there* *is a first quadrant homological spectral sequence*

$$E^2_{pq} = H_{p+q}\{S^B_q \mathbf{L}_{B/A}\} \Rightarrow \mathrm{Tor}^A_{p+q}(B, B)$$

of bigraded algebras, anticommutative for the total degree $p+q$. *This spectral sequence has the following properties:*

Low dimensional isomorphisms:

$$D_0(B/A) = 0 \tag{6.4}$$
$$D_1(B/A) \cong \mathrm{Tor}^A_1(B, B)$$

Edge homomorphisms:

$$\mathrm{Tor}^A_n(B, B) \to D_n(B/A) \tag{6.5}$$

$$\Lambda^B_n D_1(B/A) \to \mathrm{Tor}^A_n(B, B), \tag{6.6}$$

where 6.5 *annihilates the decomposable elements of the algebra* $\mathrm{Tor}^A_*(B, B)$ *and where* 6.6 *is the unique anticommutative B-algebra morphism extending the isomorphism* 6.4.

Five term exact sequence:

$$\mathrm{Tor}^A_3(B, B) \to D_3(B/A) \xrightarrow{d_2} \Lambda^B_2 D_1(B/A) \to \mathrm{Tor}^A_2(B, B) \to D_2(B/A) \to 0. \tag{6.7}$$

THEOREM 6.8. *If* $B \otimes_A B \xrightarrow{\sim} B$ *and M is a B-module, then there is a first quadrant cohomological spectral sequence of B-modules*

$$E_2^{pq} = H^{p+q}\{\mathrm{Hom}_B(S^B_q \mathbf{L}_{B/A}, M)\} \Rightarrow \mathrm{Ext}^{p+q}_A(B, M).$$

The spectral sequence of Theorem 6.3 is an analogue for simplicial rings of the lower central series spectral sequence for simplicial groups of Curtis **[6]** and is derived in the following way. Let P be a simplicial cofibrant A-algebra resolution of B and let $Q = P \otimes_A B$, $J = \mathrm{Ker}\{P \otimes_A B \to B\}$. Filtering Q by the powers of the simplicial ideal J, one obtains a (possibly nonconvergent) spectral sequence

$$E^2_{pq} = H_{p+q}(J^q/J^{q+1}) \Rightarrow H_{p+q}(Q) \cong \mathrm{Tor}^A_{p+q}(B, B), \tag{6.9}$$

which converges provided the connectivity of J^q goes to infinity with q. Now by definition $J/J^2 \cong B \otimes_P \Omega_{P/A} = \mathbf{L}_{B/A}$, and moreover $J^q/J^{q+1} \simeq S^B_q(J/J^2)$; in effect P may be taken to be a polynomial ring in each dimension, whence Q is a polynomial ring and J is the ideal generated by the variables. Thus 6.9 is the spectral sequence of Theorem 6.3. The convergence of the spectral sequence follows from the formula $H_q(J^k) = 0$ if $q < k$, which may be proved either by the arguments of Curtis **[6**, §4] or by using Theorem 6.12 below, which applies since $B \otimes_A B \xrightarrow{\sim} B$ implies that $H_0(J) = 0$ and since for each k, J_k is a regular ideal of Q_k in the sense of the following definition.

DEFINITION 6.10. An ideal I in a ring A will be called *quasiregular* (resp. *regular*) if I/I^2 is a flat (resp. projective) A/I-module and if the canonical morphism of anticommutative algebras

$$\Lambda_*^{A/I}(I/I^2) \to \mathrm{Tor}_*^A(A/I, A/I)$$

is an isomorphism.

It may be proved that if I is quasiregular, then the canonical algebra morphism

$$S_*^{A/I}(I/I^2) \to \oplus\, I^n/I^{n+1} \tag{6.11}$$

is an isomorphism. Furthermore if A is noetherian then I is regular iff it is quasiregular iff it is generated locally by an A-regular sequence. For nonnoetherian rings in general the terminology used here differs from that of [**11**, IV], but seems better adapted for the cohomology theory of commutative rings. For example we have the following connectivity result.

THEOREM 6.12. *Let J be a simplicial ideal in a simplicial ring Q such that J_k is quasiregular in Q_k for each $k \geq 0$. If $H_0(J) = 0$, then $H_k(J^n) = 0$ for $k < n$.*

We now give some applications of the spectral sequence 6.3. If M is a B-module and q is an integer, we denote by $M[q]$ the object of the derived category of B-modules determined up to isomorphism by the condition that its homology groups are M in degree q and zero elsewhere. Here is a characterization of regular and quasiregular ideals using the cotangent complex.

THEOREM 6.13. *Suppose $B = A/I$. Then the following conditions are equivalent:*
(i) *I is a quasiregular ideal of A*
(ii) *$D_q(B/A, M) = 0$ for all $q \geq 2$ and all B-modules M*
(iii) *I/I^2 is a flat B-module and $\mathbf{L}_{B/A} \simeq I/I^2[1]$.*

COROLLARY 6.14. *If $B = A/I$, the following are equivalent:*
(i) *I is a regular ideal of A*
(ii) *$D^q(B/A, M) = 0$ for all $q \geq 2$ and all B-modules M*
(iii) *I/I^2 is a projective B-module and $\mathbf{L}_{B/A} \simeq I/I^2$ [1].*

Another application of the spectral sequence is the following Artin-Rees type theorem and its corollaries which was suggested by the fact that $D^*(B/A, M)$ is the sheaf cohomology of a site (§2).

THEOREM 6.15. *Let A be a noetherian ring, let I be an ideal in A, and let $A_n = A/I^{n+1}$, $n \geq 0$. Then for each $q \geq 0$, there exists an N such that the natural homomorphism*

$$D_q(A_{n+N}/A, A_0) \to D_q(A_n/A, A_0)$$

is zero for all $n \geq 0$.

COROLLARY 6.16. *Let A be a noetherian ring, let B be the localization of an A-algebra of finite type C with respect to a multiplicative system in C, and let M be a B-module (not necessarily finitely generated). Then given $u \in D^q(B/A, M)$ with $q > 0$, there exists a surjective morphism $p: B' \to B$ of A-algebras, where the kernel of p is a finitely generated nilpotent ideal of B', such that $p^*u = 0$, where p^* is the natural homomorphism $D^q(B/A, M) \to D^q(B'/A, M)$.*

COROLLARY 6.17. *Suppose A is noetherian, B is an A-algebra of finite type and M is a B-module. Let T be the category of finite type A-algebras over B endowed with the topology* **[26]** *associated to the pretopology in which the covering families are single morphisms $B'' \to B'$ which are surjective and have nilpotent kernel. Then $D^q(B/A, M)$ is isomorphic to the cohomology of the sheaf $B' \mapsto \operatorname{Der}(B'/A, M)$ on the site T.*

For $q \geq 2$ the noetherian hypotheses of 6.15–6.17 are essential. 6.17 was originally proved in order to define as sheaf cohomology global groups $D^*(X/Y, F)$ generalizing the $D^*(B/A, M)$ for any finite type morphism $f: X \to Y$ of noetherian schemes and any quasi-coherent sheaf F on X. However Illusie **[14]** has recently found a general definition of the cotangent complex for any morphism of ring objects in a topos, rendering this method of little interest.

7. **Degeneracy of the fundamental spectral sequence in characteristic zero.** Suppose that B is an A-algebra such that $B \otimes_A B \xrightarrow{\sim} B$ and such that $\operatorname{Tor}_*^A(B, B)$ is flat as a B-module. Then **[2]** $\operatorname{Tor}_*^A(B, B)$ has a natural structure of a commutative Hopf algebra over B. In particular if B is of characteristic zero, that is, an algebra over the rational numbers, then there is a canonical Poincaré-Birkhoff-Witt isomorphism

$$\tilde{S}^B\{Q_*\} \simeq \operatorname{Tor}_*^A(B, B) \tag{7.1}$$

of graded anticommutative B-algebras, where Q_* is the indecomposable quotient B-module of $\operatorname{Tor}_*^A(B, B)$, and where $\tilde{S}^B = \oplus\, \tilde{S}_n^B$ is the anticommutative symmetric algebra functor on graded B-modules. This isomorphism is not in general compatible with the coalgebra structures.

Also when B is of characteristic zero, the functor $M \mapsto S_m^B M$ on the category of B-modules is a direct summand of the functor $M \mapsto M^{\otimes m}$. Consequently by the Künneth spectral sequences 5.2, the natural map

$$\tilde{S}_m^B\{H_*(K)\} \to H_*\{S_m^B K\} \tag{7.2}$$

is an isomorphism in dimensions $\leq q$ if K is a simplicial B-module such that K_n and $H_n(K)$ are flat B-modules for $n < q$. Using this fact and the isomorphism 7.1, one may analyze the fundamental spectral sequence and obtain the following results.

THEOREM 7.3. *Suppose that B is an A-algebra such that $B \otimes_A B \xrightarrow{\sim} B$. Assume that B is of characteristic zero and that $\operatorname{Tor}_*^A(B, B)$ is flat as a B-module. Then the spectral sequence* 6.3 *degenerates and yields an isomorphism of $D_*(B/A)$ with the indecomposable quotient B-module of $\operatorname{Tor}_*^A(B, B)$. Furthermore there is a canonical B-algebra isomorphism*

$$\tilde{S}^B\{D_*(B/A)\} \cong \operatorname{Tor}_*^A(B, B),$$

$D_*(B/A)$ is a flat B-module, and $D_*(B/A)$ has a natural (anticommutative) graded Lie coalgebra structure over B.

COROLLARY 7.4. *Let A be a local noetherian ring with residue field k of characteristic zero. Then $D^*(k/A, k)$ is isomorphic to the subspace of primitive elements of the cocommutative Hopf algebra* $\mathrm{Ext}_A^*(k, k)$. *In particular $D^*(k/A, k)$ has a natural structure as a (anticommutative) graded Lie algebra.*

COROLLARY 7.5. *Let A and B be local noetherian rings with the same residue field k of characteristic zero, and let $A \times_k B$ be their fibre product over k (geometrically* Spec $(A \times_k B)$ *is the wedge of* Spec A *and* Spec B). *Then*

$$D^*(k/A, k) \vee D^*(k/B, k) \xrightarrow{\sim} D^*(k/A \times_k B, k)$$

where $\vee$ denotes the direct sum in the category of (anticommutative) graded Lie algebras.

EXAMPLE 7.6. Let $A = k \oplus m$ where $m^2 = 0$. Then $\mathrm{Ext}_A^*(k, k)$ is the tensor algebra over k generated by $\mathrm{Ext}_A^1(k, k) = m'$, so $D^*(k/A, k)$ is the free graded Lie algebra generated by m' in degree 1.

EXAMPLE 7.7. Let A, B be regular local rings of dimensions p and q respectively with common residue field k. Then $D^*(k/A \times_k B, k)$ is the graded *Lie* algebra with generators $x_1, \ldots, x_p, y_1, \ldots, y_q$ of degree 1 and subject to the relations $[x_i, x_j] = 0$ for $1 \leq i, j \leq p$, $[y_i, y_j] = 0$ for $1 \leq i, j \leq q$.

When $\mathrm{Tor}_*^A(B, B)$ fails to be flat over B, it is easy to find examples where the conclusions of 7.3 fail to hold, e.g. $B = A/I$ where A is a regular local ring and I is not a regular ideal. However we have the following degeneracy theorem proved by the techniques of §9.

THEOREM 7.8. *Suppose that A is an augmented B-algebra and that B is of characteristic zero. Then the spectral sequences* 6.3 *and* 6.8 *are degenerate; in fact there are canonical isomorphisms*

$$\bigoplus_{q=0}^{n} H_n\{S_q^B \mathbf{L}_{B/A}\} \cong \mathrm{Tor}_n^A(B, B)$$

$$\bigoplus_{q=0}^{n} H^n\{\mathrm{Hom}_B(S_q^B \mathbf{L}_{B/A}, M)\} \cong \mathrm{Ext}_A^n(B, M).$$

8. **Relations with associative algebra cohomology.** Let $A \to B$ be a morphism of (commutative) rings and let M be a B-module. If R is a simplicial A-algebra resolution of B which in every dimension is flat as an A-module, then by the Künneth spectral sequences 5.2. the weak homotopy type of the simplicial B-algebra $R \otimes_A B$ is independent of R. This means that if R' is another such resolution of B, then $R \otimes_A B$ and $R' \otimes_A B$ are canonically isomorphic in the homotopy category Ho (sA-alg) of simplicial A-algebras, which is the category obtained from the category of simplicial A-algebras by formally inverting the morphisms which induce isomorphisms on homology. We denote by $B \overset{\mathbf{L}}{\otimes}_A B$ the object $R \otimes_A B$, unique up to canonical isomorphism, of the category Ho (sA-alg). According to 3.6, the cohomology of B as just an associative A-algebra with values in M

may be calculated as certain Ext groups of morphisms in the homotopy category of simplicial modules over $B \overset{\mathbf{L}}{\otimes}_A B$. Therefore the following spectral sequences may be regarded as a relation between the cohomologies of B as a commutative A-algebra and as just an associative A-algebra.

THEOREM 8.1. *There is a spectral sequence of B-algebras*

$$(8.2)\qquad E^2_{pq} = H_p\{\Lambda^B_q \mathbf{L}_{B/A}\} \Rightarrow \operatorname{Tor}_{p+q}{}^{B \overset{\mathbf{L}}{\otimes}_A B}(B, B)$$

anticommutative for the total degree $p + q$, where Λ is the exterior algebra functor. There is also a spectral sequence of B-modules

$$(8.3)\qquad E_2^{pq} = H^p\{\operatorname{Hom}_B(\Lambda^B_q \mathbf{L}_{B/A}, M)\} \Rightarrow \operatorname{Ext}_{B \overset{\mathbf{L}}{\otimes}_A B}{}^{p+q}(B, M)$$

Furthermore if $\operatorname{Tor}^A_q(B, B) = 0$ for $q > 0$, then $B \overset{\mathbf{L}}{\otimes}_A B$ may be replaced by $B \otimes_A B$ in these spectral sequences.

We sketch the proof. If R is an augmented B-algebra, let $\mathscr{B}(R)$ be the total bar construction of R, and let $\bar{\mathscr{B}}(R) = B \otimes_R \mathscr{B}(R)$. Let P be a cofibrant simplicial A-algebra resolution of B so that $P \otimes_A B = B \overset{\mathbf{L}}{\otimes}_A B$. Let $\mathscr{B}(P \otimes_A B)$ be the result of applying the functor $\mathscr{B}$ dimension-wise to the simplicial augmented B-algebra $P \otimes_A B$. It is a chain complex of simplicial $P \otimes_A B$ modules. Let us think of the simplicial structure of $\mathscr{B}(P \otimes_A B)$ as being in the vertical direction and the differentials of the bar construction as running horizontally, and let $(N^h)^{-1}\mathscr{B}(P \otimes_A B)$ be the simplicial object in the category of simplicial $P \otimes_A B$ modules obtained by applying the inverse of the normalization functor in the horizontal direction. Then the diagonal $\Delta(N^h)^{-1}\mathscr{B}(P \otimes_A B)$ of this double simplicial object is a flat $P \otimes_A B$-module resolution of B, hence $\Delta(N^h)^{-1}\bar{\mathscr{B}}(P \otimes_A B)$ is isomorphic to $B \overset{\mathbf{L}}{\otimes}_{B \overset{\mathbf{L}}{\otimes}_A B} B$. The double simplicial object $(N^h)^{-1}\bar{\mathscr{B}}(P \otimes_A B)$ gives rise to a spectral sequence with abutment

$$H_n(B \overset{\mathbf{L}}{\otimes}_{B \overset{\mathbf{L}}{\otimes}_A B} B) = \operatorname{Tor}_n{}^{B \overset{\mathbf{L}}{\otimes}_A B}(B, B)$$

and whose E^2 term is

$$E^2_{pq} = H^v_p H^h_q \bar{\mathscr{B}}(P \otimes_A B) = H_p \operatorname{Tor}_q{}^{P \otimes_A B}(B, B) = H_p\{\Lambda^B_q \mathbf{L}_{B/A}\},$$

thus giving the spectral sequence 8.2. The other is derived by applying the functor $\operatorname{Hom}_B(?, M)$ to $\Delta(N^h)^{-1}\mathscr{B}(P \otimes_A B)$.

The bar construction $\bar{\mathscr{B}}(R)$ has the structure of a commutative differential graded Hopf algebra over B when R is a commutative augmented B-algebra, hence if B is of characteristic 0, there is a canonical Poincaré-Birkhoff-Witt isomorphism of differential graded algebras

$$(8.4)\qquad \bar{\mathscr{B}}(R) \simeq \tilde{S}^B(Q),$$

where Q is the indecomposable part. Using this fact one can show that there is a canonical isomorphism

$$S^B\{\mathbf{L}_{B/A}\} \simeq B \overset{\mathbf{L}}{\otimes}_{B \overset{\mathbf{L}}{\otimes}_A B} B \tag{8.5}$$

in the homotopy category of simplicial B-algebras. In slightly more concrete terms this amounts to the following result.

THEOREM 8.6. *If B is of characteristic zero, then the spectral sequences of* 8.1 *are degenerate and in fact there are canonical isomorphisms*

$$\bigoplus_{p+q=n} H_p\{\Lambda_q^B \mathbf{L}_{B/A}\} \simeq \operatorname{Tor}_n^{B \overset{\mathbf{L}}{\otimes}_A B}(B, B)$$

$$\bigoplus_{p+q=n} H^p\{\operatorname{Hom}_B(\Lambda_q^B \mathbf{L}_{B/A}, M)\} \simeq \operatorname{Ext}_{B \overset{\mathbf{L}}{\otimes}_A B}{}^n(B, B).$$

COROLLARY 8.7. *When B is of characteristic zero, then commutative algebra cohomology $D^q(B/A, M)$ is canonically a direct summand of the associative algebra cohomology* $\operatorname{Ext}_{B \overset{\mathbf{L}}{\otimes}_A B}{}^{q+1}(B, M)$.

The corollary in the case where A is a field is due to Barr [**4**]. His proof uses an explicit formula for the projection operator of $\bar{\mathscr{B}}(R)$ onto Q in 8.4.

9. **Differential graded algebra resolutions in characteristic zero.** In this section we give an example of what appears to be a rather general phenomon, that in characteristic zero simplicial objects can be replaced by differential graded objects (see also [**20**]).

Let $\mathscr{D}$ be the abelian category of chain complexes of abelian groups. The tensor product of chain complexes is a coherently associative and commutative tensor product operation in the sense of MacLane [**17**], where the isomorphism of commutativity $T: X \otimes Y \simeq Y \otimes X$ is that of Koszul: $T(x \otimes y) = (-1)^{pq} y \otimes x$ if x and y have degrees p and q respectively. Much of the following discussion may be generalized to a class of such abelian categories with tensor product.

A ring R in $\mathscr{D}$ with respect to the tensor structure is just a differential graded (DG) ring in the usual sense. If M is an object of $\mathscr{D}$ which is a left R-module, i.e. a left *DG* module over R, and if N is a right R-module, then the tensor product $N \otimes_R M$ and its derived functors are also objects of $\mathscr{D}$. If R is a commutative ring in $\mathscr{D}$, that is, an anticommutative *DG* ring in the usual sense, and if M is an R-module, then the symmetric algebra $S^R\{M\} = \bigoplus_{q=0}^{\infty} S_q^R M$ is defined by the standard universal mapping property relative to commutative R-algebras. (When R is the *DG* ring which is the ring B concentrated in degree zero and M is a complex of B-modules, then this symmetric algebra is also denoted by $\tilde{S}$ in this paper.) We may also define the exterior algebra $\Lambda^R M = \bigoplus_{q=0}^{\infty} \Lambda_q^R M$ by means of the standard universal mapping property relative to graded anticommutative R-algebras in $\mathscr{D}$ (thus $\Lambda^R M = \bigoplus (\Lambda_q^R M)_n$ is a bigraded ring anticommutative for the degree $q + n$).

One may prove that if $A \to B$ is a morphism of commutative rings in $\mathscr{D}$, then $\operatorname{Tor}_*^A(B, B)$ is a graded anticommutative ring in $\mathscr{D}$. Hence if $B = A/I$, there is a canonical isomorphism of graded anticommutative B-algebras in $\mathscr{D}$

$$\Lambda_*^B(I/I^2) \to \mathrm{Tor}_*^A(B, B). \tag{9.3}$$

We now carry over the definition 6.10 to this context and say that I is quasiregular if I/I^2 is B-flat and if 9.3 is an isomorphism.

PROPOSITION 9.4. *Suppose that B is a commutative ring in $\mathscr{D}$ and M is a B-module. Assume that M is B-flat and that $n: M \to M$ is an isomorphism for each integer $n \neq 0$. Then the augmentation ideal of $S^B M$ is quasiregular.*

In effect the Koszul complex $S^B M \otimes_B \Lambda_*^B M$ is a flat resolution of B as an $S^B M$-algebra and hence can be used to calculate the Tor in 9.3.

Let $u: A \to B$ be a morphism of (ordinary) commutative rings and regard u as a morphism of DG rings concentrated in degree zero. By a *DG anticommutative A-algebra resolution of B* we mean a factorization $u = pi$, $i: A \to R$, $p: R \to B$ in $\mathscr{D}$ such that p induces isomorphisms on homology.

THEOREM 9.5. *Suppose $A \to B$ is a morphism of commutative rings and that B is of characteristic zero. Let R be a DG anticommutative A-algebra resolution of B such that R is A-flat and such that the augmentation ideal J of the DG ring $R \otimes_A B$ is quasiregular. Then there is a canonical isomorphism $J/J^2 \simeq \mathbf{L}_{B/A}$ in the derived category of B-modules.*

If R is a free DG anticommutative A-algebra, i.e. ignoring differentials R is isomorphic to $\tilde{S}^A\{N\}$ where N is a free complex of A-modules, then the augmentation ideal of $R \otimes_A B$ is quasiregular because of 9.4 and the fact that Tor's in $\mathscr{D}$ do not depend on the differentials. Therefore Theorem 9.5 implies that the cotangent complex can be calculated by using free DG anticommutative A-algebra resolutions of B, provided B is of characteristic zero. One may show that this is equivalent to the method used by Lichtenbaum-Schlessinger **[16]** to define $D^q(B/A, M)$ for $q \leq 2$.

We now apply this theorem to the bar construction of an augmented B-algebra A where B is of characteristic zero. Let $\mathscr{B}(A)$ be the total bar construction and $\overline{\mathscr{B}}(A) = B \otimes_A \mathscr{B}(A)$. If A is B-flat, then $\mathscr{B}(A)$ is a flat DG anticommutative A-algebra resolution of B. Moreover $\overline{\mathscr{B}}(A)$ is a commutative DG Hopf algebra over B so by the Poincaré-Birkhoff-Witt isomorphism 8.4 it is isomorphic as a DG B-algebra to $\tilde{S}Q$, where Q is the indecomposable part. Using 9.4 it follows that the augmentation ideal of $\overline{\mathscr{B}}(A)$ is quasiregular. By the theorem the cotangent complex $\mathbf{L}_{B/A}$ is Q. Thus we have

COROLLARY 9.6. *If A is a flat augmented B-algebra and if B is of characteristic zero, then $\mathbf{L}_{B/A} \simeq J/J^2$ where J is the augmentation ideal in the bar construction $\overline{\mathscr{B}}(A)$. Therefore*

$$\mathbf{L}_{B/A} \simeq L'(\bar{A}[1]) \tag{9.7}$$

where $\bar{A}[1]$ *is the augmentation ideal of* A *situated in degree* 1, *where* L' *is the free anticommutative graded coLie algebra functor, and where the differential on the right hand side of* 9.7 *is the unique degree* -1 *coderivation which in dimension* 2 *is the map* $S_2^B \bar{A} \to \bar{A}$ *given by the multiplication of* A.

Suppose now that B is a flat A-algebra of characteristic zero and consider $B \otimes_A B$ as an augmented B-algebra where the maps $B \to B \otimes_A B \to B$ are given by $b \mapsto b \otimes 1, x \otimes y \mapsto xy$. Applying 9.6 one obtains the following result allowing one to calculate the commutative algebra cohomology by the cochain complex of Harrison **[13]**.

COROLLARY 9.8. *If* B *is a flat* A*-algebra of characteristic zero, then there is an isomorphism* $\mathbf{L}_{B/A} \simeq \Sigma^{-1}(J/J^2)$ *in the derived category of* B*-modules, where* Σ *is the suspension functor, and where* J *is the augmentation ideal of the bar construction* $\bar{\mathscr{B}}(B \otimes_A B)$. *Moreover if* B *is projective as an* A*-module, then*

$$(9.9) \qquad D^q(B/A, M) = H^{q+1}\{\mathrm{Harr}^*(B/A, M)\}, \qquad \textit{where}$$

$$\mathrm{Harr}^q(B/A, M) = \{f \in \mathrm{Hom}_A(\otimes_A^q B, M), f \textit{ vanishes on decomposable elements for the shuffle product}\}$$

$$(df)(b_1, \ldots, b_{q+1}) = b_1 f(b_2, \ldots, b_{q+1}) + \sum_{i=1}^{q} (-1)^i f(b_1, \ldots, b_i b_{i+1}, \ldots, b_{q+1}) + (-1)^{q+1} b_{q+1} f(b_1, \ldots, b_q).$$

When A is a field, the formula 9.9 is due to Barr **[4]**.

10. **Local noetherian rings.** Let A be a local noetherian ring with residue field k and maximal ideal m_A. Suppose that A is the quotient of a regular local ring R, for example if A is complete. R may be chosen to be of minimal dimension, in which case $m_R/m_R^2 \xrightarrow{\sim} m_A/m_A^2 \simeq D_1(k/A)$. Moreover the minimal number of generators for the ideal $\mathrm{Ker}\{R \to A\}$ is then $\dim_k D_2(k/A)$. These results may be generalized as follows.

PROPOSITION 10.1. (i) *If* A *is a local ring which is the quotient of a regular local ring, then there exists a simplicial resolution* P *of* A *such that each* P_q *is a regular local ring and such that each simplicial operation* $P_q \to P_r$ *is a local homomorphism. Moreover there is a canonical isomorphism*

$$(10.2) \qquad m_P/m_P^2 \simeq \Sigma^{-1} \mathbf{L}_{k/A}$$

in the derived category of k*-modules, where* Σ *is the suspension functor and* m_P *is the simplicial maximal ideal of* P.

(ii) *It is possible to choose* P *to be minimal in the sense that all the differentials of the normalized subcomplex of* m_P/m_P^2 *are zero. The completion* $\hat{P}$ *of such a minimal* P *is unique up to noncanonical isomorphism.*

Such a P as in (i) will be called a simplicial regular local ring resolution of A. It follows from 10.2 that the number of nondegenerate parameters in dimension $q-1$ of P is at least $\dim_k D_q(k/A)$ with equality if P is minimal.

If A is an arbitrary local noetherian ring, then its completion $\hat{A}$ is flat as an A-module, so by the flat base change theorem we have that $\mathbf{L}_{k/A} \simeq \mathbf{L}_{k/\hat{A}}$. Choosing a simplicial regular local ring resolution P of $\hat{A}$ and filtering P by the ideals m_P^n, we obtain a spectral sequence:

PROPOSITION 10.3. *If A is a local noetherian ring with maximal ideal m_A and residue field k, there is a spectral sequence of anticommutative k-algebras with*

$$E^2_{pq} = H_{p+q}\{S^k_q(\Sigma^{-1}\mathbf{L}_{k/A})\}$$

and

$$\begin{aligned} E^\infty_{pq} &= 0 && \text{if } p+q \neq 0 \\ &= m^q_A/m^{q+1}_A && \text{if } p+q = 0. \end{aligned}$$

This spectral sequence is not a first quadrant spectral sequence. To remedy this defect, define the Koszul homology $K_*(A)$ of A to be the homology of the Koszul complex $K.(\mathbf{x}, A)$, where $\mathbf{x} = (x_1, \ldots, x_n)$ is a minimal system of generators for m_A. This Koszul homology depends only on A and is isomorphic to $\operatorname{Tor}^R_*(k, A)$ if A is the quotient of a regular local ring R such that $m_R/m_R^2 \simeq m_A/m_A^2$. If P is a simplicial regular local ring resolution of $\hat{A}$, then filtering $P \otimes_{P_0} k$ by powers of its augmentation ideal yields a spectral sequence relating the cotangent complex and Koszul homology.

PROPOSITION 10.4. *There is a first quadrant spectral sequence of k-algebras*

$$E^2_{pq} = H_{p+q}S^k_q\{\Sigma^{-1}T_{\geq 2}\mathbf{L}_{k/A}\} \Rightarrow K_{p+q}(A)$$

where $T_{\geq 2}\mathbf{L}_{k/A}$ is the "Postnikov" subcomplex of $\mathbf{L}_{k/A}$ with the same homology groups in dimensions ≥ 2 and zero homology groups in dimensions ≤ 1. This spectral sequence has the following properties:

Low dimensional isomorphism:

$$K_1(A) \simeq D_2(k/A) \tag{10.5}$$

Edge homomorphisms:

$$K_n(A) \to D_{n+1}(k/A) \tag{10.6}$$

$$\Lambda^k_n D_2(k/A) \to K_n(A) \tag{10.7}$$

where 10.6 *annihilates the decomposable elements of $K_*(A)$ and where* 10.7 *is the unique algebra homomorphism extending the isomorphism* 10.5.

Five term exact sequence:

$$K_3(A) \to D_4(k/A) \to \Lambda^2 K_1(A) \to K_2(A) \to D_3(k/A) \to 0.$$

COROLLARY 10.8. *A is a "local-complete-intersection" local ring* **[24]** *if and only if the map $\Lambda^2 K_1(A) \to K_2(A)$ induced by the algebra structure of $K_*(A)$ is surjective.*

11. **Euler characteristics for graded rings.** Let $A = \bigoplus_{n=0}^{\infty} A_n$ be a graded ring and let B be a graded A-algebra. Then the cotangent complex has a natural structure as a simplicial graded B-module, for it may be calculated using simplicial cofibrant resolutions in the category of graded A-algebras. Consequently

$$D_q(B/A) = \bigoplus_{n=0}^{\infty} D_q(B/A)_n$$

is a graded B-module.

THEOREM 11.1. *Let $A = \bigoplus_{n=0}^{\infty} A_n$ be a graded ring such that A_0 is a field k and such that each A_n is a finite-dimensional vector space over k. Then $D_q(k/A)_n$ is finite-dimensional for each q and n and*

$$D_q(k/A)_n = 0 \quad \text{for } n < q, \tag{11.2}$$

so that the integers

$$\chi_n = \sum_q (-1)^q \dim_k D_q(k/A)_n$$

are defined. Furthermore there is an identity of formal power series

$$\sum_{n \geq 0} (\dim_k A_n) t^n = \prod_{n \geq 1} (1 - t^n)^{\chi_n}. \tag{11.3}$$

This is proved by analyzing the fundamental spectral sequence 6.3 with respect to the grading induced by that of A. The key point is the following connectivity assertion for the symmetric algebra functor.

LEMMA 11.4. *If M is a simplicial k-module such that $H_q(M) = 0$ for $q > n$, then $H_q(S_m^k M) = 0$ for $q > mn$.*

When A is a graded local noetherian ring with residue field k and $D_q(k/A) = 0$ for q sufficiently large, 11.3 becomes an identity of rational functions of t. Equating orders of the pole at $t = 1$ we have the

COROLLARY 11.5. *Let A be a graded local noetherian ring with residue field k. Suppose that $D_q(k/A) = 0$ for q sufficiently large or equivalently that $\mathbf{L}_{k/A}$ is of finite projective dimension as a complex of k-modules. Then*

$$-\dim A = \sum_q (-1)^q \dim_k D_q(k/A)$$

where dim A *is the dimension of* A **[22]**.

REMARK 11.6. We do not know if the corollary remains valid without the hypothesis that A be graded. This would follow from the following special case of the conjecture 5.6:

Conjecture 11.7. If A is a local noetherian ring with residue field k and if $D_q(k/A) = 0$ for q sufficiently large, then A is a local-complete-intersection.

EXAMPLE 11.8. Let A be the graded Artin ring $A = k \oplus V \oplus k \oplus 0 \ldots$, where the multiplication is given by a nondegenerate quadratic form on the k-vector space V. Then A is a Gorenstein ring, however the Poincaré series of A,

$1 + (\dim V)t + t^2$, for $\dim V \geq 3$ has roots which are not roots of unity. Thus by 11.3 there are infinitely many nonzero χ_n and so $D_q(k/A) \neq 0$ for infinitely many q.

Since previous work in local algebra has not produced a class of local noetherian rings intermediate between the local-complete-intersection and Gorenstein rings, we consider the above example evidence for the conjecture 11.7.

BIBLIOGRAPHY

1. M. André, *Method simpliciale en algèbre homologique et algèbre commutative*, Lecture Notes in Mathematics no. 32, Springer, Berlin, 1967.

2. E. F. Assmus, *On the homology of local rings*, Illinois J. Math. **3** (1959), 187–199.

3. M. Barr, *Shukla cohomology and triples*, J. Algebra **5** (1967), 222–231.

4. ———, *Harrison homology, Hochschild homology and triples*, J. Algebra **8** (1963), 314–323.

5. M. Barr and J. Beck, *Acyclic models and triples*, Proceedings of the Conference on Categorical Algebra, La Jolla, Springer, Berlin, 1966.

6. E. Curtis, *Some relations between homotopy and homology*, Ann. of Math. (2) **83** (1965), 386–413.

7. A. Dold and D. Puppe, *Homologie nicht-additiver Funktoren*, Ann. Inst. Fourier **11** (1961), 201–312.

8. M. Gerstenhaber, *On the deformation of rings and algebras*. II, Ann. of Math. (2) **84** (1966), 1–19.

9. A. Grothendieck, *Categories cofibrées additives et complex cotangent relatif*, Lecture Notes in Mathematics no. 79, Springer, Berlin, 1968.

10. A. Grothendieck et al., *Séminaire de Géometrie Algèbrique* 6, Exposé VII (to appear).

11. A. Grothendieck and J. Dieudonné, *Élements de Géometrie Algèbrique*, Inst. Hautes Études Sci. Publ. Math.

12. A. Grothendieck and M. Demazure, *Séminaire de Géometrie Algèbrique* 1963–64, *Schémas en groupes* (to appear).

13. D. K. Harrison, *Commutative algebras and cohomology*, Trans. Amer. Math. Soc. **104** (1962), 191–204.

14. L. Illusie, *Complexe cotangent relatif d'un faisceau d'algèbres*, Comptes Rendus **268**, Série A (1969), 278 and 323.

15. F. W. Lawvere, *Functorial semantics of algebraic theories*, Proc. Nat. Acad. Sci. USA **50** (1963), 869–872.

16. S. Lichtenbaum and M. Schlessinger, *The cotangent complex of a morphism*, Trans. Amer. Math. Soc. **128** (1967), 41–70.

17. S. MacLane, *Categorical algebra*, Bull. Amer. Math. Soc. **71** (1965), 40–106.

18. D. Quillen, *Homotopical algebra*, Lecture Notes in Mathematics, Springer, Berlin, 1967.

19. ———, *On the homology of commutative rings*, Mimeographed Notes, M.I.T.

20. ———, *Rational homotopy theory*, Ann. of Math. (to appear).

21. M. Schlessinger, Thesis, Harvard, 1965.

22. J.-P. Serre, *Algebra locale. Multiplicities*, Lecture Notes in Mathematics no. 11, Springer 1965.

23. U. Shukla, *Cohomologie des algèbres associatives*, Ann. Sci. École Norm. Sup. **78** (1961), 163–209.

24. J. Tate, *Homology of noetherian rings and local rings*, Illinois J. Math. **1** (1957), 14–27.

25. J.-L. Verdier, *Categories derivées, quelques resultats* (*état* 0), Mimeographed Notes, Inst. Hautes Études Sci. Publ. Math.

26. ———, *Seminaire de géometrie algèbrique* (1963–64), *Cohomology étale des schemas* Exposés I—V (to appear).

MASSACHUSETTS INSTITUTE OF TECHNOLOGY

Nonabelian homological algebra and K-theory

Richard G. Swan

The main purpose of this paper is to propose a possible definition of the functors $K_n(R)$ for arbitrary rings R and all integers $n \geq 0$. It is as yet too early to tell if this theory will prove to be useful and whether it should be considered to give the "correct" generalization of the functors K_0 and K_1. It does seem to have some useful properties and the methods used to define it may prove useful in other situations. Recently, a different definition of $K_n(R)$ was proposed by O. Villamayor. His theory is obtained in quite a different way and his K_1 differs somewhat from that of Bass [**6**]. In the theory I will present here, the K_1 is the same as that of Bass and the K_2 is closely related to (and probably the same as) the one defined by Milnor [**30**], [**36**].

The method used to define this theory is that of simplicial resolutions. It has been known for some time [**11**] that ordinary chain complexes and projective resolutions do not behave properly in the nonabelian case. In fact, the reason they work so well in the abelian case is that in this case, the category of chain complexes is equivalent to the category of simplicial abelian groups [**11**]. In the general case, it is always the simplicial object which behaves correctly.

The method I will use to construct these resolutions was directly inspired by the work of André [**1**], [**2**]. André shows that his cohomology is independent of the choice of resolution by using a standard complex and a spectral sequence argument to compare the cohomology obtained from any resolution with that obtained from the standard complex. This does not seem to work in the general case where there is no abelian structure. Instead, I will prove a comparison theorem analogous to the usual one [**7**], [**26**] in the abelian case. This theory is developed in §§1 to 4. In §§5 and 6 we apply it to generalize the theory of acyclic models to the nonabelian case. In §7 we give some topological motivation for the definition of K_n. This definition is then given in §8. In §§9 and 10 we give some remarks and conjectures concerning these functors.

One remark should be made about the notation. If $\mathscr{C}$ is a category, I will denote the set of morphisms from A to B in $\mathscr{C}$ by $\mathscr{C}(A, B)$. However, if the category in question has not been assigned any symbol, I will just write Hom (A, B) for this set. The category intended should be clear from the context.

1. **Simplicial sets and complexes.** In this section we will collect some results, mostly well known, which will enable us to avoid making long calculations with face and degeneracy operators.

Let $\mathscr{S}$ be the category whose objects are the sets $[n] = \{0, 1, \ldots, n\}$ for integers $n \geq 0$ and whose morphisms are the monotone nondecreasing functions. If $\mathscr{C}$ is any category, a simplicial object in $\mathscr{C}$ is, by definition, a contravariant functor A from $\mathscr{S}$ to $\mathscr{C}$, i.e. $A: \mathscr{S}^0 \to \mathscr{C}$. We write $A_n = A([n])$. Let $\varepsilon_i: [n-1] \to [n]$ be the unique monotone map with image $[n] - \{i\}$, $i = 0, \ldots, n$ and $\delta_i: [n+1] \to [n]$ the unique monotone map with i as the only element with two pre-images, $i = 0, \ldots, n$. Let $\partial_i = A(\varepsilon_i)$, $s_i = A(\delta_i)$. It is a standard result [29] that there is a one-to-one correspondence between simplicial objects in $\mathscr{C}$ and sequences A_0, A_1, $A_2, \ldots$ of objects with maps $\partial_i: A_n \to A_{n-1}$, $s_i: A_n \to A_{n+1}$ satisfying the relations

$$(1)\qquad \begin{gathered}\partial_i\partial_j = \partial_{j-1}\partial_i \text{ for } i < j, \quad s_is_j = s_{j+1}s_i \text{ for } i \leq j,\\ \partial_is_j = s_{j-1}\partial_i \text{ for } i < j, \quad \partial_js_j = \partial_{j+1}s_j = \mathrm{id}, \quad \partial_is_j = s_j\partial_{i-1} \text{ for } i > j+1.\end{gathered}$$

A morphism of simplicial objects is a natural transformation of functors. Equivalently, it is a sequence of maps $f_n: A_n \to B_n$ which commute with the face and degeneracy operators ∂_i and s_j.

If $\mathscr{C}$ is the category of sets, groups, etc., we refer to a simplicial object in $\mathscr{C}$ as a simplicial set (semisimplicial complex), simplicial group, etc.. We shall assume the reader is familiar with the properties of such objects. A good exposition may be found in [29].

A simplicial complex (in the classical sense) is a collection K of nonempty finite subsets of some set X such that $\sigma \in K$, $\tau \subset \sigma$, $\tau \neq \varnothing$ implies $\tau \in K$. Let K_n be the set of those $\sigma \in K$ with $|\sigma| = n + 1$, i.e. σ has $n + 1$ elements.

We say L is a subcomplex of K if $L \subset K$ and L is itself a simplicial complex. Clearly unions and intersections of subcomplexes are again subcomplexes.

We shall consider here only *ordered* simplicial complexes [17]. By definition, this means that each set $\sigma \in K$ is totally ordered and that if $\tau \subset \sigma$, the ordering of τ agrees with that of σ. To every such ordered complex we associate a simplicial set $S(K)$ as follows: Let $S_n(K)$ consist of all $(n+1)$-tuples $(v_0, \ldots, v_n)$ such that $\{v_0, \ldots, v_n\} = \sigma \in K$ and $v_0 \leq v_1 \leq \cdots \leq v_n$ in the ordering of σ. The v_i need not be distinct. If $\alpha: [m] \to [n]$ in $\mathscr{S}$, define $\hat{\alpha}: S_n(K) \to S_m(K)$ by $\hat{\alpha}(v_0, \ldots, v_n) = (v_{\alpha(0)}, \ldots, v_{\alpha(m)})$. It is trivial to verify that this makes $S(K)$ a simplicial set. Note that

$$(2)\qquad \begin{aligned}\partial_i(v_0, \ldots, v_n) &= (v_0, \ldots, v_{i-1}, v_{i+1}, \ldots, v_n),\\ s_i(v_0, \ldots, v_n) &= (v_0, \ldots, v_i, v_i, \ldots, v_n).\end{aligned}$$

If $L \subset K$, clearly $S(L) \subset S(K)$ and $S(L)$ is a subsimplicial set of $S(K)$.

Let K be an ordered simplicial complex. If $\sigma \in K_n$ then $\sigma = \{v_0, v_1, \ldots, v_n\}$ where $v_0 < v_1 < \cdots < v_n$. Define $\partial_i\sigma = \{v_0, \ldots, v_{i-1}, v_{i+1}, \ldots, v_n\}$, i.e. omit v_i.

Then $\partial_i \sigma \in K_{n-1}$, $i = 0, 1, \ldots, n$. Let A be a simplicial set. We define a *map* f from K to A to be a collection of maps $f_n : K_n \to A_n$ such that $\partial_i f_n = f_{n-1} \partial_i$ for $i = 0, \ldots, n$, $n \geq 0$. Let Map (K, A) be the set of such maps. We have an obvious map $u = u_K : K \to S(K)$ by $u(\{v_0, \ldots, v_n\}) = (v_0, \ldots, v_n)$. We will show that this is a *universal* such map. In other words, if $f: K \to B$ is any map into a simplicial set B, then there is a unique $g: S(K) \to B$ with $f = gu$. Let Hom (A, B) denote the set of morphisms $A \to B$ of simplicial sets. Define θ: Hom $(S(K), B) \to$ Map (K, B) by $\theta(g) = gu$.

PROPOSITION 1.1. *θ: Hom $(S(K), B) \to$ Map (K, B) is a bijection.*

PROOF. It is clear that there is some map $w: K \to U$ which is universal, i.e. Hom $(U, B) \to$ Map (K, B) by $g \to gw$ is a bijection. This follows from the adjoint functor theorem **[18]** or by general results of universal algebra **[8]**. Let $h: U \to S(K)$ be the unique map with $hw = u$. We claim that h is an isomorphism. Any element of $S(K)$ can be obtained by applying degeneracy operations to the nondegenerate elements $(v_0, \ldots, v_n)$ with $v_0 < \cdots < v_n$. These elements have the form $u(\{v_0, \ldots, v_n\})$ and so lie in the image of h. Since h is a map of simplicial sets, this shows that h is onto. If we let U' be the subsimplicial set of U generated (with respect to the operations ∂_i, s_i) by $w(K)$, then U' clearly has the same universal property as U, so $U' = U$. Using the relations (1) we can put any composition of ∂_i's and s_i's in the form $s_{j_1} \cdots s_{j_m} \partial_{i_1} \cdots \partial_{i_m}$. But $w(K)$ is stable under the ∂_i's since $w: K \to U$ is a map. Therefore any element of U has the form $s_{j_1} \cdots s_{j_m} w(\sigma)$ where $\sigma \in K$. Using the relations (1) we can assume $j_1 > j_2 > \cdots > j_m$. Now the elements of $S(K)$ which are nondegenerate, i.e. not of the form $s_i x$ for some i, are precisely the elements of the form $u(\sigma)$ with $\sigma \in K$. Since $h(s_{j_1} \cdots s_{j_m} w(\sigma)) = s_{j_1} \cdots s_{j_m} u(\sigma)$, the fact that h is an injection follows from the following lemma of Eilenberg-Zilber **[21]**.

LEMMA 1.2. *Let A be a simplicial set. Then any element $x \in A$ can be put uniquely in the form $x = s_{j_1} \cdots s_{j_m} y$, where $j_1 > j_2 > \cdots > j_m$ and y is nondegenerate.*

PROOF. It is clear that x has this form. The only problem is to show the uniqueness. Suppose we apply a face operator ∂_i to x. The relations (1) show that we can commute the ∂_i past each s_j (with changes of index) except that at some point the ∂_i may cancel some s_j. Thus $\partial_i x$ is either of the form $s_{k_1} \cdots s_{k_m} \partial_l y$ or $s_{k_1} \cdots s_{k_{m-1}} y$. The latter case will occur, e.g. if $i = j_1$. From this it is clear that we cannot reach a nondegenerate element by applying fewer than m face operators to x but we can get a nondegenerate element by applying m face operators. This shows that m is unique. Write $m = \text{dgn}(x)$. If $m = 0$ there is nothing to prove. We proceed by induction on m. If $i < j_m$, relations (1) show that ∂_i will commute past all the s_j and so $\partial_i x = s_{k_1} \cdots s_{k_m} \partial_i y$. Thus dgn $(\partial_i x) \geq m$ for $i < j_m$. If $i = j_m$, however, $\partial_i x = s_{k_1} \cdots s_{k_{m-1}} \partial_i s_{j_m} y = s_{k_1} \cdots s_{k_{m-1}} y$, so dgn $(\partial_i x) = m - 1$. Therefore j_n is characterized as the first i for which dgn $(\partial_i x) = m - 1$. By induction, y and the remaining j_ν are uniquely determined by $\partial_{j_m} x$. (Note that $k_\nu = j_\nu - 1$.)

We now apply Proposition 1.1 to the following situation. Let K be an ordered simplicial complex and let L, M be subcomplexes of K (with the induced ordering) such that $L \cup M = K$. If A is simplicial set and $h: K \to A$ is a map, then $f = h \mid L$, $g = h \mid M$ are maps such that $f \mid L \cap M = g \mid L \cap M$. Conversely, if we are given $f: L \to A$, $g: M \to B$ with $f \mid L \cap M = g \mid L \cap M$ then it is trivial to verify that we get a unique map $h: K \to A$ by setting $h(\sigma) = f(\sigma)$ or $g(\sigma)$, whichever is defined. Applying Proposition 1.1 we get

COROLLARY 1.3. *If $K = L \cup M$, the diagram*

$$\begin{array}{ccc} S(L \cap M) & \longrightarrow & S(L) \\ \downarrow & & \downarrow \\ S(M) & \longrightarrow & S(K) \end{array}$$

is co-cartesian, i.e. a pushout diagram in the category of simplicial sets.

We can reformulate Proposition 1.1 as follows. To give a map $f: K \to A$, we give maps $f_n: K_n \to A_n$ with $\partial_i f_n = f_{n-1}\partial_i$. Since $K \to S(K)$ is a universal such map we have

COROLLARY 1.4. *$S(K)$ is the simplicial set generated by symbols $\bar{\sigma}$ corresponding to the simplexes of K with the relations*

$$\partial_i \bar{\sigma} = \overline{\partial_i \sigma} \text{ and } \deg \bar{\sigma} = n \text{ if } \sigma \in K_n.$$

A precise definition of these terms may be found in [8]. We must keep in mind the fact that a simplicial set is a *graded* algebra of a certain type.

If K is finite-dimensional we can find a much smaller presentation of $S(K)$. If $\dim K = N$, each simplex σ of K is contained in a maximal one τ since $|\tau| \le N + 1$ for all $\tau \in K$. Clearly $\sigma = \partial_{i_1} \cdots \partial_{i_s} \tau$ for some $i_1, \ldots, i_s$. Therefore $S(K)$ is generated by the maximal simplexes of K.

PROPOSITION 1.5. *Let K be a finite-dimensional ordered simplicial complex. Then $S(K)$ is generated by symbols $\bar{\sigma}$, one for each maximal simplex σ of K, with the following relations: For each pair of maximal simplexes σ, τ, choose expressions for $\sigma \cap \tau$ of the form $\sigma \cap \tau = \partial_{i_1} \cdots \partial_{i_s} \sigma = \partial_{j_1} \cdots \partial_{j_t} \tau$. Then the relations $\partial_{i_1} \cdots \partial_{i_s} \bar{\sigma} = \partial_{j_1} \cdots \partial_{j_t} \bar{\tau}$ for each pair σ, τ suffice to define $S(K)$.*

PROOF. We have already observed that these σ generate $S(K)$. It is clear that the given relations hold. Suppose A is a simplicial set and f is a map from the maximal simplexes of K to A, preserving degree, such that if σ, τ are two maximal simplexes and we express $\tau \cap \sigma$ as Proposition 1.5 then

$$\partial_{i_1} \cdots \partial_{i_s} f(\sigma) = \partial_{j_1} \cdots \partial_{j_t} f(\tau).$$

We must show that f extends to a morphism $S(K) \to A$. The uniqueness of this extension is clear since the $\bar{\sigma}$ generate $S(K)$. By Proposition 1.1 it will suffice to extend f to a map $g: K \to A$. Let $\sigma \in K$ and let τ be a maximal simplex containing σ. Write $\sigma = \partial_{k_1} \cdots \partial_{k_r} \tau$ and define $g(\sigma) = \partial_{k_1} \cdots \partial_{k_r} f(\tau)$. We must show that this is independent of the choice of τ and $k_1, \ldots, k_r$. It will then be clear that $g(\partial_i \sigma) = \partial_i g(\sigma)$. Let $\alpha = \varepsilon_{k_r} \cdots \varepsilon_{k_1}: [m] \to [n]$ where $\sigma \in K_m$ and $\tau \in K_n$. Then α is injective and if $\tau = \{v_0, \ldots, v_n\}$, $v_0 < \cdots < v_n$, then $\sigma = \{v_{\alpha(0)}, \ldots, v_{\alpha(m)}\}$. Therefore the map α is unique and does not depend on the choice of $k_1, \ldots, k_r$. On A, $\partial_{k_1} \cdots \partial_{k_r} = A(\alpha)$ so $g(\sigma) = A(\alpha) f(\tau)$ does not depend on the choice of $k_1, \ldots, k_r$. Suppose now $\sigma \subset \tau'$ where τ' is another maximal simplex. Let $\tau \cap \tau' = \partial_{i_1} \cdots \partial_{i_s} \tau = \partial_{j_1} \cdots \partial_{j_t} \tau'$ be the expressions chosen in Proposition 1.5. Since $\sigma \subset \tau \cap \tau'$ we can write $\sigma = \partial_{k_1} \cdots \partial_{k_r} (\tau \cap \tau')$ so $\sigma = \partial_{k_1} \cdots \partial_{k_r} \partial_{i_1} \cdots \partial_{i_s} \tau = \partial_{k_1} \cdots \partial_{k_r} \partial_{j_1} \cdots \partial_{j_t} \tau'$. The values for $g(\sigma)$ obtained are $\partial_{k_1} \cdots \partial_{k_r} \partial_{i_1} \cdots \partial_{i_s} f(\tau)$ and $\partial_{k_1} \cdots \partial_{k_r} \partial_{j_1} \cdots \partial_{j_t} f(\tau')$, but these are equal by the hypothesis on f.

We now apply this result to some special cases. Let $\bar{\sigma}^n$ be the n-simplex, consisting of all subsets of a set with $n + 1$ elements. Let $\Delta^n = S(\bar{\sigma}^n)$. Since $\bar{\sigma}^n$ has a unique maximal simplex, we see that Δ^n is the free simplicial set on one element ι^n of dimension n. Therefore if A is any simplicial set, Hom $(\Delta^n, A) = A_n$ where $f: \Delta^n \to A$ corresponds to $f(\iota^n) \in A_n$. By Yoneda's lemma **[18]** the representable functor Hom (, $[n]$) on S has the same property. Therefore we can identify Δ^n with the functor Hom (, $[n]$)$: S^0 \to$ Sets.

Let $\dot{\sigma}^n$ be the boundary of $\bar{\sigma}^n$, i.e. $\dot{\sigma}^n = \bar{\sigma}^n - \{\sigma^n\}$. Then $\dot{\sigma}^n$ has as maximal simplexes the $\partial_i \sigma^n$, $i = 0, \ldots, n$. By the relations (1) we have $\partial_i \partial_j \sigma^n = \partial_{j-1} \partial_i \sigma^n$ for $i < j$. This clearly is the simplex $\partial_i \sigma^n \cap \partial_j \sigma^n$ since it lies in $\partial_i \sigma^n$ and $\partial_j \sigma^n$ and has dimension $n - 2$. Let $\dot{\Delta}^n = S(\dot{\sigma}^n) \subset \Delta^n$. Then $\dot{\Delta}^n$ is generated by the elements $\iota_i = \partial_i \iota$, $i = 0, \ldots, n$ with the relations $\partial_i \iota_j = \partial_{j-1} \iota_i$ for $i < j$.

Let K and L be ordered simplicial complexes formed by certain subsets of sets X and Y respectively. The product $K \times L$ is defined to be the collection of subsets of $X \times Y$ of the form $\{(v_0, w_0), \ldots, (v_n, w_n)\}$ where

$$\{v_0, \ldots, v_n\} \in K_n, \qquad \{w_0, \ldots, w_n\} \in L_n, \qquad (v_i, w_i) \neq (v_{i+1}, w_{i+1})$$

and

$$v_0 \leq \cdots \leq v_n, \qquad w_0 \leq \cdots \leq w_n.$$

We order this subset by $(v_0, w_0) < \cdots < (v_n, w_n)$. It is clear from the definition that $S(K \times L) = S(K) \times S(L)$. Let $I = \Delta^1$ and consider $\Delta^n \times I = S(\bar{\sigma}^n \times \bar{\sigma}^1)$. This has maximal simplexes $\tau_i = \{(0, 0), \ldots, (i, 0), (i, 1), \ldots, (n, 1)\}$ **[17]** where we denote the vertices of $\bar{\sigma}^n$ by $0, 1, \ldots, n$ and those of $\bar{\sigma}^1$ by 0, 1. Here i runs from 0 to n.

Corollary 1.6. *$\Delta^n \times I$ is generated by the elements $\tau_0, \ldots, \tau_n$ of degree $n + 1$ with the relations $\partial_{i+1} \tau_i = \partial_{i+1} \tau_{i+1}$ for $i = 0, \ldots, n - 1$.*

Proof. It is easier to apply the proof of Proposition 1.5 here. Only the very last part needs to be modified. We first note that $\partial_{i+1} \tau_i = \partial_{i+1} \tau_{i+1}$ is the intersection $\tau_i \cap \tau_{i+1}$. Suppose σ is some simplex of $\bar{\sigma}^n \times \bar{\sigma}^1$ and $\sigma \subset \tau_i \cap \tau_j$ where $i < j$. We must show that $g(\sigma)$ defined in the proof of Proposition 1.5 is the same for

τ_i and τ_j. This is clear for $j = i + 1$. Now for $i < j$, $\tau_i \cap \tau_j = \{(0, 0), \ldots, (i, 0), (j, 1), \ldots, (j, n)\}$. This lies in all τ_k for $i \le k \le j$. Therefore if τ_k, τ_{k+1} give the same value of $g(\sigma)$ for all k we see that τ_i and τ_j will give the same $g(\sigma)$.

This result is often useful in constructing homotopies.

We conclude this section by pointing out a consequence of Corollary 1.3. Let K be a finite ordered simplicial complex and let L be a subcomplex of K. Let σ be maximal among the simplexes of K not in L. Let $M = K - \{\sigma\}$. This is clearly a subcomplex of K and $L \subset M$. Let $\bar{\sigma}$ be the subcomplex of K consisting of all faces of σ. Then $K = M \cup \bar{\sigma}$ while $M \cap \bar{\sigma} = \dot{\sigma}$, the subcomplex consisting of all proper faces of σ. Suppose $\sigma \in K_n$ so $S(\bar{\sigma}) = \Delta^n$, $S(\dot{\sigma}) = \dot{\Delta}^n$. By Corollary 1.3, the diagram

$$\begin{array}{ccc} \dot{\Delta}^n & \longrightarrow & \Delta^n \\ \downarrow & & \downarrow \\ S(M) & \longrightarrow & S(K) \end{array}$$

is co-cartesian. By induction on the number of simplexes in $K - L$ we see that we can obtain $S(K)$ from $S(L)$ by a finite number of such pushouts. These pushouts are the simplicial version of the familiar process of attaching cells to a space. They will form the basis for our construction of simplicial resolutions.

2. **Simplicial homotopies.** Let $\mathscr{C}$ be a category having finite direct sums which we denote by $A \amalg B$ or $\coprod A_i$. Let $\mathscr{F}$ be the category of finite sets and all functions. If $A \in \mathscr{C}$ and $S \in \mathscr{F}$ define $S \times A = \coprod_S A$, the direct sum of $|S|$ copies of A indexed by S [**31**]. Let $i_s: A \to S \times A$ be the sth injection. Clearly $S \times A$ is a functor in A. If $f: S \to T$, define $f \times A: S \times A \to T \times A$ by $(f \times A)i_s = i_{f(s)}$. This gives us a functor $\mathscr{F} \times \mathscr{C} \to \mathscr{C}$ characterized by the adjunction property

$$(1) \qquad \mathscr{C}(S \times A, B) = \operatorname{Hom}(S, \mathscr{C}(A, B))$$

where Hom denotes maps of sets here. This property shows that if we regard $S \times A$ as a functor in S, holding A fixed, it will preserve direct limits.

Suppose S is a simplicial object of $\mathscr{F}$, i.e. $S: \mathscr{S}^0 \to \mathscr{F}$. If $Q \in \mathscr{C}$ the composition $\mathscr{S}^0 \xrightarrow{S} \mathscr{F} \xrightarrow{\times Q} \mathscr{C}$ gives us a simplicial object $S \times Q$ of $\mathscr{C}$. If B is a simplicial object of $\mathscr{C}$, the naturality of (1) shows that

$$(2) \qquad \operatorname{Hom}(S \times Q, B) = \operatorname{Hom}(S, \mathscr{C}(Q, B))$$

where the Hom's now denote maps of simplicial objects and $\mathscr{C}(Q, B)$ is the simplicial set defined by the composition $\mathscr{S}^0 \xrightarrow{B} \mathscr{C} \xrightarrow{\mathscr{C}(Q,-)}$ Sets. Again it follows that the functor $- \times Q$ preserves direct limits.

If S is a simplicial object of $\mathscr{F}$ and A is a simplicial object of $\mathscr{C}$, the composition

$$\mathscr{S}^0 \xrightarrow{S \times A} \mathscr{F} \times \mathscr{C} \longrightarrow \mathscr{C}$$

gives us a simplicial object $S \times A$ of $\mathscr{C}$. We leave the formulation of the appropriate adjunction property to the reader since it will not be needed here. We use this construction here to formulate the notion of homotopy. Using the notation of §1, let $\bar{\sigma}^1$ be a 1-simplex, and $\bar{\sigma}^0$ a 0-simplex. Let $I = \Delta^1 = S(\bar{\sigma}^1)$ and $* = \Delta^0 = S(\bar{\sigma}^0)$. There are two inclusions $i_0, i_1 : \bar{\sigma}^0 \to \bar{\sigma}^1$ which induce $i_0, i_1 : * \to I$. By the definition (or by (1)), $* \times A$ is canonically isomorphic to A. We will identify them. The two maps $* \to I$ give $i_0, i_1 : A = * \times A \to I \times A$.

DEFINITION. Let A, B be simplicial objects of $\mathscr{C}$. We say two morphisms $f, g : A \to B$ are homotopic if there is a morphism $h : I \times A \to B$ such that $f = hi_0$, $g = hi_1$.

This agrees with the usual definition for simplicial sets. However, even for simplicial groups it looks quite different because if G is a simplicial group and G' its underlying set then clearly $(I \times G)' \neq I \times G'$. Therefore we must show that this definition agrees with the usual one.

PROPOSITION 2.1. *Let A and B be simplicial objects in $\mathscr{C}$. Let $f, g : A \to B$. Then there is a one-one correspondence between homotopies $h : I \times A \to B$ between f and g and collections of maps $h_i^{(n)} : A_n \to B_n$ for $i = -1, 0, \ldots, n$, $n \geq 0$ satisfying*

(a) $f \mid A_n = h_{-1}^{(n)}, g \mid A_n = h_n^{(n)}$,

(b) $\partial_i h_j^{(n)} = h_{j-1}^{(n-1)} \partial_i$ *if* $i \leq j$, $\partial_i h_j^{(n)} = h_j^{(n-1)} \partial_i$ *if* $i > j$,

(c) $s_i h_j^{(n)} = h_{j+1}^{(n+1)} s_i$ *if* $i \leq j$, $s_i h_j^{(n)} = h_j^{(n+1)} s_i$ *if* $i > j$.

PROOF. The 1-simplex $\bar{\sigma}^1$ has two vertices 0 and 1. We order it by $0 < 1$. Now I_n consists of all sequences $(0, \ldots, 1, \ldots, 1)$. We number these $x_{-1}^{(n)}, \ldots, x_n^{(n)}$ where $x_i^{(n)}$ has exactly $i + 1$ entries 0. The simplicial set $*$ has one element $*_n$ in each dimension n and $i_0(*_n) = x_{-1}^{(n)}$, $i_1(*_n) = x_n^{(n)}$. If K is a simplicial set, a map $h : K \times A \to B$ is given by a collection of maps $h^{(n)} : K_n \times A_n \to B_n$ commuting with ∂_i, s_i. By (1) a map $h^{(n)} : K_n \times A_n \to B_n$ is equivalent to a collection of maps $h_x^{(n)} : A_n \to B_n$ indexed by the elements $x \in K_n$. The condition $\partial_i h^{(n)} = h^{(n-1)} \partial_i$ is expressed by the diagram

$$\begin{array}{ccc} A_n \times K_n & \longrightarrow & B_n \\ \downarrow{\scriptstyle \partial_i} \qquad \downarrow{\scriptstyle \partial_i} & & \downarrow{\scriptstyle \partial_i} \\ A_{n-1} \times K_{n-1} & \longrightarrow & B_{n-1} \end{array}$$

and so is clearly equivalent to the set of conditions $\partial_i h_x = h_{\partial_i x} \partial_i$. Similarly the condition $s_i h^{(n)} = h^{(n+1)} s_i$ is equivalent to $s_i h_x = h_{s_i x} s_i$. If $K = I$, define $h_i^{(n)} = h_x^{(n)}$ for $x = x_i^{(n)}$. A trivial computation shows that our conditions translate into (b) and (c) and that the conditions $f = hi_0$, $g = hi_1$ translate into (a).

The conditions of Proposition 2.1 make sense in any category even without direct sums. We may use them as the definition of homotopy in the general case. It should be remarked that homotopy in this sense is not an equivalence relation even for simplicial sets [**29**]. However this will not cause any difficulty.

If $F:\mathscr{C} \to \mathscr{D}$ is a (covariant) functor and A is a simplicial object of $\mathscr{C}$, the composition $\mathscr{S}^0 \xrightarrow{A} \mathscr{C} \xrightarrow{F} \mathscr{D}$ defines a simplicial object $F(A)$ of $\mathscr{D}$. Since the conditions of Proposition 2.1 are preserved by F we get

COROLLARY 2.2. *If $f, g: A \to B$ are homotopic, so are $F(f), F(g): F(A) \to F(B)$.*

The conditions of Proposition 2.1 are not the usual ones [**29**]. However, it is easy to see they are equivalent.

PROPOSITION 2.3. *Let A and B be simplicial objects of $\mathscr{C}$. Let $f, g: A \to B$. Then there is a one-one correspondence between collections of maps $h_i^{(n)}$ satisfying the conditions of Proposition 2.1 and collections of maps $k_i^{(n)}: A_n \to B_{n+1}$ for $i = 0, \ldots, n$, $n \geq 0$ satisfying the conditions*

(a) $\partial_0 k_0^{(n)} = f \mid A_n$, $\partial_{n+1} k_n^{(n)} = g \mid A_n$,

(b) $\partial_i k_j^{(n)} = k_{j-1}^{(n-1)} \partial_i$ *if* $i < j$, $\partial_{j+1} k_{j+1}^{(n)} = \partial_{j+1} k_j^{(n)}$ *and* $\partial_i k_j^{(n)} = k_j^{(n-1)} \partial_{i-1}$ *for* $i > j+1$,

(c) $s_i k_j^{(n)} = k_{j+1}^{(n+1)} s_i$ *if* $i \leq j$, $s_i k_j^{(n)} = k_j^{(n+1)} s_{i-1}$ *if* $i > j$.

This correspondence is given by $k_i^{(n)} = h_i^{(n)} s_i$ and $k_i^{(n)} = \partial_{i+1} h_i^{(n)}$ or $\partial_{i+1} h_{i+1}^{(n)}$, whichever is defined.

The proof consists of verifying all the assertions by direct (and very easy) computations.

3. **Simplicial cofibrations.** We now define the analogue of the process of attaching cells to a space. We begin with a few simple lemmas.

LEMMA 3.1. *Let $Q \in \mathscr{C}$ and let A be a simplicial object of $\mathscr{C}$. Then there is a natural isomorphism* $\mathrm{Hom}\,(\Delta^n \times Q, A) = \mathscr{C}(Q, A_n)$.

PROOF. By (2), $\mathrm{Hom}\,(\Delta^n \times Q, A) = \mathrm{Hom}\,(\Delta^n, \mathscr{C}(Q, A))$. By §1, Δ^n is free on one generator ι of dimension n so $\mathrm{Hom}\,(\Delta^n, X) = X_n$ for any simplicial set X.

By a finite system $\mathbf{D}$ in $\mathscr{C}$ we will mean an indexed collection of objects D_α and morphisms $\varphi_{\alpha\beta}^\gamma: D_\alpha \to D_\beta$. If $\mathscr{C}$ has finite inverse limits we can define $D = \varprojlim \mathbf{D}$. However, we shall avoid this assumption by simply defining $\mathscr{C}(Q, \mathbf{D})$ for $Q \in \mathscr{C}$ to be the set of all collections (f_α) such that $f_\alpha: Q \to D_\alpha$ and $\varphi_{\alpha\beta}^\gamma f_\alpha = f_\beta$ for all α, β, γ. Therefore if $D = \varprojlim \mathbf{D}$ exists we have $\mathscr{C}(Q, D) = \mathscr{C}(Q, \mathbf{D})$ by definition.

DEFINITION. Let A be a simplicial object of $\mathscr{C}$. By $\mathbf{Z}_n(A)$ we will mean the finite system with objects D_i for $i = 0, \ldots, n+1$ and D_{ij} for $0 \leq i < j \leq n+1$, where $D_i = A_n$ for all i and $D_{ij} = A_{n-1}$ for all i, j, with maps $\partial_i: D_j \to D_{ij}$, $\partial_{j-1}: D_i \to D_{ij}$ given by $\partial_i: A_n \to A_{n-1}$, $\partial_{j-1}: A_n \to A_{n-1}$.

Thus a map $Q \to \mathbf{D}$ is equivalent to a collection of maps $f_i: Q \to A_n$, $i = 0, \ldots, n+1$ satisfying the conditions $\partial_i f_i = \partial_{j-1} f_j$ for $i < j$. In particular, we have a canonical map $\partial: A_{n+1} \to \mathbf{Z}_n(A)$ given by the maps $\partial_i: A_{n+1} \to A_n$. If $\mathscr{C}$ has finite inverse limits we may define $Z_n(A) = \varprojlim \mathbf{Z}_n(A)$. For simplicial sets, $Z_n(A)$ is the set of all sequences $(x_0, \ldots, x_{n+1})$ with $x_i \in A_n$ and $\partial_i x_j = \partial_{j-1} x_i$ for $i < j$.

LEMMA 3.2. *Let* $Q \in \mathscr{C}$ *and let* A *be a simplicial object of* $\mathscr{C}$. *Then there is a natural isomorphism* $\mathrm{Hom}\,(\dot{\Delta}^n \times Q, A) = \mathscr{C}(Q, \mathbf{Z}_{n-1}(A))$. *Also, the map*

$$\mathrm{Hom}\,(\Delta^n \times Q, A) \to \mathrm{Hom}\,(\dot{\Delta}^n \times Q, A)$$

given by $\dot{\Delta}^n \subset \Delta^n$ *coincides with the map* $\mathscr{C}(Q, \partial): \mathscr{C}(Q, A_n) \to \mathscr{C}(Q, \mathbf{Z}_{n-1}(A))$.

PROOF. By (2), $\mathrm{Hom}\,(\dot{\Delta}^n \times Q, A) = \mathrm{Hom}\,(\dot{\Delta}^n, \mathscr{C}(Q, A))$. In §1 we showed that $\dot{\Delta}^n$ is generated by $\iota_0, \ldots, \iota_n \in (\dot{\Delta}^n)_{n-1}$ with relations $\partial_i \iota_j = \partial_{j-1}\iota_i$ for $i < j$. Thus an element of $\mathrm{Hom}\,(\dot{\Delta}^n, \mathscr{C}(Q, A))$ is a collection of maps $f_0, \ldots, f_n \in \mathscr{C}(Q, A_{n-1})$ satisfying the corresponding relations, i.e. a map $Q \to \mathbf{Z}_{n-1}(A)$. Since $\partial_i \iota = \iota_i$ where ι generates Δ^n, the last part follows immediately.

We now give a simple condition for the existence of pushouts.

LEMMA 3.3. *Let* $f: A \to B$, $g: A \to C$ *be maps of simplicial objects in* $\mathscr{C}$. *Suppose for every* n, *the pushout* D_n *of the following diagram exists:*

$$(1)_n \qquad \begin{array}{ccc} A_n & \xrightarrow{f_n} & B_n \\ {\scriptstyle g_n}\downarrow & & \downarrow{\scriptstyle b_n} \\ C_n & \xrightarrow[c_n]{} & D_n. \end{array}$$

Then the pushout D *of the diagram*

$$(2) \qquad \begin{array}{ccc} A & \xrightarrow{f} & B \\ {\scriptstyle g}\downarrow & & \downarrow{\scriptstyle b} \\ C & \xrightarrow{c} & D \end{array}$$

in the category of simplicial objects of $\mathscr{C}$ *exists, and in dimension* n *the diagram* (2) *coincides with the diagram* $(1)_n$. *The same holds for more general direct limits.*

PROOF. If $\alpha: [m] \to [n]$ in $\mathscr{C}$, it induces a map $\hat{\alpha}: (1)_n \to (1)_m$ of diagrams and hence a map $\hat{\alpha}: D_n \to D_m$. It is trivial to verify that this assignment of maps makes (D_n) a simplicial object and yield a co-cartesian diagram (2).

We will call a co-cartesian diagram (or pushout) (2) normal if each resulting diagram $(1)_n$ is co-cartesian. If $\mathscr{C}$ has pushouts, this will always be the case by Lemma 3.3. In particular, it is true for simplicial sets.

COROLLARY 3.4. *The functor* $S \times A$, *where* S *is a simplicial object in* $\mathscr{F}$ *and* A *is a simplicial object in* $\mathscr{C}$, *preserves normal pushouts in either variable. In particular, a pushout* $\mathscr{F}$ *becomes a normal pushout in* $\mathscr{C}$ *under the functor* $- \times A$. *The same holds for more general direct limits.*

PROOF. The functor $\mathscr{F} \times \mathscr{C} \to \mathscr{C}$ preserves pushouts by the adjunction property (1), §2. Apply Lemma 3.3.

LEMMA 3.5. *Let T be a finite set, $S \subset T$, and $U = T - S$. Let $Q, C \in \mathscr{C}$. Then the diagram*

$$\begin{array}{ccc} S \times Q & \longrightarrow & T \times Q \\ {\scriptstyle f}\Big\downarrow & & \\ C & & \end{array}$$

(with $S \times Q \to T \times Q$ by $S \subset T$ and any f) has a pushout in $\mathscr{C}$ given by

$$\begin{array}{ccc} S \times Q & \longrightarrow & T \times Q \\ {\scriptstyle f}\Big\downarrow & & \Big\downarrow{\scriptstyle p} \\ C & \xrightarrow[i]{} & C \amalg (U \times Q) \end{array} \tag{4}$$

where i is the injection of C in the direct sum, $p \mid S \times Q = if$, and $p \mid U \times Q$ is the injection of $U \times Q$ in the direct sum.

PROOF. The functor $- \times Q$ preserves direct sums so $T \times Q = (S \times Q) \amalg (U \times Q)$. Let $A = S \times Q$, $B = U \times Q$. Then (4) is just

$$\begin{array}{ccc} A & \longrightarrow & A \amalg B \\ \Big\downarrow & & \Big\downarrow \\ C & \longrightarrow & C \amalg B. \end{array}$$

This is trivially co-cartesian.

We now come to the main construction. Let A be a simplicial object in $\mathscr{C}$, let $Q \in \mathscr{C}$, and let $\eta : Q \to \mathbf{Z}_{n-1}(A)$ be given. By Lemma 3.2, η gives us a map $\dot{\Delta}^n \times Q \to A$. Consider the diagram

$$\begin{array}{ccc} \dot{\Delta}^n \times Q & \longrightarrow & \Delta^n \times Q \\ \Big\downarrow & & \\ A. & & \end{array} \tag{5}$$

By Lemma 3.5, this has a pushout in each dimension. By Lemma 3.4, it therefore has a pushout

$$\begin{array}{ccc} \dot{\Delta}^n \times Q & \longrightarrow & \Delta^n \times Q \\ \Big\downarrow & & \Big\downarrow \\ A & \longrightarrow & A * Q. \end{array}$$

DEFINITION. The map $A \to A * Q$ is called an elementary cofibration obtained by attaching Q to A by means of the map η.

It is easy to determine the structure of $(A * Q)_m$ for each m. Clearly $(\Delta^n)_m - (\dot{\Delta}^n)_m$ consists of all elements $s_{j_1} \cdots s_{j_r}\iota$ where $j_1 > \cdots > j_r, m = r + n$. Referring to Lemmas 2.7 and 2.8 we see that $(A * Q)_m$ is the direct sum of A_m and one copy of Q for each $s_{j_1} \cdots s_{j_r}, j_1 > \cdots > j_r$, $m = r + n$. Therefore $A * Q$ agrees with the complex defined by André [**1**]. In particular, $(A * Q)_n = A_n \amalg Q$, $(A * Q)_m = A_m$ for $m < n$.

We can now iterate the above construction.

DEFINITION. A map $f: A \to B$ of simplicial objects in $\mathscr{C}$ will be called a cofibration[1] if B has a filtration $B^{(0)} \subset B^{(1)} \subset \cdots$ by simplicial subobjects such that

(1) $f: A \to B^{(0)}$ is an isomorphism

(2) $B^{(i)} \subset B^{(i+1)}$ is an elementary cofibration obtained by attaching an object $Q^{(i+1)}$ to $B^{(i)}$ by a map $\eta_{i+1}: Q_{i+1} \to \mathbf{Z}_{n_i}(B^{(i)})$

(3) For each m, $B_m^{(i)} = B_m$ for sufficiently large i.

Clearly any cofibration can be constructed by starting with $B^{(0)} = A$, forming $B^{(i+1)}$ from $B^{(i)}$ as in (2) and insuring that (3) holds by making sure only a finite number of attachments are made in any given dimension, i.e. $n_i \to \infty$.

Suppose now that $\mathscr{P}$ is some class of objects of $\mathscr{C}$. We say that a cofibration is a $\mathscr{P}$-cofibration if each attached Q_i lies in $\mathscr{P}$.

DEFINITION. Let $\mathscr{P}$ be a class of objects of $\mathscr{C}$. Let $\mathbf{D}$ be a finite system in $\mathscr{C}$ and $f \in \mathscr{C}(X, \mathbf{D})$. We say that f is a $\mathscr{P}$-epimorphism if $(1, f): \mathscr{C}(P, X) \to \mathscr{C}(P, \mathbf{D})$ is onto for all $P \in \mathscr{P}$.

Note that a $\mathscr{P}$-epimorphism need not be an epimorphism in general [**16**].

DEFINITION. Let $\mathscr{P}$ be a class of objects in $\mathscr{C}$. A simplicial object C of $\mathscr{C}$ is said to be $\mathscr{P}$-aspherical in dimension n if $\partial: C_{n+1} \to \mathbf{Z}_n(C)$ is a $\mathscr{P}$-epimorphism.

REMARK. For each $Q \in \mathscr{P}$, $X(Q) = \mathscr{C}(Q, C)$ is a simplicial set. Clearly C is $\mathscr{P}$-aspherical in dimension n if and only if for all $Q \in \mathscr{P}$, $X(Q)$ is aspherical in dimension n.

Using these notions we can now formulate the basic comparison theorem.

THEOREM 3.6. *Let $A \to A * Q$ be the elementary cofibration obtained by attaching Q to A in dimension n by a map $\eta: Q \to \mathbf{Z}_{n-1}(A)$. Let C be a simplicial object of $\mathscr{C}$. Let $\mathscr{P}$ be a class of objects of $\mathscr{C}$ with $Q \in \mathscr{P}$. Then*

(1) *If C is $\mathscr{P}$-aspherical in dimension $n - 1$, then any map $A \to C$ extends to a map $A * Q \to C$.*

(2) *Let C be $\mathscr{P}$-aspherical in dimensions n and $n - 1$. Let $f, g: A * Q \to C$. If $f \mid A \simeq g \mid A$, then $f \simeq g$ and we can choose the homotopy $f \simeq g$ to extend the homotopy $f \mid A \simeq g \mid A$.*

---

[1] It is not clear whether the theory presented here can be considered part of a homotopy theory in the sense of (**33**). If this is so it is not clear that the present definition will be the correct choice for cofibrations. If not, the concept defined here should probably be given another name.

PROOF. Since A $*$ Q is defined as a pushout, we can prove (1) by producing a map $\Delta^n \times Q \to C$ making the diagram

$$\begin{array}{ccc} \dot{\Delta}^n \times Q & \longrightarrow & \Delta^n \times Q \\ \downarrow & & \downarrow \\ A & \longrightarrow & C \end{array}$$

commute. In other words we must extend the map $\dot{\Delta}^n \times Q \to C$ to $\Delta^n \times Q$. We must show that $\mathrm{Hom}\,(\Delta^n \times Q, C) \to \mathrm{Hom}\,(\dot{\Delta}^n \times Q, C)$ is onto. By Lemmas 3.1 and 3.2, this is just the map $(1, \partial):\mathscr{C}(Q, C_n) \to \mathscr{C}(Q, \mathbf{Z}_{n-1}(C))$. This is onto by the definition of $\mathscr{P}$-asphericity.

To prove (2) we make use of the obvious natural isomorphism $S \times (T \times A) = (S \times T) \times A$. By Corollary 3.4 and the definition of $A * Q$ we see that the diagram

$$\begin{array}{ccc} I \times \dot{\Delta}^n \times Q & \longrightarrow & I \times \Delta^n \times Q \\ \downarrow & & \downarrow \\ I \times A & \longrightarrow & I \times (A * Q) \end{array}$$

is co-cartesian. Since $f \mid A \simeq g \mid A$, we have $h:I \times A \to C$ with $hi_0 = f \mid A$, $hi_1 = g \mid A$. Composing with the first vertical map in (9) we get $I \times \dot{\Delta}^n \times Q \to C$. Suppose $k:I \times \Delta^n \times Q \to C$ extends this. Since (9) is co-cartesian, we get $r:I \times (A * Q) \to C$ with $r \mid I \times A = h$. Let

$$f' = ri_0 : A * Q \to C, \; g' = ri_1 : A * Q \to C.$$

Clearly $f' \mid A = f \mid A, g' \mid A = g \mid A$. The definition of $A * Q$ as a pushout and Lemma 3.1 show that a map of $A * Q$ is completely determined by its restrictions to A and $\Delta^n \times Q$. Therefore to have $f' = f$, $g' = g$ we must choose k so $ki_0:\Delta^n \times Q \to C$ agrees with $\Delta^n \times Q \to A * Q \xrightarrow{i} C$ and $ki_1:\Delta^n \times Q \to C$ agrees with $\Delta^n \times Q \to A * Q \xrightarrow{g} C$. Call these latter maps f_0 and g_0. Therefore we have the maps shown in the following diagram:

$$(10) \qquad \begin{array}{ccc} \dot{\Delta}^n \times Q & \longrightarrow & \Delta^n \times Q \\ {\scriptstyle i_0}\downarrow & & \downarrow{\scriptstyle f_0} \\ I \times \dot{\Delta}^n \times Q & \longrightarrow & C \\ {\scriptstyle i_1}\uparrow & & \uparrow{\scriptstyle g_0} \\ \dot{\Delta}^n \times Q & \longrightarrow & \Delta^n \times Q. \end{array}$$

We must find a map of $I \times \Delta^n \times Q$ into C factoring these maps as follows:

$$(11)\qquad \begin{array}{ccccc} \dot\Delta^n \times Q & \longrightarrow & \Delta^n \times Q & & \\ \big\downarrow{\scriptstyle i_0} & & \big\downarrow{\scriptstyle i_0} & & \\ I \times \dot\Delta^n \times Q & \longrightarrow & I \times \Delta^n \times Q & \longrightarrow & C \\ \big\uparrow{\scriptstyle i_1} & & \big\uparrow{\scriptstyle i_0} & & \\ \dot\Delta^n \times Q & \longrightarrow & \Delta^n \times Q. & & \end{array}$$

Now let σ^n be an n-simplex and $\dot\sigma^n$ its boundary. Let J be the 1-simplex with vertices 0, 1. Let Σ be the complex $(J \times \dot\sigma^n) \cup (\{0\} \times \sigma^n) \cup (\{1\} \times \sigma^n)$. Let $S = S(\Sigma)$ be the associated simplicial set. By Corollary 1.3, S is the direct limit of the diagram

$$\begin{array}{ccc} \dot\Delta^n & \longrightarrow & \Delta^n \\ \big\downarrow & & \big\downarrow \\ I \times \dot\Delta^n & \longrightarrow & S \\ \big\uparrow & & \big\uparrow \\ \dot\Delta^n & \longrightarrow & \Delta^n \end{array}$$

By Corollary 3.4, $S \times Q$ is the direct limit of the diagram (10) (with C omitted) so we get a unique map $S \times Q \to C$. We must now extend this to $I \times \Delta^n \times Q \to C$. Now every simplex of $J \times \sigma^n$ of dimension $\leq n - 1$ lies in $J \times \dot\sigma^n$. In fact the vertices of $J \times \sigma^n$ are $(0, v)$, $(1, v)$ where v runs over the vertices of σ^n. A simplex of dimension $\leq n - 1$ can involve at most n different v. These v lie in a simplex of $\dot\sigma^n$. Thus $J \times \sigma^n$ is obtained from Σ by adding n and $n + 1$ simplexes as at the end of §1. Therefore we can find subsimplicial sets

$$S = S^{(0)} \subset S^{(1)} \subset \cdots \subset S^{(k)} = I \times \Delta^n$$

where we pass from $S^{(i)}$ to $S^{(i+1)}$ by forming pushouts (with $v = n$ or $n + 1$)

$$(12)\qquad \begin{array}{ccc} \dot\Delta^v & \longrightarrow & \Delta^v \\ \big\downarrow & & \big\downarrow \\ S^{(i)} & \longrightarrow & S^{(i+1)}. \end{array}$$

By Corollary 3.4, the product of (12) with Q is co-cartesian, so $S^{(i)} \times Q \to S^{(i+1)} \times Q$ is an elementary cofibration obtained by adjoining Q in dimension n or $n + 1$. By part (1) of the theorem, the map $S \times Q \to C$ extends to $I \times \Delta^n \times Q \to C$.

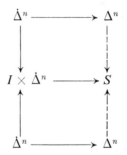

COROLLARY 3.7. *Let $A \to B$ be a $\mathscr{P}$-cofibration in $\mathscr{C}$ and let C be a simplicial object of $\mathscr{C}$ which is $\mathscr{P}$-aspherical in all dimensions. Then*

(1) *Any map $f: A \to C$ extends to $f': B \to C$.*

(2) *If $f, g: A \to C$ extend to $f', g': B \to C$ and $f \simeq g$ then $f' \simeq g'$.*

PROOF. We have $A \xrightarrow{\simeq} B^{(0)} \subset B^{(1)} \subset \cdots \subset B$. By Theorem 3.6 we can extend maps and homotopies one stage at a time. Since $B_n = B_n^{(i)}$ for large i, the maps and homotopies so obtained clearly give the required maps and homotopies on B.

4. **Simplicial resolutions.** Let A be an object of $\mathscr{C}$. By a simplicial resolution of A we mean a simplicial object X of $\mathscr{C}$ together with a map $\varepsilon: X_0 \to A$ such that $\varepsilon\partial_0 = \varepsilon\partial_1$. We write $\varepsilon: X \to A$ by abuse of notation. Such a resolution may also be regarded as an augmented simplicial object $\tilde{X}$ by setting $\tilde{X}_n = X_n$ for $n \geq 0$, $X_{-1} = A$ and $\partial_0 = \varepsilon: X_0 \to X_{-1}$. We say that the resolution is $\mathscr{P}$-aspherical if X is $\mathscr{P}$-aspherical in all dimensions > 0 and in addition the diagram

$$(1) \qquad X_1 \underset{\partial_1}{\overset{\partial_0}{\rightrightarrows}} X_0 \xrightarrow{\varepsilon} A$$

is $\mathscr{P}$-exact, i.e. for all $P \in \mathscr{P}$, the diagram

$$(2) \qquad (P, X_1) \underset{(1,\partial_1)}{\overset{(1,\partial_0)}{\rightrightarrows}} (P, X_0) \xrightarrow{(1,\varepsilon)} (P, A)$$

is exact ($(1, \varepsilon)$ is the difference cokernel of $(1, \partial_0)$ and $(1, \partial_1)$). This is equivalent to the condition that $\partial: \tilde{X}_n \to \mathbf{Z}_{n-1}(\tilde{X})$ is a $\mathscr{P}$-epimorphism for all $n \geq 0$, where $\mathbf{Z}_0(\tilde{X})$ is the finite system $X_0 \xrightarrow{\varepsilon} A \xleftarrow{\varepsilon} X_0$, where $\partial: X_1 \to \mathbf{Z}_0(\tilde{X})$ is given by

$$\begin{array}{ccc} X_1 & \xrightarrow{\partial_0} & X_0 \\ {\scriptstyle \partial_1}\downarrow & & \downarrow{\scriptstyle \varepsilon} \\ X_0 & \xrightarrow{\varepsilon} & A \end{array}$$

and where $\partial: X_0 \to Z_{-1}(\tilde{X})$ is $\partial_0: X_0 \to X_{-1}$.

REMARK. Clearly $\tilde{X}$ is $\mathscr{P}$-aspherical if and only if the simplicial set Hom $(Q, \tilde{X})$ is aspherical for all $Q \in \mathscr{P}$. An augmented simplicial set K is aspherical if and only if for each set $x_0, \ldots, x_{n+1} \in K_n$ with $\partial_i x_j = \partial_{j-1} x_i$ for $i < j$, there is some $y \in K_{n+1}$ with $\partial_i y = x_i$, $n \geq -1$. This is equivalent to the requirement that K satisfies the Kan condition **[29]** and $\pi_i K = 0$, $i \geq -1$. In fact, given n faces of a potential n-simplex, we fill in the $(n + 1)$st face by asphericity in dimension $n - 1$ and then fill in the n-simplex by asphericity in dimension n.

DEFINITION. The resolution $\varepsilon: X \to A$ is called $\mathscr{P}$-cofibrant if X has a filtration $X^{(0)} \subset X^{(1)} \subset \cdots$ such that

(1) $X^{(0)} = \Delta^0 \times Q^{(0)}$ with $Q^{(0)} \in \mathscr{P}$,

(2) For each $i > 0$ either (a) $X^{(i)} = X^{(i-1)} \amalg (\Delta^0 \times Q^{(i)})$, $Q^{(i)} \in \mathscr{P}$, or (b) $X^{(i-1)} \to X^{(i)}$ is an elementary cofibration obtained by attaching some $Q^{(i)} \in \mathscr{P}$ to $X^{(i-1)}$ by a map $\eta: Q^{(i)} \to \mathbf{Z}_{n_i}(X^{(i-1)})$, $n_i \geq 0$.

(3) Only a finite number of attachments occur in any given dimension, i.e. case (a) occurs only a finite number of times and in case (b), $n_i \to \infty$.

We can now state the main comparison theorem.

THEOREM 4.1. *Let $A, B \in \mathscr{C}$. Let $\mathscr{P}$ be a class of objects of $\mathscr{C}$. Let $\varepsilon: X \to A$ be a $\mathscr{P}$-cofibrant simplicial resolution of A and let $\varepsilon': Y \to B$ be a $\mathscr{P}$-aspherical simplicial resolution of B. Let $f: A \to B$ in $\mathscr{C}$. Then*

(1) *There is a map $g: X \to Y$ of simplicial objects such that the diagram*

$$\begin{array}{ccc} X_0 & \xrightarrow{\varepsilon} & A \\ {\scriptstyle g_0}\downarrow & & \downarrow{\scriptstyle f} \\ Y_0 & \xrightarrow{\varepsilon'} & B \end{array}$$

commutes (i.e. g extends f).

(2) *If $g, h: X \to Y$ satisfy* (1) (*for the same f*), *then $g \simeq h$.*

PROOF. Let $X^{(0)} \subset X^{(1)} \subset \cdots$ be the filtration given by the definition of $\mathscr{P}$-cofibrant. Then $X^{(0)} = \Delta^0 \times Q^{(0)}$. Clearly $X_0^{(0)} = Q^{(0)}$. To define $g^{(0)}: X^{(0)} \to Y$ we must give a map $\gamma: Q^{(0)} \to Y_0$. This satisfies the condition that the diagram

$$\begin{array}{ccc} Q^{(0)} & \xrightarrow{\varepsilon} & A \\ {\scriptstyle \gamma}\downarrow & & \downarrow{\scriptstyle f} \\ Y_0 & \xrightarrow{\varepsilon'} & B \end{array}$$

commutes. Since ε' is a $\mathscr{P}$-epimorphism and $Q^{(0)} \in \mathscr{P}$, such a map γ will exist. Suppose now that $g^{(i-1)}: X^{(i-1)} \to Y$ is defined. In case (a), $X^{(i)} = X^{(i-1)} \amalg (\Delta^0 \times Q^{(i)})$. As in the case of $X^{(0)}$ there is a map $\Delta^0 \times Q^{(i)} \to Y$ such that the diagram

$$\begin{array}{ccc} \Delta^0 \times Q^{(i)} & \xrightarrow{\varepsilon} & A \\ \downarrow & & \downarrow \\ Y & \xrightarrow{\varepsilon'} & B \end{array}$$

commutes. This map together with $g^{(i-1)}$ defines $g^{(i)}$. In case (b), we apply Theorem 3.6. Since $Q^{(i)}$ is attached in a dimension > 0, $X_0^{(i)} = X_0^{(i-1)}$ in this case and the commutativity of the diagram is automatic.

Suppose now we have $g, h: X \to Y$. Let $g^{(n)} = g \mid X^{(n)}$ and $h^{(n)} = h \mid X^{(n)}$. To show $g^{(0)} \simeq h^{(0)}$ we must find a map $k: I \times \Delta^0 \times Q^{(0)} \to Y$ with $g^{(0)} = ki_0$, $h^{(0)} = ki_1$. Let σ^1 be a 1-simplex with vertices 0, 1. Then $I = S(\sigma^1)$. Let $E = S(\{0, 1\})$. Then $E = \Delta^0 \amalg \Delta^0$ and $k \mid E \times \Delta^0 \times Q^{(0)}$ is uniquely defined by the conditions $g^{(0)} = ki_0$, $h^{(0)} = ki_1$. We must extend this to $I \times \Delta^0 \times Q^{(0)}$. But $I \times \Delta^0 \simeq I$ is obtained from $E \times \Delta^0 \simeq E$ by attaching one simplex of dimension 1.

Therefore, as in the proof of Theorem 3.6, $E \times \Delta^0 \times Q^{(0)} \to I \times \Delta^0 \times Q^{(0)}$ is an elementary cofibration. Theorem 3.6 now gives the necessary extension.

Next we must extend a homotopy between $g^{(i-1)}$ and $h^{(i-1)}$ to one between $g^{(i)}$ and $h^{(i)}$. In case (a) we have $X^{(i)} = X^{(i-1)} \amalg (\Delta^0 \times Q^{(i)})$. In this case $I \times X^{(i)} = (I \times X^{(i-1)}) \amalg (I \times \Delta^0 \times Q^{(i)})$ since $I \times -$ preserves direct limits. Take the homotopy $g^{(i-1)} \simeq h^{(i-1)}$ on the first summand and define the homotopy on the second by the same method used for $g^{(0)} \simeq h^{(0)}$. In case (b), apply Theorem 3.6.

Finally, once we have the map or homotopy defined for all $X^{(n)}$, we get the required map or homotopy on X by $X = \varinjlim X^{(n)}$.

Corollary 4.2. *Let $A \in \mathscr{C}$. Let $\varepsilon: X \to A$ and $\varepsilon': Y \to A$ be $\mathscr{P}$-aspherical, $\mathscr{P}$-cofibrant resolutions. Then there are maps $f: X \to Y$, $g: Y \to X$ extending* id_A, *and* $fg \simeq \mathrm{id}_Y$, $gf \simeq \mathrm{id}_X$.

We can now define derived functors as in the abelian case, getting homotopy classes of simplicial objects [**2**]. It follows directly from the definition or Proposition 2.1 that $f \simeq f'$ implies $gf \simeq gf'$, $fh \simeq f'h$ for maps of simplicial objects. Using the equivalence relation generated by the relation of homotopy, we can define homotopy classes of maps. Let $\mathbf{H}(\mathscr{C})$ be the category of simplicial objects of $\mathscr{C}$ and homotopy classes of maps. Suppose each object A of $\mathscr{C}$ has a $\mathscr{P}$-aspherical, $\mathscr{P}$-cofibrant resolution $\varepsilon: X \to A$. Choose one for each A and define a functor $L_{\mathscr{P}}: \mathscr{C} \to \mathbf{H}(\mathscr{C})$ by $L_{\mathscr{P}}(A) = X$. If $f: A \to B$, lift it to $g: X \to Y$ (where $Y = L_{\mathscr{P}}(B)$) and define $L_{\mathscr{P}}(f)$ to be the homotopy class of g. By Theorem 4.1 this is a well-defined covariant functor. By Corollary 4.2 it is unique up to *unique* isomorphism. Now if $F: \mathscr{C} \to \mathscr{D}$ is any covariant functor, it defines a functor $\mathbf{H}(F): \mathbf{H}(\mathscr{C}) \to \mathbf{H}(\mathscr{D})$ by $(\mathbf{H}(F)(X))_n = F(X_n)$. By Proposition 2.1 this preserves homotopy classes of maps.

Definition. $\mathbf{L}_{\mathscr{P}}F = \mathbf{H}(F) \cdot L_{\mathscr{P}}: \mathscr{C} \to \mathbf{H}(\mathscr{D})$.

If D is the category of sets, groups, abelian groups, etc. we can define homotopy groups and sets $L_nF = \pi_n\mathbf{L}F$.

To apply this construction we need to show the existence of resolutions.

Definition. Let $\mathscr{P}$ be a class of objects of $\mathscr{C}$. We say that $\mathscr{P}$ is adequate if for each finite system $\mathbf{D}$ in $\mathscr{C}$, there is a $\mathscr{P}$-epimorphism $Q \to \mathbf{D}$ with $Q \in \mathscr{P}$.

If $\mathscr{C}$ has finite inverse limits, it will clearly suffice to check this condition for objects $D \in \mathscr{C}$.

Proposition 4.3. *If $\mathscr{P}$ is adequate every object of $\mathscr{C}$ has a $\mathscr{P}$-aspherical, $\mathscr{P}$-cofibrant resolution.*

Proof. Let $A \in \mathscr{C}$. Let $Q^{(0)} \to A$ be a $\mathscr{P}$-epimorphism with $Q^{(0)} \in \mathscr{P}$. Let $X^{(0)} = \Delta^0 \times Q^{(0)}$ with $\varepsilon: X_0^{(0)} = Q^{(0)} \to A$ the given map. Let $\mathbf{D}$ be the finite system $X_0 \xrightarrow{\varepsilon} A \xleftarrow{\varepsilon} X_0$. Let $\gamma: Q^{(1)} \to \mathbf{D}$ be a $\mathscr{P}$-epimorphism with $Q^{(1)} \in \mathscr{P}$. Then γ is determined by a pair of maps $\eta_0, \eta_1: Q^{(1)} \to X_0$, i.e. a map $\eta: Q^{(1)} \to \mathbf{Z}_0(X)$. Attach $Q^{(1)}$ to $X^{(0)}$ by η, getting $X^{(1)}$. The sequence

$$X_1^{(1)} \underset{\partial_1}{\overset{\partial_0}{\rightrightarrows}} X_0^{(1)} \xrightarrow{\varepsilon} A$$

is now clearly $\mathscr{P}$-exact. Suppose $X^{(n)}$ has been defined such that $\varepsilon: X^{(n)} \to A$ is $\mathscr{P}$-aspherical in dimensions $< n$. Let $\eta: Q^{(n+1)} \to \mathbf{Z}_n(X^{(n)})$ be a $\mathscr{P}$-epimorphism with $Q^{(n+1)} \in \mathscr{P}$. Attach $Q^{(n+1)}$ to $X^{(n)}$ by η getting $X^{(n+1)}$. Finally, let $X = \cup X^{(n)}$.

REMARK. Suppose $\mathscr{E}$ is a class of morphisms in $\mathscr{C}$. Define $P \in \mathscr{C}$ to be $\mathscr{E}$-projective if $\mathscr{C}(P, f)$ is onto for all $f \in \mathscr{E}$. The relation "$\mathscr{C}(P, f)$ is onto" sets up a "galois connection" [8] between objects and maps in $\mathscr{C}$ as in [16]. Therefore if $\mathscr{P}^*$ is the class of $\mathscr{P}$-epimorphisms and $\mathscr{E}^*$ the class of $\mathscr{E}$-projectives, we have contravariance: $\mathscr{P}_1 \subset \mathscr{P}_2$ implies $\mathscr{P}_1^* \supset \mathscr{P}_2^*$, $\mathscr{E}_1 \subset \mathscr{E}_2$ implies $\mathscr{E}_1^* \supset \mathscr{E}_2^*$; and we have the adjunction: $\mathscr{P} \subset \mathscr{E}^*$ if and only if $\mathscr{E} \subset \mathscr{P}^*$. Thus $\mathscr{E}^* = \mathscr{E}^{***}$, $\mathscr{P}^* = \mathscr{P}^{***}$ just as in [16]. It is often convenient to choose $\mathscr{E}$ first and then set $\mathscr{P} = \mathscr{E}^*$.

REMARK. Suppose, for convenience, that $\mathscr{C}$ has finite inverse limits. We say that a class of maps $\mathscr{E}$ is nearly adequate if for any $A \in \mathscr{C}$ and finite set of maps $f_1, \ldots, f_n \in \mathscr{E}$, there is a map $g: Q \to A$ such that $g \in \mathscr{E}$ and $\mathscr{C}(Q, f_1), \ldots, \mathscr{C}(Q, f_n)$ are all onto. For such an $\mathscr{E}$, we can construct resolutions which are $\mathscr{E}$-aspherical and cofibrant, the attached $Q^{(i)}$'s being $\mathscr{E}^{(i)}$-projective, where the $\mathscr{E}^{(i)}$ are any preassigned finite subsets of $\mathscr{E}$. The arguments above show that these resolutions form an inverse system up to homotopy. Thus we obtain a functor $L: \mathscr{C} \to$ pro $\mathbf{H}(\mathscr{C})$. This is clearly very closely related to Verdier's theory of hypercoverings [3], [37]. I have not yet checked to see whether the two theories yield the same results.

5. **Aspherical models.** We first consider the abelian case (acyclic models) obtain some motivation. Suppose $\mathscr{A}$ and $\mathscr{B}$ are abelian categories and $S: \mathscr{A} \to \mathscr{B}$, $T: \mathscr{B} \to \mathscr{A}$ are adjoint functors, $\mathscr{B}(SA, B) \approx \mathscr{A}(A, TB)$. Suppose also that T is exact. Since S is right exact it is reasonable to look at its left derived functors $S_n = L_n S$. We can then define a homology theory on $\mathscr{B}$ by setting $H_n = S_n \circ T$: $\mathscr{B} \to \mathscr{B}$. Similarly if S is exact we let $T^n = R^n T$ and define a cohomology theory on $\mathscr{A}$ by $H^n = T^n \circ S: \mathscr{A} \to \mathscr{A}$. The idea of defining a cohomology theory in this way is suggested by the following remark due to P. Freyd. Let $\mathscr{C}$ be an abelian category. Let $\mathscr{A}$ consist of all additive covariant functors $\mathscr{C} \to \mathscr{A}\mathscr{b}$ and let $\mathscr{B}$ be the full subcategory of left exact functors. Let $i: \mathscr{B} \to \mathscr{A}$ be the inclusion. This has an exact left adjoint $R^0: \mathscr{A} \to \mathscr{B}$ [18], [20]. Freyd's remark is that we may calculate the higher derived functors of some $F \in \mathscr{A}$ by $R^n F = i^n(R^0 F)$ where $i^n = R^n(i)$.

We now apply the above idea to the following special case. Let $\mathscr{M}$, $\mathscr{C}$ be categories and $i: \mathscr{M} \to \mathscr{C}$ a covariant functor. Define res: $\mathscr{A}\mathscr{b}^{\mathscr{C}} \to \mathscr{A}^{\mathscr{M}}$ by res $F = F \circ i$, i.e. if i is an inclusion, res is just restriction to $\mathscr{M}$. This functor is clearly exact. It has a left adjoint adj: $\mathscr{A}\mathscr{b}^{\mathscr{M}} \to \mathscr{A}^{\mathscr{C}}$ [24]. We shall see below that $\mathscr{A}\mathscr{b}^{\mathscr{M}}$ has enough projectives so we can consider the derived functors $\text{adj}_n = L_n(\text{adj})$ and define $H_n: \mathscr{A}\mathscr{b}^{\mathscr{C}} \to \mathscr{A}\mathscr{b}^{\mathscr{C}}$ to be $\text{adj}_n \circ \text{res}$. If we are given some $F: \mathscr{C} \to \mathscr{A}\mathscr{b}$, we may define $H_n(C, F) = H_n(F)(C)$ for $C \in \mathscr{C}$. This point of view has been noted independently by Oberst [32], who has shown that most of the standard cohomology theories can be obtained in this way.

The usefulness of this point of view for the present work is that it leads very directly to the theory of acyclic models. This has been observed independently by Ulmer and will appear in a set of notes which he is preparing. We think of $\mathcal{M}$ as the category of models. Let $F \in \mathcal{A}b^{\mathcal{C}}$. To calculate $H_n(F)$, we choose a projective resolution $P_* \to \operatorname{res} F$ in $\mathcal{A}b^{\mathcal{M}}$, apply adj getting a chain complex $C_* = \operatorname{adj} P_*$ in $\mathcal{A}b^{\mathcal{C}}$, and then take homology. Now it is easy to see that res adj is the identity on $\mathcal{A}b^{\mathcal{M}}$. Therefore $\operatorname{res} C_* = P_*$. Thus the condition that C_* be acyclic on models is equivalent to the exactness of the resolution P_*. Now, assuming $\mathcal{M}$ is a full subcategory of $\mathcal{C}$, the condition that a functor $C \in \mathcal{A}b^{\mathcal{C}}$ be "representable" in the sense of acyclic models turns out to be equivalent to the condition that $C \approx \operatorname{adj} P$ where P is projective in $\mathcal{A}b^{\mathcal{M}}$. This in turn is equivalent to the conditions (a) res C is projective in $\mathcal{A}b^{\mathcal{M}}$, and (b) $C \approx \operatorname{adj} \operatorname{res} C$. The proof is the exact analogue of that given below for the nonabelian case. We now obtain the comparison theorem for acyclic model theory [**14**] as follows. Let C_*, D_* be chain complexes in $\mathcal{A}b^{\mathcal{C}}$ augmented by $C_* \to A$, $D_* \to B$. Assume C_* is "representable" and $D_* \to B$ is acyclic on models. Given $A \to B$, apply the functor res and invoke the usual projective-acyclic comparison theorem getting $\operatorname{res} C_* \to \operatorname{res} D_*$ unique up to homotopy. Then apply the functor adj and get $C_* = \operatorname{adj} \operatorname{res} C_* \to \operatorname{adj} \operatorname{res} D_* \to D_*$, unique up to homotopy. Any two maps $C_* \to D_*$ over the given $A \to B$ are homotopic by the same argument.

We now generalize to the nonabelian case. Let $\mathcal{G}$ be a right complete category. Let $i: \mathcal{M} \to \mathcal{C}$ be as above. Define $\operatorname{res}: \mathcal{G}^{\mathcal{C}} \to \mathcal{G}^{\mathcal{M}}$ by $\operatorname{res} F = F \circ i$. This again has a right adjoint $\operatorname{adj}: \mathcal{G}^{\mathcal{M}} \to \mathcal{G}^{\mathcal{C}}$ given by $\operatorname{adj} G(C) = \varinjlim G \mid \mathcal{M}_C$ where $\mathcal{M}_C$ is the category whose objects are pairs (M, f) with $f: i(M) \to C$ and the obvious morphisms [**24**]. We want to define left derived functors $L_{\mathcal{P}}$ adj as in §4. This requires us to choose some class $\mathcal{P}$ and show that it is adequate. One standard way to do this is as follows. The notation introduced here will be used constantly throughout the rest of this paper.

For simplicity, we assume $\mathcal{G}$ has finite inverse limits. This implies the same property for all functor categories $\mathcal{G}^{\mathcal{M}}$, $\mathcal{G}^{\mathcal{C}}$. With this assumption, a class $\mathcal{P}$ will be adequate if and only if each object admits a $\mathcal{P}$-epimorphism $Q \to X$ with $Q \in \mathcal{P}$.

Let $\mathcal{P}$ be a class of objects of $\mathcal{G}$ and $\mathcal{E}$ a class of maps of $\mathcal{G}$ such that $\mathcal{P}^* = \mathcal{E}$, $\mathcal{E}^* = \mathcal{P}$ as in §4.

DEFINITION. The class $\mathcal{E}'$ in $\mathcal{G}^{\mathcal{M}}$ consists of all maps $\eta: F \to G$ such that for each $M \in \mathcal{M}$, the map $\eta_M: F(M) \to G(M)$ is in $\mathcal{E}$. Define $\mathcal{P}' = \mathcal{E}'^*$.

PROPOSITION 5.1. *We have* $\mathcal{P}'^* = \mathcal{E}'$. *If* $\mathcal{P}$ *is adequate, so is* $\mathcal{P}'$.

PROOF. We use the following general construction. Let $\mathcal{M}_d$ be the discrete category consisting of all objects of $\mathcal{M}$ and all identity maps. The inclusion $\mathcal{M}_d \subset \mathcal{M}$ gives a functor $\mathcal{G}^{\mathcal{M}} \to \mathcal{G}^{\mathcal{M}_d}$ which has a right adjoint as above. Explictly, this functor can be described as follows. An object of $\mathcal{G}^{\mathcal{M}_d}$ is just an indexed collection $(Q_M)_{M \in \mathcal{M}}$ of objects of $\mathcal{G}$. Under the adjoint, this is sent into an object $P \in \mathcal{G}^{\mathcal{M}}$ where $P(X) = \coprod_{M \to X} Q_M$. This follows immediately from Kan's general construction. The adjunction property is

$$(1) \qquad \operatorname{Hom}(P, F) = \prod_{M \in \mathscr{M}} \operatorname{Hom}(Q_M, F(M)).$$

It should be noted that if $\mathscr{M}$ is a full subcategory of $\mathscr{C}$ and i is the inclusion, then the inclusion $\mathscr{M}_d \subset \mathscr{M} \subset \mathscr{C}$ gives us a functor $\mathscr{G}^{\mathscr{C}} \to \mathscr{G}^{\mathscr{M}_d}$ whose right adjoint is given by exactly the same formula as P above (using all $X \in \mathscr{C}$). The composition $\mathscr{G}^{\mathscr{M}_d} \to \mathscr{G}^{\mathscr{C}} \to \mathscr{G}^{\mathscr{M}}$ is thus the same right adjoint of $\mathscr{G}^{\mathscr{M}} \to \mathscr{G}^{\mathscr{M}_d}$.

Suppose now all $Q_M \in \mathscr{P}$. If $\eta: F \to G$ lies in $\mathscr{E}'$, then each $F(M) \to G(M)$ lies in $\mathscr{E}$, so $\operatorname{Hom}(Q_M, F(M)) \to \operatorname{Hom}(Q_M, G(M))$ is onto. Therefore, by (1), $\operatorname{Hom}(P, F) \to \operatorname{Hom}(P, G)$ is onto. Thus $P \in \mathscr{P}'$ by definition. Conversely, if $F \to G$ lies in $\mathscr{P}'^*$, then (1) shows that $\operatorname{Hom}(Q_M, F(M)) \to \operatorname{Hom}(Q_M, G(M))$ is onto for all $Q_M \in \mathscr{P}$ so $F(M) \to G(M)$ lies in $\mathscr{E}$ for all M. Thus $\mathscr{P}'^* = \mathscr{E}'$.

Finally, suppose $\mathscr{P}$ is adequate. Given $F \in \mathscr{G}^{\mathscr{M}}$, choose $\nu_M: Q_M \to F(M)$ in $\mathscr{E}$ for each M, with $Q_M \in \mathscr{P}$. Construct P as above. By (1), there is some $\nu: P \to F$ so that ν induces $\nu_M: Q_M \to F(M)$. We have just seen that $\nu \in \mathscr{E}'$ and $P \in \mathscr{P}'$.

As an example we may let $\mathscr{G}$ be an abelian category with enough projectives, let $\mathscr{E}$ consist of all epimorphisms, and let $\mathscr{P}$ consist of all projectives. This gives us the usual theory of acyclic models by virtue of the equivalence between chain complexes and simplicial abelian groups [**11**].

We now examine the case where $\mathscr{M}$ is a full subcategory of $\mathscr{C}$. In this case we can follow the method of [**10**]. As above, let $\mathscr{P}$ be a class of objects of $\mathscr{G}$ and $\mathscr{E}$ a class of maps of $\mathscr{G}$ with $\mathscr{E}^* = \mathscr{P}$, $\mathscr{P}^* = \mathscr{E}$. We continue to assume that $\mathscr{G}$ has finite inverse limits. The notation introduced in the following definition will also be used constantly throughout the rest of this paper.

DEFINITION. Let $\mathscr{E}''$ be the class of maps $F \to G$ in $\mathscr{G}^{\mathscr{C}}$ such that for each object $M \in \mathscr{M}$, the map $F(M) \to G(M)$ is in $\mathscr{E}$. Let $\mathscr{P}'' = \mathscr{E}''^*$.

PROPOSITION 5.2. *We have* $\mathscr{P}''^* = \mathscr{E}''$. *If* $\mathscr{P}$ *is adequate, so is* $\mathscr{P}''$.

PROOF. Define the functor $P(X)$ for $X \in \mathscr{C}$ as in the proof of Proposition 5.1. The formula (1) continues to hold. The rest of the proof of Proposition 5.1. goes through without change.

Suppose $F \in \mathscr{G}^{\mathscr{C}}$. Assume $\mathscr{P}$ is adequate. Using the method first considered, we form $\operatorname{res} F \in \mathscr{G}^{\mathscr{M}}$, choose a $\mathscr{P}'$-aspherical, $\mathscr{P}'$-cofibrant resolution $\varepsilon: P' \to \operatorname{res} F$. We then apply the adjoint $\operatorname{adj}: \mathscr{G}^{\mathscr{M}} \to \mathscr{G}^{\mathscr{C}}$ getting $\operatorname{adj} P' \to \operatorname{adj} \operatorname{res} F \to F$. This leads to the same result as the second method.

PROPOSITION 5.3. *The resolution* $\operatorname{adj} P' \to F$ *is* $\mathscr{P}''$*-aspherical and* $\mathscr{P}''$*-cofibrant.*

PROOF. Since $\mathscr{M}$ is a full subcategory of $\mathscr{C}$, it is trivial to check that $\operatorname{res} \operatorname{adj} = \operatorname{id}$. Now, by the definition of $\mathscr{E}''$, a resolution $P'' \to F$ in $\mathscr{G}^{\mathscr{C}}$ is $\mathscr{P}''$-aspherical if and only if $\operatorname{res} P'' \to \operatorname{res} F$ is $\mathscr{P}'$-aspherical. But $\operatorname{res} \operatorname{adj} P' \to \operatorname{res} F$ is just $P' \to \operatorname{res} F$. Since adj is a right adjoint, it preserves all direct limits. Now P' is obtained by starting with some $\Delta^0 \times Q^{(0)}$ and adjoining various $Q^{(n)} \in \mathscr{P}'$ successively, either by forming $P^{(n)} = P^{(n-1)} \coprod (\Delta^0 \times Q^{(n)})$ or letting $P^{(n)}$ be the pushout of

$$\begin{array}{ccc} \dot{\Delta}^r \times Q^{(n)} & \longrightarrow & \Delta^r \times Q^{(n)} \\ \downarrow & & \downarrow \\ P^{(n-1)} & \dashrightarrow & P^{(n)}. \end{array}$$

Now if $Q \in \mathcal{G}^{\mathcal{M}}$ and K is a simplicial set, we have adj $(K \times Q) = K \times$ adj Q. This follows from Corollary 3.4. To show that adj P' is $\mathcal{P}''$-cofibrant we need only show all adj $Q^{(n)} \in \mathcal{P}''$.

LEMMA 5.3. *Let $G \in \mathcal{G}^{\mathcal{C}}$. Then $G \in \mathcal{P}''$ if and only if G has the form $G =$* adj *Q where $Q \in \mathcal{G}^{\mathcal{M}}$ and $Q \in \mathcal{P}'$ or equivalently, if* res *$G \in \mathcal{P}'$ and* adj res *$G \to G$ is an isomorphism.*

PROOF. If $Q \in \mathcal{P}'$ and $f: F \to F'$ in $\mathcal{P}^{\mathcal{C}}$, we have Hom (adj Q, f) $=$ Hom (Q, res f). But $f \in \mathcal{E}''$ if and only if res $f \in \mathcal{E}$. Thus adj $Q \in \mathcal{P}''$ if $Q \in \mathcal{P}'$. Conversely, suppose $G \in \mathcal{P}''$. Construct $\eta: P \to G$ as in the proof of Proposition 5.2. Since $G \in \mathcal{P}''$ and $\eta \in \mathcal{E}''$, η splits, i.e. there is some $\nu: G \to P$ such that $\eta\nu = \mathrm{id}_G$. Now P is obtained from $(Q_n) \in \mathcal{G}^{\mathcal{M}_d}$ by applying the adjoint $\mathcal{G}^{\mathcal{M}_d} \to \mathcal{G}^{\mathcal{C}}$. This is the composition of the two adjoints $\mathcal{G}^{\mathcal{M}_d} \to \mathcal{G}^{\mathcal{M}} \to \mathcal{G}^{\mathcal{C}}$. The first of these, applied to (Q_M), yields the P of Proposition 5.1, but this is just res P. Therefore adj res $P = P$. Now applying adj res to the maps ν, η gives

$$\begin{array}{ccccc} \text{adj res } G & \longrightarrow & \text{adj res } P & \longrightarrow & \text{adj res } G \\ \downarrow & & \downarrow & & \downarrow \\ G & \xrightarrow{\ \nu\ } & P & \xrightarrow{\ \eta\ } & G. \end{array}$$

Both horizontal compositions are the identity. Therefore the composition $G \to P =$ adj res $P \to$ adj res G is an inverse for adj res $G \to G$, so this map is an isomorphism. Also res G is a retract of res $P \in \mathcal{P}'$, so res $G \in \mathcal{P}'$.

A similar argument works in the other direction. Let $F \in \mathcal{G}^{\mathcal{C}}$ and let $\varepsilon: \mathcal{P}'' \to F$ be a $\mathcal{P}''$-cofibrant, $\mathcal{P}''$-aspherical resolution of F.

PROPOSITION 5.4. *The resolution* res $P'' \to$ res F *is $\mathcal{P}'$-cofibrant and $\mathcal{P}'$-aspherical, and* adj res $P'' = P''$.

PROOF. Clearly $f \in \mathcal{E}''$ implies res $f \in \mathcal{E}'$ and Lemma 5.3 shows that $Q \in \mathcal{P}''$ implies res $Q \in \mathcal{P}'$. Also res preserves all direct and inverse limits. It is clear from this that res $P'' \to$ res F is $\mathcal{P}'$-cofibrant and $\mathcal{P}'$-aspherical. Now, in each dimension, P'' is a direct sum of various $Q_\alpha \in \mathcal{P}''$. Therefore, the definition $\mathcal{P}'' = \mathcal{E}''*$ shows that $P''_n \in \mathcal{P}''$. Lemma 5.3 shows that adj res $P'' \to P''$ is an isomorphism.

Therefore the simplicial resolutions $P'' \to F$ obtained by applying $\mathbf{L}_{\mathcal{P}'}$ adj to res F are the same as the $\mathcal{P}''$-aspherical, $\mathcal{P}''$-cofibrant resolutions of F.

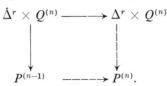

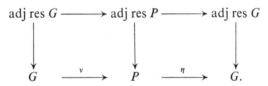

REMARK. Suppose we consider the case where $\mathscr{P}$ consists of all objects of $\mathscr{G}$ and $\mathscr{E}$ consists of all split epimorphisms. Clearly $\mathscr{P}$ is adequate. This case is closely related to the work of André **[2]**. In fact the standard resolution used in **[2]** is easily shown to be $\mathscr{P}$-cofibrant. If $\mathscr{G}$ is abelian, it is also very easy to show that the standard resolution is $\mathscr{P}$-aspherical. This shows that André's method leads to the same cohomology as the method used here. However, if $\mathscr{G}$ is not abelian, it is not clear whether the standard resolution will be $\mathscr{P}$-aspherical.

6. **Group valued functors.** We now assume $\mathscr{G} = \mathscr{G}p$, the category of all groups. The results obtained here will also hold for the case $\mathscr{G} = \mathscr{A}b$, the category of abelian groups, but in this case we merely recover the classical theory of acyclic models.

LEMMA 6.1. *A map $\eta : F \to G$ in $\mathscr{G}p^{\mathscr{M}}$ is an epimorphism if and only if $\eta_M : F(M) \to G(M)$ is onto for all $M \in \mathscr{M}$.*

PROOF. Let $F'(M) = im\eta_M \subset G(M)$. Let $H(M)$ be the pushout of the diagram.

$$\begin{array}{ccc} F'(M) & \longrightarrow & G(M) \\ \Big\downarrow & & \Big\downarrow \mu \\ G(M) & \underset{\nu}{\dashrightarrow} & H(M). \end{array}$$

If $M \to M'$ in $\mathscr{M}$ we get a map of diagrams and hence a map of their pushouts. Thus H is a functor, $H \in \mathscr{G}^{\mathscr{M}}$ and $\mu, \nu : G \to H$, $\mu\eta = \nu\eta$. If η is an epimorphism, this implies $\mu = \nu$. But $H(M)$ is the free product of two copies of $G(M)$ with amalgamated subgroup $F'(M)$. Therefore $\mu = \nu$ can only happen if $F'(M) = G(M)$. The converse is trivial.

It follows from this that if $\mathscr{E}$ is the class of epimorphisms in $\mathscr{G}p$, then $\mathscr{E}'$ defined in §5, is the class of epimorphisms in $\mathscr{G}p^{\mathscr{M}}$. Thus $\mathscr{P}'$ is the class of projectives in $\mathscr{G}p^{\mathscr{M}}$. The existence of free groups shows that $\mathscr{P}$ is adequate. Therefore so is $\mathscr{P}'$. This class $\mathscr{E}$ will be the only one we will use for the rest of this paper.

Now if G is a simplicial object of $\mathscr{G}p^{\mathscr{C}}$, we have a simplicial group $G(C)$ for each $C \in \mathscr{C}$. Therefore we can consider the groups $\pi_n(G(C))$. This gives a sequence of functors $\pi_n G \in \mathscr{G}p^{\mathscr{C}}$. Given $F \in \mathscr{G}p^{\mathscr{M}}$, let $P' \to F$ be an $\mathscr{P}'$-aspherical, $\mathscr{P}'$-cofibrant resolution in $\mathscr{G}^{\mathscr{M}}$. Then adj P' is a simplicial object of $\mathscr{G}p^{\mathscr{C}}$.

DEFINITION. Let $\mathrm{adj}_n\,(F) = \pi_n(\mathrm{adj}\ P')$. Clearly $\mathrm{adj}_n : \mathscr{G}p^{\mathscr{M}} \to \mathscr{G}p^{\mathscr{C}}$ is a covariant factor.

DEFINITION. Let $H_n : \mathscr{G}^{\mathscr{C}} \to \mathscr{G}^{\mathscr{C}}$ be given by $H_n = \mathrm{adj}_n \circ \mathrm{res}$.

Since π_n of a simplicial group is abelian for $n \neq 0$, $\mathrm{adj}_n\,F$ and $H_n(F)$ lie in the subcategory $\mathscr{A}b^{\mathscr{C}} \subset \mathscr{G}p^{\mathscr{C}}$ for $n \neq 0$.

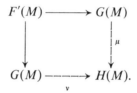

We can also define augmented homology groups as follows. We have $\varepsilon : P' \to \mathrm{res}\ F$ so $\mathrm{adj}\ P' \to \mathrm{adj}\ \mathrm{res}\ F \to F$. Define $\tilde{H}_n(F)$ to be π_n of the augmented simplicial object obtained from $\operatorname{adj} P'$ by adding F in dimension -1 and letting $\partial_0 : \operatorname{adj} P'_0 \to F$ be the map induced by $\operatorname{adj} P' \to F$. Clearly $\tilde{H}_n(F) = H_n(F)$ for $n \geq 1$, while $\tilde{H}_0(F)$, $\tilde{H}_{-1}(F)$ are the kernel and cokernel of the map $\varepsilon : H_0(F) \to F$. Of course $\tilde{H}_{-1}(F)$ will only be defined if the image of ε is normal in F.

We now show how to obtain a long exact sequence for each map $C \to C'$ in $\mathscr{C}$. If G is a simplicial group, the Moore subcomplex of G is defined by letting $M_n(G)$ be the subgroup of G_n consisting of all $x \in G_n$ such that $\partial_i x = 1$ for $i < n$. The map $\partial : M_n(G) \to M_{n-1}(G)$ is given by ∂_n. It is well known **[29]** that $M_*(G)$ is a nonabelian chain complex whose homology gives the homotopy $\pi_n(G)$.

LEMMA 6.2. *Let $f: G \to H$ be a map of simplicial groups. Let $N = \ker f$ i.e. $N_n = \ker f_n : G_n \to H_n$ for each n. Then N is a simplicial subgroup of G and*

$$0 \to M_*(N) \to M_*(G) \to M_*(H)$$

is exact. If $f_n : G_n \to H_n$ is onto, then

$$1 \to M_n(N) \to M_n(G) \to M_n(H) \to 1$$

is exact.

PROOF. Only the last assertion is nontrivial. Let $h \in M_n(H)$. There is an element $g \in G_n$ with $f(g) = h$. Suppose we have such an element with $\partial_i g = 1$ for $i < m$ where $m < n$. This is clear if $m = 0$. Let $g' = g(s_m \partial_m g)^{-1}$. Then $\partial_i g' = 1$ for $i \leq m$ and $f(g') = h(s_m \partial_m h)^{-1} = h$. Repeating this we eventually find an element $g \in M_n(G)$ with $f(g) = h$. Note that $s_m \partial_m$ is only defined for $m < n$, since there is no s_n on G_{n-1}.

COROLLARY 6.3. *Let $f: G \to H$ be a map of simplicial groups which is onto. Let $N = \ker f$. Then there is along exact sequence*

$$\cdots \to \pi_n N \to \pi_n G \to \pi_n H \to \pi_{n-1} N \to \cdots \to \pi_0 G \to \pi_0 H \to 0$$

The proof is the same as that in the abelian case.

To get a long exact sequence when f is not onto we must use an auxiliary construction.

DEFINITION. Let G be a simplicial group. Define PG to be the simplicial group given by $(PG)_n = \ker \partial_0^{n+1} : G_{n+1} \to G_0$ with the same ∂_i and s_i as for G. Define $p: PG \to G$ by $p_n = \partial_{n+1} : G_{n+1} \to G_n$. Let $s_* : (PG)_n \to (PG)_{n+1}$ be $s_{n+1} : G_{n+1} \to G_{n+2}$.

LEMMA 6.4. *PG is a simplicial group, p is a map of simplicial groups. The map $s_* : (PG)_n \to (PG)_{n+1}$ satisfies the conditions $\partial_i s_* = s_* \partial_i$ for $i \leq n$, $\partial_{n+1} s_* = \mathrm{id}$ if $n > 0$. If $n = 0$, we have $\partial_0 s_*(x) = 1$ for all x and $\partial_1 s_* = \mathrm{id}$.*

This follows by easy calculations. We also have $s_i s_* = s_* s_i$ for $i \leq n$, $s_{n+1} s_* = s_* s_*$.

LEMMA 6.5. *Let G be a simplicial group. Suppose there is a homomorphism $S: G_n \to G_{n+1}$ for a given $n > 0$ satisfying $\partial_i S = S \partial_i$, $i = 0, \ldots, n$, $\partial_{n+1} S = \mathrm{id}$.*

Then G is aspherical in dimension n. The same is true in dimension $n = 0$ if we have $\partial_0 Sx = 1$ for all x.

PROOF. Let $x \in M_n(G)$, $\partial x = 1$. Then $\partial_i x = 1$ for $i = 0, \ldots, n$. Therefore $Sx \in M_{n+1}(G)$ but $\partial S(x) = \partial_{n+1} S(x) = x$.

LEMMA 6.6. *For any simplicial group G, the simplicial group PG is contractible. The map $M_n(PG) \to M_n(G)$ induced by $p: PG \to G$ is onto for $n > 0$.*

PROOF. By Lemma 6.5, PG is aspherical in all dimensions. Since it satisfies the Kan condition [**29**], it is contractible [**29**]. If $x \in M_n(G)$, $n > 0$, then $\partial_0 x = 1$. Let $y = s_n x$. Then $\partial_0^{n+1} y = \partial_0^n \partial_n y = \partial_0^n x = 1$, so $y \in (PG)_n$. We have $p(y) = \partial_{n+1} y = x$. But $\partial_i y = \partial_i s_n x = s_{n-1} \partial_i x = 1$ for $i < n$, so $y \in M_n(PG)$.

PROPOSITION 6.7. *To each map of simplicial groups $f: G \to H$, there is associated, in a functorial way, a long exact sequence*

$$\cdots \to \pi_n(G \to H) \to \pi_n(G) \to \pi_n(H) \to \pi_{n-1}(G \to H) \to \cdots \to \pi_0(G) \to \pi_0(H).$$

PROOF. Consider the direct sum (free product) $G \amalg PH$. Map this into H by f on G and p on PH, getting $1 \to N \to G \amalg PH \to H$. Since PH is contractible, $G \amalg PH$ has the homotopy type of $G \amalg 1 = G$ so the inclusion $G \to G \amalg PH$ induces isomorphisms $\pi_n G \xrightarrow{\approx} \pi_n(G \amalg PH)$. We have

$$1 \to M_n(N) \to M_n(G \amalg HP) \to M_n(H) \to 1,$$

the exactness at the right being a consequence of Lemma 6.6. The long exact sequence follows as in Corollary 6.3, except that there is no 0 at the right since $M_0(G \amalg PH) \to M_0(H)$ need not be onto.

To conclude this section, we observe that there is a somewhat simpler construction for cofibrant aspherical resolutions in the case of group valued functors. Suppose G is a simplicial group. Then $Z_{n-1}(G) = \{(x_0, \ldots, x_n) \mid x_i \in G_{n-1}, \partial_i x_j = \partial_{j-1} x_i \text{ for } i < j\}$. We can embed $ZM_{n-1}(G) = \{x \in G_{n-1} \mid \partial_i x = 1, i \leq n-1\}$ in $Z_{n-1}(G)$ by sending $x \in ZM_{n-1}(G)$ to $(1, 1, \ldots, 1, x) \in Z_n(G)$. This extends immediately to simplicial objects in $\mathcal{G}p^{\mathcal{M}}$ (or $\mathcal{G}p^{\mathcal{C}}$). Suppose now G is such an object, Q is an object of $\mathcal{G}p^{\mathcal{M}}$ and $\eta: Q \to ZM_{n-1}(G) \subset Z_{n-1}(G)$. We may use η to attach Q to G getting $G * Q$. As always $(G * Q)_i = G_i$ for $i < n$ and in particular $M_{n-1}(G * Q) = M_{n-1}(G)$. Now $\partial_i \eta(Q) = 1$ for $i < n$, so $Q \subset M_n(G * Q)$. If we choose $\eta: Q \to ZM_{n-1}(G)$ to be in $\mathcal{E}'$(or $\mathcal{E}''$), then $Q(M) \to ZM_{n-1}(G)(M)$ is onto for each $M \in \mathcal{M}$. Therefore $\pi_{n-1}(G * Q(M)) = 0$ for $M \in \mathcal{M}$, so $G * Q$ is aspherical in dimension $n - 1$. Thus it suffices to attach our Q's by maps $\eta: Q \to ZM_{n-1}(G)$ rather than by the more complicated maps $Q \to Z_{n-1}(G)$. Similarly, in starting our resolution, when we attach a $Q^{(1)}$ to $\Delta^0 \times Q^{(0)}$ in dimension 1, our aim is to make

$$Q^{(1)} \underset{\eta_1}{\overset{\eta_0}{\rightrightarrows}} Q^{(0)} \longrightarrow F$$

$\mathscr{P}'$-exact, i.e. for each $M \in \mathscr{M}$, we want $F(M)$ to be the difference cokernel of $Q^{(1)}(M) \rightrightarrows Q^{(0)}(M)$. For this it will clearly suffice to let N be the kernel of $Q^{(0)} \to F$, find $Q^{(1)} \in \mathscr{P}'$ and a $\mathscr{P}'$-epimorphism $\eta: Q^{(1)} \to N$ and define $\eta_0(Q^{(1)}) = 1$, $\eta_1 = \eta$. In other words, we find $\eta: Q^{(1)} \to ZM_0(\Delta^0 \times Q^{(0)}) = M_0(\Delta^0 \times Q^{(0)}) = Q^{(0)}$ such that $Q^{(1)} \in \mathscr{P}'$ and $Q^{(1)} \to \ker [Q^0 \to F]$ is in $\mathscr{E}'$.

It is quite easy to determine π_0 of the resulting simplicial resolution. In fact, $\pi_0(G_*)$ is the difference cokernel of $G_1 \rightrightarrows G_0$. Now, after attaching $Q^{(1)}$, we get $G_0 = Q^{(0)}$, $G_1 = Q^{(0)} \amalg Q^{(1)}$. Since ∂_0, ∂_1 are the same on the summand $Q^{(0)}$ of G_1, $\pi_0(G_*)$ will be the difference cokernel of $Q^{(1)} \rightrightarrows Q^{(0)}$. Since we chose $Q^{(1)}$ so that $\partial_0 \mid Q^{(1)}$ is trivial and $\partial_1 \mid Q^{(1)}$ is η, this shows that $\pi_0(G_*(C))$ is the quotient of $Q^{(0)}(C)$ by the normal subgroup generated by the image of $\eta: Q^{(1)}(C) \to Q^{(0)}(C)$.

7. **Topological *K*-theory.** In this section we consider topological K-theory in order to get some motivation for the definition given in §8. The reader who is not familiar with topological K-theory may skip this section with no essential loss.

Let X be a finite CW-complex with base point. Let $G = GL(\mathbf{C}) = \varinjlim GL(n, \mathbf{C})$ be the infinite general linear group over $\mathbf{C}$ with the usual topology. Then $\tilde{K}^{-1}(X) = [X, G]$ and $\tilde{K}^{-n} = [S^{n-1}X, G]$ for $n \geq 1$ **[4]**. We can easily obtain these groups as the cohomology groups of a nonabelian cochain complex. Let PX denote the path space of X. Let $C^{-n}(X)$ be the set of all maps (not homotopy classes) of X into $P\Omega^n G$. This has a natural group structure obtained from that of G. Define $\partial: C^{-n}(X) \to C^{-n+1}(X)$ to be the map induced by $P\Omega^{n+1}G \to P\Omega^n G$, this latter map being obtained from the diagram

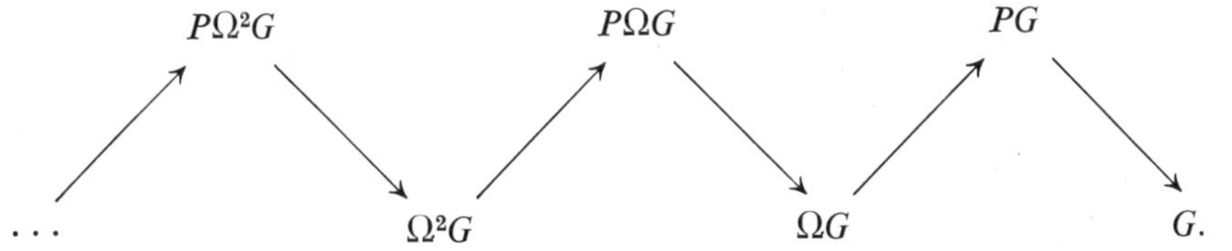

If $f \in C^n(X)$ and $\partial f = 1$, then the composition $X \to P\Omega^n G \to \Omega^n G$ must be trivial, since $\Omega^n G \to P\Omega^{n-1} G$ is $1 - 1$. Thus f maps X into $\Omega^{n+1}G$. If $f, g: X \to \Omega^{n+1}G$ and $f \simeq g$, then $f^{-1}g$ lifts to a map $X \to P\Omega^{n+1}G$, since $P\Omega^{n+1}G \to \Omega^{n+1}G$ is a fibration. Conversely if $f^{-1}g$ lifts then $f \simeq g$, since $P\Omega^{n+1}G$ is contractible. It follows that $\tilde{K}^{-n}(X) = H^{-n+2}(C^*(X))$ for $n \geq 2$. To get the same result for $n = 1$ we must augment the complex, defining $C^1(X)$ to consist of all maps of X into G.

If we repeat the above construction replacing G by an Eilenberg-MacLane space $K(\pi, n)$, we get a portion of the ordinary cohomology of X. This suggests that the main differences between K-theory and ordinary cohomology theory are: (a) the use of nonabelian groups, and (b) the choice of $G = GL(\mathbf{C})$ as the initial group in place of $K(\pi, n)$.

This construction is easily modified so as to agree with our simplicial point of view. The functor P is defined to be $P(X) = \{\omega: I \to X \mid \omega(0) = *\}$ where $*$ is the basepoint. Let 1 be the identity functor and define $P \to 1$ by $\omega \to \omega(1)$. Define $P \to P \circ P$ by sending ω to $f: I \times I \to X$ where $f(x, t) = \omega(st)$. It is trivial to verify that these two maps make P into a comonad [15], [27] which is right adjoint to the usual monad structure on the cone functor. Associated with this comonad, there is a simplicial functor P_* defined in the usual way. Applying this functor to $G = GL(\mathbf{C})$ gives us a simplicial topological group $P_*(G)$. If X is any space, let $G_n(X)$ be all maps of X into $P_n(G)$. Then $G_* = (G_n)$ is a functor from spaces to simplicial groups. Now if A is a simplicial group, the Moore subcomplex $M(A)$ is a nonabelian chain complex and $\pi_n(A) = H_n(M_*(A))$. If X is a topological space, it is trivial to verify that $M_n(G_*(X)) = C^{-n}(X)$ as defined above. Therefore $\tilde{K}^{-n}(X) = \pi_{n-2}(G_*(X))$. To get $\tilde{K}^{-2}$ and $\tilde{K}^{-1}$ we must assume $G_*(X)$ is augmented, having $GL(\mathbf{C})^X$ in dimension -1.

I will now try to give an analogue of this in the algebraic case. As mentioned above the main difference between K-theory and cohomology seems to lie in replacing abelian cochain complexes by nonabelian ones and starting with $GL(\mathbf{C})$.

8. **Algebraic *K*-theory.** Let us first recall briefly the definitions of K_1 and K_2 given by Bass and Milnor respectively. Let R be a ring with unit. Let $GL(R) = \varinjlim GL_n(R)$ be the infinite general linear group over R and let $E(R)$ be the subgroup generated by all $1 + re_{ij}$ where $r \in R$ and e_{ij} is the matrix with 1 in the (i, j)-place and 0 elsewhere. Then $E(R) = [GL(R), GL(R)]$. Bass defines $K_1(R) = GL(R)/E(R)$. Let $ST(R)$ be the group generated by symbols $e_{ij}(r)$ for $r \in R$, $i, j \geq 1$ integers, $i \neq j$, with the Steinberg relations [35]

(1) $e_{ij}(r + s) = e_{ij}(r)e_{ij}(s)$,
(2) $[e_{ij}(r), e_{km}(s)] = 1$ if $i \neq m, j \neq k$,
(3) $[e_{ij}(r), e_{jm}(s)] = e_{im}(rs)$ if $i \neq m$,
(4) $[e_{ij}(r), e_{ki}(s)] = e_{kj}(-sr)$ if $j \neq k$.

Note that (4) is a consequence of (3). Define $ST(R) \to GL(R)$ by sending $e_{ij}(r)$ to $1 + re_{ij}$. The image is $E(R)$. The kernel turns out to be the center of $ST(R)$ [30], [36] and we define $K_2(R) = \ker [ST(R) \to GL(R)]$.

In each case we take the obvious generators or relations ($1 + re_{ij}$ for generators of $GL(R)$, the Steinberg relations for relations in $E(R)$) and let $K_i(R)$, $i = 1, 2$ be whatever is left over. This is very close to the well-known prescription of Eilenberg-MacLane for defining cohomology [13]. This suggests defining $K_n(R)$ by the same procedure. Since we are using nonabelian chain groups, this agrees with the point of view of §7.

The main difficulty with this approach is that it is not at all obvious what the "obvious" generators and relations should be. One way of deciding this in ordinary cohomology theory is to choose some category of models and define the "obvious" relations to be those which hold for models. This leads to the theory of acyclic models. Therefore we will define K_n using the theory developed in §5 and §6.

The only problem remaining is the choice of models. In the topological case, the usual choice is the class of contractible spaces. The algebraic analogue is the class of free rings. However this is still not quite correct. If X is a contractible space with basepoint x, we do not have $\tilde{K}^{-n}(X) = 0$ for $n \neq 0$ but only $K^{-n}(X, x) = 0$. The algebraic analogue of a basepoint is an augmentation. This suggests considering the category of augmented $\mathbf{Z}$-algebras. We use instead an equivalent category. Let $\mathscr{R}$ be the category of all associative rings, not necessarily with unit and ring homomorphisms (which need not preserve units). Let $\mathscr{A}$ be the category of augmented $\mathbf{Z}$-algebras. An object of $\mathscr{A}$ consists of a ring A with unit and a unit preserving ring homomorphism $\varepsilon: A \to \mathbf{Z}$. Define a functor from $\mathscr{A}$ to $\mathscr{R}$ by sending (A, ε) to $\ker \varepsilon$. If $R \in \mathscr{R}$ we can adjoin a unit to R by letting $R_u = \mathbf{Z} \oplus R$ as abelian group, with multiplication given by $(m, r)(n, s) = (mn, ms + nr + rs)$. Define $\varepsilon: R_u \to \mathbf{Z}$ by $\varepsilon(m, r) = m$. Then (R_u, ε) is an augmented $\mathbf{Z}$-algebra. Clearly the functors $(A, \varepsilon) \to \ker \varepsilon$, $R \to (R_u, \varepsilon)$ define an equivalence between $\mathscr{A}$ and $\mathscr{R}$.

Let $\mathscr{R}_1$ be the subcategory of $\mathscr{R}$ consisting of rings with unit and unit preserving homomorphisms. Note that $\mathscr{R}_1$ is *not* a *full* subcategory of $\mathscr{R}$. The inclusion $i: \mathscr{R}_1 \subset \mathscr{R}$ has a left adjoint which sends R to R_u. The fact that $R_u \neq R$ even for $R \in \mathscr{R}_1$ shows again that $\mathscr{R}_1$ is not full in $\mathscr{R}$. Let $F: \mathscr{R}_1 \to \mathscr{A}\mathscr{b}$ be a covariant functor from $\mathscr{R}_1$ to the category of abelian groups. We define a functor $\tilde{F}: \mathscr{R} \to \mathscr{A}\mathscr{b}$ by setting $\tilde{F}(R) = \ker [F(R_u) \to F(\mathbf{Z})]$.

LEMMA 8.1. *For $R \in \mathscr{R}_1$, there is a natural isomorphism $R_u \approx R \times \mathbf{Z}$.*

PROOF. Send R_u to R by $(m, r) \to m1 + r$ and to $\mathbf{Z}$ by $\varepsilon(m, r) = m$.

COROLLARY 8.2. *If $F: R_1 \to \mathscr{A}\mathscr{b}$ preserves finite direct products then $\tilde{F} \mid R_1 = F$.*

In this case we can write F for $\tilde{F}$ without any confusion.

LEMMA 8.3. *The functors $K_1, K_2: \mathscr{R}_1 \to \mathscr{A}\mathscr{b}$ preserve finite direct products.*

PROOF. Clearly $GL(R \times R') = GL(R) \times GL(R')$. We must show the same for ST. We have a map $ST(R \times R') \to ST(R) \times ST(R')$. Now identify $R \times R'$ with $R \oplus R'$ so R and R' are embedded as $R \times 0$, $0 \times R'$. Then for $r \in R$, $r' \in R'$ we have $e_{ij}(r, r') = e_{ij}(r)e_{ij}(r')$. The $e_{ij}(r)$ for $r \in R$ satisfy all the relations for $ST(R)$ and no more, since the subgroup they generate maps onto $ST(R)$ under $ST(R \times R') \to ST(R) \times ST(R')$. Thus these elements generate a subgroup $ST(R) \subset ST(R \times R')$. Similarly we have a subgroup $ST(R') \subset ST(R \times R')$. If we can show that these subgroups commute we will have a map $ST(R) \times ST(R') \to ST(R \times R')$ which is clearly an inverse for $ST(R \times R') \to ST(R) \times ST(R')$. Now $e_{ij}(r)$ and $e_{km}(r')$ commute by the Steinberg relations except when $k = j$, $m = i$. In this case, choose $n \neq i, j$. Let e' be the unit of R'. Then $e_{ji}(r') = [e_{jn}(e'), e_{ni}(r')]$. But both terms of this commutator commute with $e_{ij}(r)$ by the Steinberg relations.

It follows from this that we can extend the functors K_1 and K_2 to $\mathscr{R}$ by setting $K_i(R) = \ker [K_i(R_u) \to K_i(\mathbf{Z})]$. Similarly we can extend GL and ST. It is very easy to describe $GL(R)$ directly for $R \in \mathscr{R}$. It is, by definition, the kernel of $GL(R_u) \to GL(\mathbf{Z})$ and so consists of all invertible matrices $I + Q$ where Q has entries in R. Consider all matrices $Q = (q_{ij})$, $i, j \geq 1$ integers, with only a finite number of nonzero entries. Define an operation by $P \circ Q = P + Q + PQ$. Let

$GL(R_u)$ consist of all such matrices Q for which there is a P with $P \circ Q = Q \circ P = 0$. The operation $P \circ Q$ makes $GL(R)$ into a group. It is isomorphic to ker $[GL(R_u) \to GL(\mathbf{Z})]$ by the map sending P to $I + P$. In fact, if $A \to B$ is any map in $\mathscr{R}$ with kernel R then $GL(R) \approx \ker [GL(A) \to GL(B)]$. If R has a unit, the map $P \mapsto I + P$ identifies the present $GL(R)$ with the usual one. Since $GL(R_u)$ is a split extension with kernel $GL(R)$ and quotient $GL(\mathbf{Z})$, we see that $GL(\mathbf{Z})$ acts naturally on $GL(R)$. In terms of the above representation, this is simply given by $A \cdot Q = AQA^{-1}$ for $A \in GL(\mathbf{Z})$, $Q \in GL(R)$.

The group $ST(R)$ is not as easy to describe. As in the case of $GL(R)$, the group $ST(\mathbf{Z})$ acts naturally on $ST(R)$. We can describe $ST(R)$ in terms of this action.

LEMMA 8.4. *For any $R \in \mathscr{R}$, the group $ST(R)$ can be presented as a group having $ST(\mathbf{Z})$ as group of operators and, with respect to these operators, the generators $e_{ij}(r)$ for $i, j \geq 1$, $i \neq j$, $r \in R$. The relations are*

(1) $e_{ij}(r)e_{ij}(s) = e_{ij}(r + s)$
(2) $[e_{ij}(r), e_{km}(s)] = 1$ *if* $j \neq k$, $i \neq m$
(3) $[e_{ij}(r), e_{jk}(s)] = e_{ik}(rs)$
(4) $e_{ij}(z) \cdot e_{ij}(r) = e_{ij}(r)$ *for* $z \in \mathbf{Z}$
(5) $e_{ij}(z) \cdot e_{km}(s) = e_{km}(s)$ *for* $z \in \mathbf{Z}$, $i \neq m, j \neq k$
(6) $e_{ij}(z) \cdot e_{jk}(s) = e_{ik}(zs)e_{jk}(s)$ *for* $z \in \mathbf{Z}$, $i \neq k$
(7) $e_{ij}(z) \cdot e_{ki}(s) = e_{kj}(-zs)e_{ki}(s)$ *for* $z \in \mathbf{Z}, j \neq k$.

Here $a \cdot b$ denotes aba^{-1} for $a \in ST(\mathbf{Z})$, $b \in ST(R)$.

PROOF. This is an easy consequence of the Reidemeister-Schreier theorem [**28**]. The only difficulty is to make sure that no relations are overlooked. As a check, we may verify the result as follows. Let $ST'(R)$ be the group defined by the above relations. Since it is given with an action of $ST(\mathbf{Z})$ we may form the semi-direct product $ST(\mathbf{Z}) \tilde{\times} ST'(R)$. Define $ST(\mathbf{Z}) \tilde{\times} ST'(R) \to ST(R_u)$ by sending $(e_{ij}(z), 1)$ to $e_{ij}((z, 0))$ and $(1, e_{ij}(s))$ to $e_{ij}(0, s)$. We check easily that all relations are satisfied. Define $ST(R_u) \to ST(\mathbf{Z}) \tilde{\times} ST'(R)$ by sending $e_{ij}((z, r))$ to $(e_{ij}(z), e_{ij}(r))$. Again all relations are satisfied. Both compositions are clearly the identity. Therefore $ST(R_u) = ST(\mathbf{Z}) \tilde{\times} ST'(R)$. This clearly identifies $ST'(R)$ with $ST(R)$.

We now define the map $\theta: ST(R) \to GL(R)$ as follows. The map $ST(\mathbf{Z}) \to GL(\mathbf{Z})$ makes $ST(\mathbf{Z})$ act on $GL(R)$. Let θ be the $ST(\mathbf{Z})$-homomorphism sending $e_{ij}(r)$ to re_{ij}, where re_{ij} is the matrix having r in position (i, j) and 0 elsewhere. We now have $K_2(R) = \ker \theta$, $K_1(R) = \operatorname{ckr} \theta$. This follows from the diagram

$$\begin{array}{ccccccccc}
1 & \longrightarrow & ST(R) & \longrightarrow & ST(R_u) & \longrightarrow & ST(\mathbf{Z}) & \longrightarrow & 1 \\
 & & \downarrow & & \downarrow & & \downarrow & & \\
1 & \longrightarrow & GL(R) & \longrightarrow & GL(R_u) & \longrightarrow & GL(\mathbf{Z}) & \longrightarrow & 1.
\end{array}$$

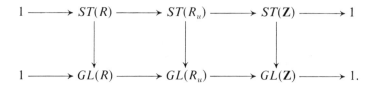

By the usual snake lemma, this gives a long exact sequence

$$0 \to \ker \theta \to K_2(R_u) \to K_2(\mathbf{Z}) \to \operatorname{ckr} \theta \to K_1(R_u) \to K_1(\mathbf{Z}) \to 0,$$

which yields our result since $R_u \to \mathbf{Z}$ splits.

For each n, let $R_u^{(n)}$ be the free associative ring with unit on n noncommuting indeterminates $x_1, \ldots, x_n$. Define $\varepsilon: R_u^{(n)} \to \mathbf{Z}$ by $\varepsilon(x_i) = 0$ for all i. This makes $R_u^{(n)}$ an augmented $\mathbf{Z}$-algebra. The kernel $R^{(n)}$ of ε consists of all noncommutative polynomials in $x_1, \ldots, x_n$ with integral coefficients and zero constant term. It is clearly the free associative ring (without unit) on $x_1, \ldots, x_n$. By **[22]**, **[34]**, ε induces an isomorphism $K_1(R_u^{(n)}) \to K_1(\mathbf{Z})$. This shows $K_1(R^{(n)}) = 0$. This is the algebraic analogue of the fact that $K^{-1}(X, x) = 0$ for a contractible space X. This makes it reasonable to expect that $K_2(R^{(n)}) = 0$. This has not yet been proved but recent work of P. M. Cohn and his students on $GL(R^n)$ should make this accessible. At any rate, it suggests again that the $R^{(n)}$ should be reasonable as models.

We therefore let $\mathscr{M}$ be the full subcategory of $\mathscr{R}$ whose objects are the $R^{(n)}$. Consider the functor $GL: \mathscr{R} \to \mathscr{G}\!p$. As in §6, we can define $\tilde{H}_n(GL): \mathscr{R} \to \mathscr{G}\!p$ for $n \geq -1$. We make the following tentative definition.

DEFINITION. Let $K_n = \tilde{H}_{n-2}(GL): \mathscr{R} \to \mathscr{G}\!p$ for $n \geq 1$.

By the remarks of §7 and those of the present section, this appears to be a reasonable choice. Further justification is given by the following result.

THEOREM 8.5. *The above definition of K_1 agrees with Bass's definition. If we denote Milnor's K_2 by K_2', there is a natural map $K_2' \to K_2$. For every $R \in \mathscr{R}$, $K_2'(R) \to K_2(R)$ is onto and the kernel is generated by the images of all $K_2'(R^{(n)})$, $n \geq 1$ under all maps $R^{(n)} \to R$.*

COROLLARY 8.6. *If $K_2'(R^{(n)}) = 0$ for all n, then $K_2' \to K_2$ is an isomorphism.*

Before giving the proof, we give a general method for constructing elements of $\mathscr{P}''$ in $\mathscr{G}\!p^{\mathscr{R}}$. Suppose we are given sets S_n for $n \geq 0$, $n \in \mathbf{Z}$. For each ring R, let $P(R)$ be the free group generated by all $R^n \times S_n$, i.e. $P(R)$ has one generator $e(r_1, \ldots, r_n, s)$ for each $n \geq 0$, $s \in S_n$, $r_1, \ldots, r_n \in R$. If $f: R \to R'$ define $P(f): P(R) \to P(R')$ by sending $e(r_1, \ldots, r_n, s)$ to $e(fr_1, \ldots, fr_n, s)$. Clearly P is a functor from $\mathscr{R}$ to $\mathscr{G}\!p$. Let $\mathscr{P}''$ be as in §6.

LEMMA 8.7. *The functor P lies in $\mathscr{P}''$.*

PROOF. Let $\eta: F \to G$ lie in $\mathscr{E}''$, i.e. $\eta: F(R^{(n)}) \to G(R^{(n)})$ is onto for each n. Let $\varphi: P \to G$. We must lift φ to $\psi: P \to F$. For $r = (r_1, \ldots, r_n)$, let $\hat{r}: R^{(n)} \to R$ be the unique map sending x_i to r_i, $i = 1, \ldots, n$. Then $P(\hat{r})$ sends $e(x_1, \ldots, x_n, s)$ to $e(r_1, \ldots, r_n, s)$. For each $n \geq 0$ and $s \in S_n$ choose an element $a_n(s) \in F(R^{(n)})$ which maps onto $\varphi(e(x_1, \ldots, x_n, s))$ under η. Define $\psi_R: P(R) \to F(R)$ by sending $e(r_1, \ldots, r_n, s)$ to $F(\hat{r})a_n(s)$. It is a routine matter to check that ψ is natural and lifts φ.

PROOF OF THEOREM 8.5. We must construct a simplicial resolution for GL. For the first step we need to find some $G_0 \in \mathscr{P}''$ and a map $G_0 \to GL$ in $\mathscr{E}''$. Let

$G_0(R)$ be the free group on generators $e_{ij}(r, \alpha)$ for $i, j \geq 1, i \neq j, r \in R, \alpha \in ST(\mathbf{Z})$. Then $G_0 \in \mathscr{P}''$ by Lemma 8.7 with $S = \{(i,j) \mid i,j \geq 1,\ i \neq j\} \times ST(\mathbf{Z})$. Define an action of $ST(\mathbf{Z})$ on $G_0(R)$ by $\beta \cdot e_{ij}(r, \alpha) = e_{ij}(r, \beta \cdot \alpha)$. Define $\varepsilon_R : G_0(R) \to GL(R)$ by sending $e_{ij}(r, \alpha)$ to $\bar{\alpha} \cdot re_{ij}$ where $\bar{\alpha}$ is the image of α in $GL(\mathbf{Z})$ and $\bar{\alpha} \cdot x$ denotes $\bar{\alpha}x\bar{\alpha}^{-1}$ (formed in $GL(R_u)$). This defines $\varepsilon : G_0 \to GL$. This map factors through ST by $G_0 \xrightarrow{\varepsilon} ST \xrightarrow{\theta} GL$ where ε'_R sends $e_{ij}(r, \alpha)$ to $\alpha \cdot e_{ij}(r)$. Clearly $G_0(R) \to ST(R)$ is onto for all R. For $R = R^{(n)}$, $ST(R) \to GL(R)$ is also onto since $K_1(R) = 0$. Therefore $\varepsilon : G_0 \to GL$ lies in $\mathscr{E}''$. We can now extend G_0 to a simplicial resolution $G_* \to GL$ without changing the 0-dimensional part G_0, i.e. we form $\Delta^0 \times G_0 \to GL$ and attach higher dimensional material. Therefore $\pi_0(G_*)$ is a quotient of G_0, so $\tilde{H}_{-1}(GL)$ is the cokernel of $G_0 \to GL$. Since $G_0(R) \to ST(R)$ is onto, $\tilde{H}_{-1}(GL) = \text{ckr}\,(ST \to GL)$, but this is just Bass's K_1.

To calulate $\tilde{H}_0(GL)$ we must now attach some $Q \in \mathscr{P}''$ to $\Delta^0 \times G_0$ in dimension 1 as the end of §6. Let N be the kernel of $G_0 \to GL$. We must find $Q \in \mathscr{P}''$ and a map $Q \to N$ such that $Q(R^{(n)}) \to N(R^{(n)})$ is onto for $n \geq 1$. Now each element $y \in N(R^{(n)})$ can be written in the form $y = \prod_1^k e_{i_\nu j_\nu}(u_\nu, \alpha_\nu)^{\varepsilon_\nu}$ where

$$u_\nu = w_\nu(x_1, \ldots, x_n)$$

is an element of $R^{(n)}$, $\alpha_\nu \in ST(\mathbf{Z})$, and $\varepsilon_\nu = \pm 1$. For each such y, $Q(R)$ will have a generator

$$e(i_1, \ldots, i_k, j_1, \ldots, j_k, \alpha_1, \ldots, \alpha_k, \varepsilon_1, \ldots, \varepsilon_k, r_1, \ldots, r_k)$$

where $r_1, \ldots, r_k \in R$. Send $Q(R) \to G_0(R)$ by sending this generator to

$$y' = \prod_1^k e_{i_\nu j_\nu}(w_\nu(r_1, \ldots, r_n), \alpha)^{\varepsilon_\nu}.$$

Now $r = (r_1, \ldots, r_k)$ defines a map $\hat{r} : R^{(n)} \to R$ sending x_i to r_i. The induced map $N(R^{(n)}) \to N(R)$ sends y to y'. Thus the image of $Q(R)$ in $G_0(R)$ is the subgroup of $G_0(R)$ generated by the images of all $N(R^{(n)})$ under all maps $R^{(n)} \to R$. This subgroup clearly lies in $N(R)$. It is also stable under $ST(\mathbf{Z})$ since all $N(R^{(n)})$ are. Now $Q \in \mathscr{P}''$ by Lemma 8.7. Also $Q \to N$ and $Q(R^{(n)}) \to N(R^{(n)})$ is onto, so $Q \to N$ is a $\mathscr{P}''$-epimorphism. Therefore attaching Q in dimension 1 gives us a simplicial resolution which is aspherical in dimension 0. We can now calculate $H_0(GL)$ by the remark at the end of §6. This shows that $H_0(GL)(R) = \pi_0(G_*(R))$ is the quotient of $G_0(R)$ by the normal subgroup $B(R)$ generated by the image of $Q(R)$ in $G_0(R)$. In other words it is the quotient of $G_0(R)$ by the normal subgroup $B(R)$ generated by the images of all $N(R^{(n)})$ in $G_0(R)$. This subgroup is clearly stable under $ST(\mathbf{Z})$. Now every relation given in Lemma 8.4 lies in this subgroup. For example, the element $(e_{ij}(z) \cdot e_{km}(s))e_{km}(s)^{-1}$ (from relation (5)) is the image of the element $(e_{ij}(z) \cdot e_{km}(x_1))e_{km}(x_1)^{-1}$ in $N(R^{(1)})$ under the map $\hat{s}$. The elements clearly lie in N since all the relations hold in $GL(R)$. Thus $H_0(GL)(R)$ is a quotient of the group obtained by reducing $G_0(R)$ modulo these relations, i.e. of $ST(R)$. It is clear that $H_0(GL)(R)$ is the quotient of $ST(R)$ by the normal subgroup generated by the images in $ST(R)$ of all the kernels ker $[ST(R^{(n)}) \to GL(R^{(n)})]$.

But this kernel is $K_2(R^{(n)})$ and its image in $ST(R)$ lies in $K_2(R)$. The theorem follows immediately from this.

We now investigate the existence of exact sequences. If $f: R \to R'$ we get $G_*(f): G_*(R) \to G_*(R')$. The results of §6 gives us an exact sequence

$$\cdots \to K_n(R \to R') \to K_n(R) \to K_n(R') \to K_{n-1}(R \to R') \to$$
$$\cdots \to K_3(R) \to K_3(R') \to ST'(R \to R') \to ST'(R) \to ST'(R'),$$

where we write $ST'(R) = H_0(GL)(R)$. It is not clear how to include K_2 and K_1 in this sequence since our method does not give us $K_1(R \to R')$. However, in case f is onto, everything works well even in low dimensions.

THEOREM 8.8. *Let I be a 2-sided ideal of R. Then there is a natural exact sequence*

$$\cdots \to K_n(R, I) \to K_n(R) \to K_n(R/I) \to K_{n-1}(R, I) \to$$
$$\cdots \to K_2(R) \to K_2/R(I) \to K_1(R, I) \to K_1(R) \to K_1(R/I).$$

Also $K_1(R, I)$ agrees with the relative K_1 defined by Bass (at least if R has a unit).

PROOF. Let G_* be a $\mathscr{P}''$-aspherical, $\mathscr{P}''$-cofibrant resolution of GL. Each G_n is a direct sum of various Q's in $\mathscr{P}''$. Therefore $G_n \in \mathscr{P}''$.

LEMMA 8.9. *Suppose $G \in \mathscr{G}/\!\!\not{p}^{\mathscr{R}}$. If $G \in \mathscr{P}''$ and $f: R \to R'$ is onto, then $G(f)$: $G(R) \to G(R')$ is onto.*

PROOF. Define $P(R)$ to be the free group with one generator $e(u, a)$ for each $u: R^{(n)} \to R$ and $a \in G(R^{(n)})$. For $f: R \to R'$ let $P(f): P(R) \to P(R')$ by sending $e(u, a)$ to $e(fu, a)$. Then $P \in \mathscr{G}/\!\!\not{p}^{\mathscr{R}}$ and $P \in \mathscr{P}''$. In fact P is the functor given by Proposition 5.1 using the free groups generated by the elements of the $G(R^{(n)})$ as the Q_m. The $\mathscr{P}''$-epimorphism $P \to G$ is given by sending $e(u, a)$ to $G(u)a \in G(R)$. Since $G \in \mathscr{P}''$, this splits; thus $G(R)$ is naturally a retract of $P(R)$. If we show that $P(R) \to P(R')$ is onto, it will follow that $G(R) \to G(R')$ is also onto. Let $u: R^{(n)} \to R'$ and $a \in G(R^{(n)})$. Since $R^{(n)}$ is free, we can lift u to $v: R^{(n)} \to R$ and $e(v, a)$ maps to $e(u, a)$. This shows $P(R) \to P(R')$ is onto.

This lemma shows that $G_*(R) \to G_*(R/I)$ is onto. Let $G_*(R,I)$ be its kernel. Corollary 6.3 gives us a long exact sequence

$$\cdots \to K_n(R, I) \to K_n(R) \to K_n(R/I) \to \cdots \to K_3(R)$$
$$\to K_3(R/I) \to ST'(R, I) \to ST'(R) \to ST'(R/I) \to 0.$$

Applying the snake lemma to the diagram

$$\begin{array}{ccccccccc} K_3(R) \to K_3(R/I) \to & ST'(R, I) & \to & ST'(R) & \to & ST'(R/I) & \to 0 \\ & \downarrow & & \downarrow & & \downarrow & \\ 0 \to & GL(R, I) & \to & GL(R) & \to & GL(R/I) & \end{array}$$

then gives us the sequence

$$K_3(R) \to K_3(R/I) \to K_2(R, I) \to K_2(R) \to K_2(R/I) \to K_1(R, I) \to K_1(R) \to K_1(R/I).$$

To identify $K_1(R, I)$ with Bass's $K_1(R, I)$ when R has a unit, note that $ST'(R)$ and $ST'(R/I)$ are quotients of $ST(R)$, $ST(R/I)$ by Theorem 8.5. The definition of ST shows that $ST(R) \to ST(R/I)$ is onto. Let N be its kernel. We have then a diagram

$$\begin{array}{ccccccccc} 1 & \longrightarrow & N & \longrightarrow & ST(R) & \longrightarrow & ST(R/I) & \longrightarrow & 1 \\ & & \downarrow & & \downarrow & & \downarrow & & \\ 1 & \longrightarrow & GL(R, I) & \longrightarrow & GL(R) & \longrightarrow & GL(R/I). & & \end{array}$$

By definition of ST we see that N is the smallest normal subgroup of $ST(R)$ containing all $e_{ij}(u)$ for $u \in I$. The image of $ST(R)$ in $GL(R)$ is just $E(R)$. Therefore the image of N in $GL(R, I)$ is the smallest normal subgroup of $E(R)$ containing all $1 + ue_{ij}$, $u \in I$. This group is $E(R, I)$ by definition. Bass defines $K_1(R, I) = GL(R, I)/E(R, I)$. Let $E'(R, I)$ be the image of $ST'(R, I)$ in $GL(R, I)$. We must show that $E'(R, I) = E(R, I)$. Now it is easy to see that $E'(R, I)$ is the image of $\ker\,[ST'(R) \to ST'(R/I)]$ in $GL(R, I)$. Consider the diagram

$$\begin{array}{ccc} ST(R) & \xrightarrow{j} & ST(R/I) \\ \downarrow{\scriptstyle \theta} & & \downarrow{\scriptstyle \theta'} \\ ST'(R) & \xrightarrow{j'} & ST'(R/I) \\ \downarrow & & \downarrow \\ GL(R) & \longrightarrow & GL(R/I). \end{array}$$

Since $E(R, I)$ is the image of $\ker j$ in $GL(R)$ and $E'(R, I)$ is the image of $\ker j'$, we see that $E(R, I) \subset E'(R, I)$. Conversely, let $x \in \ker j'$. Since $ST(R) \to ST'(R)$ is onto, we can lift x to $y \in ST(R)$. Let z be the image of y in $ST(R/I)$. Then $z \in \ker \theta'$. If there is some $t \in \ker \theta$ which maps onto z, we use $t^{-1}y$ to lift x. Since $t^{-1}y \in \ker j$, its image in $GL(R)$ lies in $E(R, I)$. Thus $E'(R, I) \subset E(R, I)$. Now let $K_2(R)_0$ be the sum of all images $K_2(R^{(n)}) \to K_2(R)$ induced by maps $R^{(n)} \to R$, $n \geq 1$. By Theorem 8.5, $K_2(R)_0 = \ker \theta$ and $K_2(R/I)_0 = \ker \theta'$. We must show $K_2(R)_0 \to K_2(R/I)_0$ is onto. If $f: R^{(n)} \to R/I$, we can lift f to $g: R^{(n)} \to R$, since $R^{(n)}$ is free. Therefore $K_2(f)$ factors as $K_2(R^{(n)}) \to K_2(R) \to K_2(R/I)$, so the image of $K_2(f)$ lies in the image of $K_2(R)_0$ in $K_2(R/I)$.

REMARK. The exact sequence of Theorem 8.8 can be continued down to $K_0(R/I)$ by **[6]**.

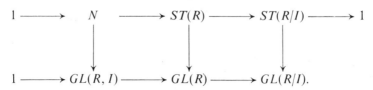

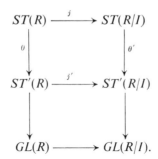

9. **Excision and Mayer-Vietoris sequence.** Milnor **[30]**, **[5]** has shown that one can define a Mayer-Vietoris sequence for K_1 and K_0. Suppose we have a Cartesian diagram in $\mathscr{R}_1$

$$(*) \qquad \begin{array}{ccc} R' & \xrightarrow{k} & R_1 \\ {\scriptstyle h}\downarrow & & \downarrow{\scriptstyle f} \\ R_2 & \xrightarrow[g]{} & R \end{array}$$

with f onto. Milnor shows there is an exact sequence

$$K_1(R') \to K_1(R_1) \oplus K_1(R_2) \to K_1(R) \to K_0(R') \to K_0(R_1) \oplus K_0(R_2) \to K_0(R).$$

In **[5]**, Bass shows that one can define K_n for all $n \leq 0$ and that this Mayer-Vietoris sequence extends to all negative dimensions. It would be very useful to have such a sequence also for positive dimensions. I have not yet been able to establish this. In this section, I will discuss some of the problems involved in proving this. Suppose first that there is such a Mayer-Vietoris sequence. Let R be a ring with unit and I a 2-sided ideal of R. As above, let I_u be the ring obtained by adjoining a unit to I. It is trivial to verify that the diagram

$$\begin{array}{ccc} I_u & \longrightarrow & R \\ \downarrow & & \downarrow \\ \mathbf{Z} & \longrightarrow & R/I \end{array}$$

is cartesian. This will then give an exact sequence

$$\cdots \to K_n(I_u) \to K_n(\mathbf{Z}) \oplus K_n(R) \to K_n(R/I) \to \cdots.$$

Since $I_u \to \mathbf{Z}$ splits, we have $K_n(I_u) = K_n(\mathbf{Z}) \oplus A_n$ where $A_n = \ker [K_n(I_u) \to K_n(\mathbf{Z})]$. If we send the summand $K_n(\mathbf{Z})$ to $K_n(\mathbf{Z}) \oplus K_n(R)$ and project on $K_n(\mathbf{Z})$, the result is clearly the identity map. Therefore we may remove the terms $K_n(\mathbf{Z})$, getting a sequence

$$\cdots \to A_n \to K_n(R) \to K_n(R/I) \to A_{n-1} \to \cdots.$$

Comparing this with the sequence of Theorem 8.8 we see that A_n and $K_n(R, I)$ can only differ by group extensions. But A_n depends only on I so we conclude that $K_n(R, I)$ depends only on I up to group extensions. The most reasonable way for this to happen is for $K_n(R, I)$ to depend only on I. A somewhat stronger formulation of this is given by the excision property. This says that if $f: R \to R'$ is a ring homomorphism and I and I' are 2-sided ideals of R and R' such that f maps I isomorphically onto I', then f maps $K_n(R, I) \to K_n(R', I')$ isomorphically. I do not know whether this property holds. If it does, the Mayer-Vietoris sequence is an easy consequence. In fact, M. G. Barratt has remarked **[12]** that it follows immediately from the diagram

$$\begin{array}{ccccc} \cdots \to K_n(R', I) & \longrightarrow & K_n(R') & \longrightarrow & K_n(R_2) \to \cdots \\ \downarrow{\scriptstyle\theta} & & \downarrow & & \downarrow \\ \cdots \to K_n(R_1, I) & \longrightarrow & K_n(R_1) & \longrightarrow & K_n(R) \to \cdots \end{array}$$

and the fact that θ is an isomorphism.

We now consider the case $n = 1$ and show that the various properties considered above are really equivalent in this case. Recall that $K_1(R, I) = GL(R, I)/E(R, I)$ where $E(R, I)$ is the smallest normal subgroup of $E(R)$ containing all elements $1 + qe_{ij}$, $q \in I$. Now it is obvious that $GL(R, I)$ depends only on I. In fact $GL(R, I) = GL(I)$. Also the elements $1 + qe_{ij}$ depend only on I. However, the normal subgroup generated by these elements and their conjugates seems to depend on R. The larger R is, the more possible conjugates there are. This makes it seem very unlikely that $E(R, I)$ and hence $K_1(R, I)$ can depend only on I. However, I do not known of any counter example and it could conceivably turn out that excision holds for K_1. If this is not the case, the following result shows that it will not be possible to extend the Mayer-Vietoris sequence without altering K_1.

PROPOSITION 9.1. *The following properties are equivalent.*

(1) *There is a functor $K_2: \mathcal{R}_1 \to \mathcal{Ab}$ such that for any cartesian diagram* (*) *in $\mathcal{R}_1$ with f onto, the following sequence is exact:*

$$K_2(R_1) \oplus K_2(R_2) \to K_2(R) \to K_1(R') \to K_1(R_1) \oplus K_1(R_2) \to K_1(R)$$

(2) *For any cartesian diagram* (*) *in $\mathcal{R}_1$ with f a split epimorphism the sequence*

$$0 \to K_1(R') \to K_1(R_1) \oplus K_1(R_2) \to K_1(R)$$

is exact.

(3) *Excision holds for K_1.*

(4) *There is a natural isomorphism $K_1(R, I) \approx K_1(I)$, for $R \in \mathcal{R}_1$.*

(5) *If $R \to R/I$ splits, the map $K_1(I_u, I) \to K_1(R, I)$ is an isomorphism.*

PROOF. If f is a split epimorphism, so is h. Therefore $K_2(R_2) \to K_2(R)$ is onto. This shows that (1) implies (2). Clearly (4) implies (3). Also (3) implies (1) by Barratt's argument using either Milnor's K_2 or the K_2 defined here. The $K_1(R)$ at the right comes from the known Mayer-Vietoris sequence [**30**], [**5**]. Given any $I \subset R$, consider the map $I_u \to R$. If (3) holds, we get an isomorphism $K_1(I_u, I) \to K_1(R, I)$. Now $I_u \to \mathbf{Z}$ splits. By the exact sequence of Theorem 8.8 or by Milnor's exact sequence for his K_2, we get an exact sequence $0 \to K_1(I_u, I) \to K_1(I_u) \to K_1(\mathbf{Z}) \to 0$. But we have defined $K_1(I)$ as the kernel of $K_1(I_u) \to K_1(\mathbf{Z})$. Therefore $K_1(I_u, I) \approx K_1(I)$. This shows that (3) implies (4). It remains to show that (2) implies (5) and (5) implies (3). Given $I \subset R$, $R \in \mathcal{R}_1$, consider the map $I_u \to R$ as above. This gives $(I_u, I) \to (R, I)$, which induces $GL(I_u, I) \to GL(R, I)$. This map is just the identity if we identify $GL(I_u, I)$ and $GL(R, I)$ with $GL(I)$. The map sends $E(I_u, I)$ into $E(R, I)$, so we have $E(I_u, I) \subset E(R, I) \subset GL(I)$. Let $F(R, I) = E(R, I)/E(U_u, I)$. To prove (3) it will suffice to show that $F(R, I) = 0$ for all R, I. In other words $E(R, I) = E(I_u, I)$ for all R, I. Let $x \in F(R, I)$. Choose a representative y for x in $E(R, I)$. By the definition of $E(R, I)$, y will be a product $\Pi z_\nu(1 + q_\nu e_{i_\nu j_\nu})z_\nu^{-1}$ where $q_\nu \in I$, and $z_\nu \in E(R)$. Write $z_\nu = \Pi(1 + r_\mu^{(\nu)} e_{k_\mu m_\mu})$. Since there are only a finite number of q_ν and $r_\mu^{(\nu)}$, we can find a map $f: R_u^{(n+p)} \to R$ such that the images of $x_1, \ldots, x_n$ include all the q_ν and the images of $x_{n+1}, \ldots, x_{n+p}$ include all the $r_\mu^{(\nu)}$. Let J be the 2-sided ideal of $R_n^{(n+p)}$ generated by $x_1, \ldots, x_n$. Clearly $f(J) \subset I$. By the choice of f, we can lift each q_ν to $q'_\nu \in J$ and each $r_\mu^{(\nu)}$ to $r_\mu^{(\nu)} \in R^{(n+p)}$. Form the product y' analogous to y using q'_ν and $r'^{(\nu)}_\mu$. Then y' represents $x' \in F(R_u^{(n+p)}, J)$ and this maps to x under $F(f)$. Therefore it will suffice to show that $F(R_u^{(n+p)}, J) = 0$. But $R_u^{(n+p)}/J \approx R_u^{(p)}$ and this is free in $\mathscr{R}$. Thus $R_u^{(n+p)} \to R_u^{(n+p)}/J$ splits. If (5) holds then $F(R_u^{(n+p)}, J) = 0$. It follows that (5) implies (3).

Finally, to see that (2) implies (5), let $R \to R/I$ be split. Form the cartesian diagram

$$\begin{array}{ccc} I_u & \longrightarrow & R \\ \downarrow & & \downarrow \\ \mathbf{Z} & \longrightarrow & R/I. \end{array}$$

By (2), we have $0 \to K_1(I_u) \to K_1(\mathbf{Z}) \oplus K_1(R) \to K_1(R/I)$. Now $K_1(I_u) \to K_1(\mathbf{Z})$ splits and its kernel is $K_1(I_u, I)$. As in the remarks about K_n above, the sequence gives $0 \to K_1(I_u, I) \to K_1(R) \to K_1(R/I)$. But $R \to R/I$ splits, so the exact sequence of Theorem 8.8 or of Milnor shows that we have $0 \to K_1(R, I) \to K_1(R) \to K_1(R/I) \to 0$. This gives a natural isomorphism $\theta: K_1(I_u, I) \approx K_1(R, I)$. Consider the map $\varphi: (I_u, I) \to (R, I)$. This gives a commutative diagram

$$\begin{array}{ccc} K_1(I_u, I) & \xrightarrow{\ \theta\ } & K_1(I_u, I) \\ \downarrow{\scriptstyle 1} & & \downarrow{\scriptstyle K_1(f)} \\ K_1(I_u, I) & \xrightarrow{\ \theta\ } & K_1(R, I), \end{array}$$

which shows $K_1(f): (I_u, I) \to K_1(R, I)$ is an isomorphism.

REMARK. As we have seen in the proof, it will suffice to require (5) to hold for pairs $(R_u^{(n+p)}, J)$ where J is the 2-sided ideal generated by $x_1, \ldots, x_n$. In this case, the sequence $0 \to K_1(R_u^{(n+p)}, J) \to K_1(R_u^{(n+p)}) \to K_1(R_u^{(p)}) \to 0$, together with the fact that $K_1(\varepsilon): K_1(R_u^{(n+p)}) \to K_1(\mathbf{Z})$ is an isomorphism [**34**], shows that $K_1(R_u^{(n+p)}, J) = 0$. Thus to prove that all the assertions of Proposition 9.1 hold, it will suffice to show that $K_1(J_u, J) = 0$ for these particular J, or, equivalently, we must show that $K_1(J_u) \to K_1(\mathbf{Z})$ is an isomorphism. Perhaps this could be done by modifying the methods of [**22**], [**34**].

If the answer to this is affirmative, one might hope to extend the result by showing that $K_n(R, I) \approx K_n(I)$ for all n. This would be a very useful result. Even if this turns out to be false, the analogue might hold in Villamayor's theory.

10. **Further problems.** In [**5**] Bass defines a functor L on $\mathscr{A}b^{\mathscr{R}_1}$ as follows. Let T be an infinite cyclic group with generator t. Let T_+ and T_- be the submonoids with unit generated by t and t^{-1}. If $F: \mathscr{R}_1 \to \mathscr{A}b$, define $LF(R)$ to be the cokernel of $F(R[T_+]) \oplus F(R[T_-]) \to F(R[T])$. Bass shows that $LK_1 = K_0$ and proposes defining $K_{-n} = L^n K_0$. It would be very interesting to know if the functors K_n defined here satisfy $LK_n = K_{n-1}$. If this is true, it might also be possible to define a simplicial spectrum [**25**] G_* in $\mathscr{G}p^{\mathscr{R}}$ such that $\pi_n G_* = K_n$ for all $n \in \mathbf{Z}$. This would give a weak analogue of Bott periodicity (with the period presumably being ∞). In the topological case, this periodicity implies that we can regard $GL(\mathbf{C})$ as an infinite loop space and so define $K^n(X)$ for all $n \in \mathbf{Z}$.

Another very important problem is to extend the definition of K_n to abelian categories and subcategories of abelian categories. This is well known for K_0 and K_1 [**6**] but does not seem to be known for K_2 or even for the relative K_1. The method used here does not seem to be very useful for this since there is no definition of $GL(\mathscr{A})$ for an abelian category $\mathscr{A}$.

It would also be interesting to compare the methods used here with other forms of nonabelian homological algebra [**9**], [**19**], [**23**]. I have not yet tried to do this.

Finally, we remark that if it turns out that Milnor's K_2 does not satisfy $K_2(R^{(n)}) = 0$, it might be better to define K_n using "generically" aspherical models (cf. [**13**]). The theory of [**13**] is easily included in that of the present paper. We use the notion of [**13**]. Let Φ be a class of maps as in [**13**]. Let $\mathscr{M}_\Phi$ be the category whose objects are the $R^{(n)}$ and whose maps are those in Φ. Then $\mathscr{M}_\Phi$ is a subcategory of $\mathscr{R}$. The corresponding theory of aspherical models yields the cohomology theory of [**13**]. To see this, let $K(\ \ )$ and $K(\ \ , \Phi)$ be as in [**13**]. For $F \in \mathscr{M}_\Phi$, we have an inclusion $K(F, \Phi) \subset K(F)$. This gives a map $K(\ \ , \Phi) \to \operatorname{res} K(\ \ )$, and so $\operatorname{adj} K(\ \ , \Phi) \to K(\ \ )$. It is an easy exercise to show that $K(\ \ , \Phi) \in \mathscr{P}'$ and that $\operatorname{adj} K(\ \ , \Phi) \to K(\ \ )$ is an isomorphism (consider the obvious map $K(\ \ ) \to \operatorname{adj} K(\ \ , \Phi)$). Therefore, if we choose a $\mathscr{P}'$-aspherical, $\mathscr{P}'$-cofibrant resolution in $\mathscr{G}p^{\mathscr{M}_\Phi}$ and apply adj, we get a Φ-acyclic resolution in $\mathscr{G}p^{\mathscr{R}}$ in the sense of [**13**].

BIBLIOGRAPHY

1. M. André, *Homology of simplicial objects*, these Proceedings, pp. 15–36.

2. ———, *Méthode simplicial en algèbre homologique et algèbre commutative*, Springer-Verlag, Berlin, 1967.

3. M. Artin and B. Mazur, *Homotopy of varieties in the etale topology*, Proceedings of a Conference on Local Fields, Springer-Verlag, Berlin, 1967.

4. M. F. Atiyah, *K-theory*, Benjamin, New York, 1967.

5. H. Bass, *Algebraic K-theory*, Benjamin, New York, 1968.

6. ———, *K-theory and stable algebra*, Publ. Math. Int. Hautes Etudes Sci. **22** (1964), 5–60.

7. H. Cartan and S. Eilenberg, *Homological algebra*, Princeton Univ. Press, Princeton, N.J., 1956.

8. P. M. Cohn, *Universal algebra*, Harper and Row, New York, 1965.

9. P. Dedeker, *Les fonteurs* Ext_π, H^2_π, *et* $H^{\underline{2}}_\pi$ *non abéliennes*, C. R. Acad. Sci. Paris **258** (1964), 4891–4894.
10. A. Dold, S. MacLane, and U. Oberst, *Projective classes and acyclic models*, Reports of the Midwest Category Seminar, Springer-Verlag, Berlin, 1967.
11. A. Dold and D. Puppe, *Homologie nicht-additiver Funktoren*, Ann. Inst. Fourier **11** (1961), 201–312.
12. E. Dyer, *Cohomology theories*, University of Chicago 1963 (Math. Lecture Notes).
13. S. Eilenberg and S. MacLane, *Homology theories for multiplicative systems*, Trans. Amer. Math. Soc. **71** (1951), 294–330.
14. ———, *Acyclic models*, Amer. J. Math. **75** (1953), 189–199.
15. S. Eilenberg and J. C. Moore, *Adjoint functors and triples*, Illinois J. Math. **9** (1965), 381–398.
16. ———, *Foundations of relative homological algebra*, Memoirs Amer. Math, Soc. No. 55, Amer. Math. Soc., Providence, R.I., 1965.
17. S. Eilenberg and N. Steenrod, *Foundations of algebraic topology*, Princeton, 1952.
18. P. Freyd, *Abelian categories*, Harper and Row, New York, 1964.
19. A. Fröhlich, *Non-abelian homological algebra*, Proc. London Math. Soc. (3) **11** (1961), 239–275.
20. P. Gabriel, *Des categories abéliennes*, Bull. Soc. Math. France **90** (1962), 323–448.
21. P. Gabriel and M. Zisman, *Calculus of fractions and homotopy theory*, Springer-Verlag, Berlin, 1966.
22. S. Gersten, *Whitehead groups of free associative algebras*, Bull. Amer. Math. Soc. **71** (1965), 157–159.
23. J. Giraud, *Cohomologie non abélienne*, C. R. Acad. Sci. Paris **260** (1965), 2392–2394, 2666–2668.
24. D. Kan, *Adjoint functors*, Trans. Amer. Math. Soc. **87** (1958), 294–329.
25. ———, *Semisimplicial spectra*, Illinois J. Math. **7** (1963), 463–478.
26. S. MacLane, *Homology*, Springer-Verlag, Berlin, 1963.
27. ———, *Categorical algebra*, Bull. Amer. Math. Soc. **71** (1965), 40–106.
28. W. Magnus, A. Karass, and D. Solitar, *Combinatorial group theory*, Interscience, New York, 1966.
29. J. P. May, *Simplicial objects in algebraic topology*, Van Nostrand, Princeton, N.J., 1967.
30. J. Milnor, *Notes on algebraic K-theory*, University of California, Los Angeles, 1968 (mimeographed notes).
31. B. Mitchell, *Theory of categories*, Academic Press, New York, 1965.
32. U. Oberst, *Homology of categories and exactness of direct limits*, Math. Z. **107** (1968), 87–115.
33. D. Quillen, *Homotopical algebra*, Springer-Verlag, Berlin, 1967.
34. J. Stallings, *Whitehead torsion of free products*, Ann. of Math. **82** (1965), 354–363.
35. R. Steinberg, *Génerateurs, relations et revètements de groupes algébriques*, Colloq. Théorie des groupes algébriques, Bruxelles 1962, Gauthier-Villars, Paris 1962, pp. 113–127.
36. R. G. Swan, *Notes on algebraic K-theory*, Springer-Verlag, Berlin, 1968.
37. J. L. Verdier, *Séminaire de Géometrie algébrique Exp V app.*, Inst. Hautes Etudes Sci., Paris 1963–4 (mimeographed).

UNIVERSITY OF CHICAGO

Groups of cohomological dimension one

John Stallings

In [**2**], we showed that a finitely generated group of cohomological dimension one is necessarily a free group. In this note we outline the ideas in that proof together with some more recent improvements.

1. **A graph-theoretic result.** Here we prove a lemma on graphs which resembles a theorem of Bergman [**1**]. Our proof was inspired by a private communication from M. Dunwoody.

A graph Γ is a set of vertices and edges, such that each edge has one or two endpoints which are vertices; local finiteness and connectedness are defined as usual. A set A of vertices of Γ is said to be connected, if some subgraph Δ, having A as its set of vertices, is connected. The complement A^* of a set of vertices A is the set of all vertices of Γ not in A; the coboundary δA is the set of those edges e having one endpoint in A and one in A^*.

1.1. (A Version of König's Theorem.) *Let Γ be a locally finite graph. Let $A_1 \supset A_2 \supset \cdots \supset A_n \supset \cdots$ be a decreasing sequence of infinite, connected sets of vertices, each containing a fixed vertex v. Then $\bigcap A_i$ is infinite.*

The proof is simple.

Suppose the graph Γ is connected and there is a subset A of vertices such that both A and A^* are infinite, but δA is finite; then we say Γ has more than one end; the definition of ends in general is the usual one. The number of edges in δA will be called the complexity of A; if this number is minimal over all A such that both A and A^* are infinite but δA finite, we call A minimal.

1.2. In a connected graph Γ with more than one end, if A is minimal, then both A and A^* are connected.

Otherwise, some infinite component of A or A^* would have less complexity.

1.3. In a connected, locally finite graph with more than one end, if $A_1 \supset \cdots \supset A_n \supset \cdots$

is a decreasing sequence of minimal subsets of vertices of Γ, each containing some fixed vertex v, then $\bigcap A_n$ is minimal and in fact is some A_k; in other words, the sequence stops decreasing at some point A_k.

We note first that $B = \bigcap A_n$ is infinite, by 1.1 and 1.2. B^* is obviously infinite. Each edge e of δB belongs to some δA_i; if e belongs to δA_i then it also belongs to δA_j for all $j > i$. Hence if c is the minimal complexity, δB contains at most c elements; in particular, δB is finite, and hence contained in some δA_k. Since δA_k is minimal, we have $\delta B = \delta A_k$. Since $B \subset A_k$, and Γ is connected, and $\delta B = \delta A_k$, it follows that $B = A_k$.

1.4. In a connected, locally finite graph Γ with more than one end, there is B minimal, such that if A is any minimal set of vertices, then some one of the sets

$$(*) \qquad A \cap B, \quad A \cap B^*, \quad A^* \cap B, \quad A^* \cap B^*$$

is finite.

This is proved by constructing B as a smallest minimal set containing a vertex v. Such a B exists by 1.3. If $\delta A \cap \delta B = \varnothing$ and all four of the sets (*) are infinite, then a simple computation would show that all the sets (*) are minimal, and then the one that contains v would be smaller than B. On the other hand, if $\delta A \cap \delta B \neq \varnothing$, a computation shows that one of the sets (*) has coboundary with fewer elements than a minimal set, and hence that particular set must be finite.

We can, in particular, apply this result to the graph of a group G with respect to a finite set T of generators. This graph Γ has for its vertices the elements of G; an edge connects every pair (g, tg) for $t \in T$. The graph Γ is connected, since T generates G; it is locally finite since T is finite. The ends of Γ can be identified with the ends of G. Finally, G acts, as a group of automorphisms, on the right of Γ.

1.5. In a finitely generated group G with more than one end, there exists a set $A \subset G$, such that A and A^* are both infinite, but (in the graph Γ) δA is finite, such that for every $g \in G$, at least one of the sets:

$$Ag \cap A, \quad Ag \cap A^*, \quad A^*g \cap A, \quad A^*g \cap A^*$$

is finite.

For, we take A to be the set B of 1.4, and notice that Ag is minimal since A is. Such an A will be called a characteristic subset of G.

2. **The bipolar structure.** Given a characteristic subset A of G, there are six possibilities as to which of the sets

$$(*) \qquad Ag \cap A, \quad Ag \cap A^*, \quad A^*g \cap A, \quad A^*g \cap A^*$$

are finite. This divides G into six subsets F, S, AA, AA^*, A^*A, A^*A^* as follows:

$g \in F$ if $Ag \cap A^*$ and $A^*g \cap A$ are finite.

$g \in S$ if $Ag \cap A$ and $A^*g \cap A^*$ are finite.

$g \in XY$ if $XAg \cap YA$ is the only finite set. In this last description, X and Y stand for A or A^* and $AQ = Q$, $A^*Q = Q^*$ for any subset $Q \subset G$.

If G has two ends, the explicit structure of G is well known: G is an extension of a finite group by either the infinite cyclic group or the infinite dihedral group. If G has more than two ends, then F is finite and the following properties can be proved:

2.1. F is a finite subgroup of G.

2.2. S may be empty. If S is nonempty, then $F \cup S$ is a group in which F has index two.

2.3. If $f \in F$, $g \in XY$, then $fg \in XY$.

2.4. If $s \in S$, $g \in XY$, then $sg \in X^* Y$.

2.5. If $g \in XY$, $h \in Y^*Z$, then $gh \in XZ$.

2.6. If $g \in XY$, then $g^{-1} \in YX$.

2.7. If $g \in G$, there is an upper bound $N(g)$ to the length n of expressions $g = g_1 g_2 \cdots g_n$ where for some $X_0, X_1, \ldots, X_n$, each $g_i \in X_{i-1}X_i^*$.

This division of G into six subsets, satisfying these seven properties, we call a bipolar structure. An element of G will be said to be decomposable if it can be written gh for some X, Y, Z, with $g \in XY$, $h \in Y^*Z$. The other elements, the indecomposable ones, form a subset P of G. In particular, by 2.7, P generates G, and, by 2.5, P contains $F \cup S$.

It is easily shown that G satisfies a certain universal property with respect to P. We generalize this circumstance in the next section.

3. **Pregroups.** A pregroup consists of a set P, an element $1 \in P$, a function $x \mapsto x^{-1}$ of P to P, a set $D \subset P \times P$, and a function $(x, y) \mapsto xy$ of D to P, satisfying these five axioms:

1. For all $x \in P$, we have $(1, x)$, $(x, 1) \in D$ and $1x = x1 = x$.
2. For all $x \in P$, we have (x, x^{-1}), $(x^{-1}, x) \in D$ and $xx^{-1} = x^{-1}x = 1$.
3. If $(x, y) \in D$, then $(y^{-1}, x^{-1}) \in D$ and $(xy)^{-1} = y^{-1}x^{-1}$.
4. If (x, y), $(y, z) \in D$, then: $(x, yz) \in D$ if and only if $(xy, z) \in D$, in which case $x(yz) = (xy)z$.
5. If (w, x), (x, y), $(y, z) \in D$, then either $(w, xy) \in D$ or $(xy, z) \in D$.

It is easily shown that the set of indecomposable elements, in a group with a bipolar structure, forms a pregroup.

Now, van der Waerden's proof [3] of the structure theorem for free products of groups can be extended, with care, to pregroups. In this proof, Axiom 5, which may seem rather ad hoc, must be applied several times. We shall just state the result.

A morphism of pregroups is a function $\varphi: P \to Q$, compatible with the structure, in particular with multiplication. Thus pregroups form a category containing the category of groups. By abstract nonsense (the Adjoint Functor Theorem), there is to each pregroup P, a universal group $U(P)$. That is, $U(P)$ is a group; there is a specific morphism $\iota: P \to U(P)$; such that, for any group X and morphism $\varphi: p \to X$, there is a unique consistent homomorphism $U(\varphi): U(P) \to X$.

It is easy to see that a group with bipolar structure is the universal group of its pregroup of indecomposable elements.

The structure theorem states that every element g of $U(P)$ can be written $g = \iota(p_1) \cdots \iota(p_n)$ where the word $(p_1, \ldots, p_n)$ satisfies the condition that $(p_i, p_{i+1}) \notin D$ for all i. Two words $(p_1, \ldots, p_n)$ and $(q_1, \ldots, q_r)$ give the same g, if and only if:

(1) $n = r$.

(2) There are $a_1, \ldots, a_{n-1} \in P$ such that $(p_1 a_1, \ldots, a_{i-1} p_i a_i, \ldots, a_{n-1}^{-1} p_n)$ is definable and equal to $(q_1, \ldots, q_n)$.

This allows us, in certain instances, to describe $U(P)$ in more classical ways, e.g. as a free product with amalgamation. Applying this analysis to groups with bipolar structure, we can show:

3.1. Let G have a bipolar structure $(F, S, AA, AA^*, A^*A, A^*A^*)$. Define $H = F \cup S$; and $G_1 = F \cup \{\text{indecomposable elements of } AA\}$; and $G_2 = F \cup \{\text{indecomposable elements of } A^*A^*\}$. These are subgroups of G.

Case 1: If $S \neq \varnothing$, then $G = H *_F G_1$.

Case 2: If $S = \varnothing$ and there is no indecomposable element of AA^*, then $G = G_1 *_F G_2$.

Case 3: If $S = \varnothing$ and there is an indecomposable element $x \in AA^*$, then $xFx^{-1} \subset G_1$; let $\varphi: F \to G_1$ denote $f \mapsto xfx^{-1}$; then $G = [G_1;\ F, \varphi]$.

In this result $A *_C B$ denotes the free product of A and B with amalgamated subgroup C. $[A;\ B, f]$ is defined, when B is a subgroup of A and f an embedding $B \to A$, to be the group obtained from A by adjoining a new generator x and relations $f(b) = xbx^{-1}$ for all $b \in B$.

4. **The characterization of infinitely-ended groups.** A few more computations can be made. In 1.5, we could add that there exists an element $g \in G$ such that $Ag \cap A^*$ is finite, but no other of $Ag \cap A$, $A^*g \cap A$, $A^*g \cap A^*$ is finite; hence in our bipolar structure, AA^* is nonempty. It is also easy to compute the number of ends of groups of the form $A *_F B$ and $[A; F, \varphi]$ for F finite. This leads to the following result:

4.1. Let G be a finitely generated group. Then G has infinitely many ends, if and only if:

Either (a): G can be written $A *_F B$, where F is a finite group, contained properly in both A and B, and of index greater than two in B.

Or (b): G can be written $[A;\ F, \varphi]$, where F is a finite, proper subgroup of A.

It should be remarked that for G to have exactly two ends, it is necessary and sufficient that either (a) $G = A *_F B$ where the finite group F has index two in both A and B, or (b) $G = [F;\ F, \varphi]$ where F is finite.

This, of course, generalizes our earlier result on torsion-free groups with infinitely many ends. From here on, we use the simple cohomological fact that a finitely generated, nontrivial group of cohomological dimension one has more than one end. Then, since we have such groups under control, we know that such a group is either infinite cyclic or a nontrivial free product. Thus, such a group can be decomposed ultimately into a free product of infinite cyclic groups; i.e., it is a free group.

REFERENCES

1. G. Bergman, *On groups acting on locally finite graphs*, Ann. of Math. **88** (1968), 335–340.
2. J. Stallings, *On torsion-free groups with infinitely many ends*, Ann. of Math. **88** (1968), 312–334.
3. B. L. van der Waerden, *Free products of groups*, Amer. J. Math. **70** (1948), 527–528.

UNIVERSITY OF CALIFORNIA, BERKELEY

Hopf fibration towers and the unstable Adams spectral sequence

Larry Smith

Let X be a Hopf space. In [**9**] we introduced the coprimitive fibre square (mod p)

$$\begin{array}{ccc} X\langle -1\rangle & \longrightarrow & L(QH_*(X;\mathbf{Z}_p)) \\ {\scriptstyle \pi\langle -1\rangle}\big\downarrow & & \big\downarrow \\ X & \xrightarrow{\phi\langle -1\rangle} & K(QH_*(X;\mathbf{Z}_p)) \end{array}$$

and indicated some of its elementary properties. By iterating this construction we obtain a tower of fibrations

$$\begin{array}{c} \vdots \\ X\langle -n\rangle \\ {\scriptstyle \pi\langle -n\rangle}\big\downarrow \\ X\langle -n+1\rangle \\ \big\downarrow \\ \vdots \\ \big\downarrow \\ X\langle -1\rangle \\ {\scriptstyle \pi\langle -1\rangle}\big\downarrow \\ X = X\langle 0\rangle \end{array}$$

over X. The exact homotopy sequences of this fibration tower may then be pasted together to form an exact couple

$$\begin{array}{ccc} A\langle X\rangle & \longrightarrow & A\langle X\rangle \\ & \nwarrow \qquad \swarrow & \\ & E\langle X\rangle . & \end{array}$$

It is reasonable to expect that this couple provides information about the p-primary part of the homotopy of X. However, a major difficulty has been the proof of a convergence theorem for the associated spectral sequence.

In the sequel we will avoid this convergence question by introducing a slight modification of the construction of the coprimitive fibre square. Under mild restrictions on a space X we will obtain a tower of fibrations

$$\begin{array}{c} \vdots \\ X(-n) \\ \Big\downarrow {\scriptstyle \pi(-n)} \\ \vdots \\ X(-1) \\ \Big\downarrow {\scriptstyle \pi(-1)} \\ X = X(0) \end{array}$$

over X called the mod-C^p-tower of X. From these fibrations we obtain in a standard way an exact couple by applying the homotopy functor with $\mathbf{Z}_p$-coefficients,

$$\begin{array}{ccc} A(X) & \longrightarrow & A(X) \\ & \nwarrow \qquad \swarrow & \\ & E(X) & \end{array}$$

The associated spectral sequence $\{E^r(X), d^r(X)\}$ behaves somewhat like an unstable Adams spectral sequence mod p. Indeed we shall show:

THEOREM. *If X is a simply connected Hopf space then the spectral sequence $\{E^r(X), d^r(X)\}$ converges in the naive sense to $\pi_*(X; \mathbf{Z}_p)$.*

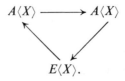

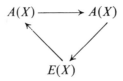

If $H_*(X; \mathbf{Z}_p)$ is coprimitive and p is an odd prime then we are able to identify $E^2(X)$. The precise result requires several preliminaries to state and may be found in §7. It is motivated by the results of **[5]**.

The spectral sequence of the modified coprimitive tower seems closely related to the unstable Adams spectral sequence constructed by Massey-Peterson in **[5]**. Indeed we identify $E^2(X)$ under the mod p analog of the conditions required to construct their spectral sequence. I am indebted to Professors Massey and Peterson for making their work available to me in prepublication form.

It would be of interest to determine what, if any, is the connection between the spectral sequence of the mod-C^p-tower and the unstable Adams spectral sequence introduced by Rector in **[12]**.

The construction of the modified coprimitive tower forces us to work with spaces more general then Hopf spaces. The precise definitions are given in §1. This enlargement of the objects under study has the advantage that the spheres, S^{2n+1}, will satisfy the requirements when p is an odd prime. Several results of a technical nature concerning constructions made in these larger categories are deferred to an appendix.

The second section introduces the mod-C^p-tower and its associated spectral sequence. The third section contains some results on homology Hopf algebras that are used in §4 to study homology properties of the mod-C^p-tower. The fifth section introduces certain functors of unstable modules over the Steenrod algebra $\mathscr{A}^*$. In the sixth section we introduce a subalgebra $\mathscr{H}^*$ of $\mathscr{A}^*$ that is important in connecting mod p homotopy and mod p-homology. This connection is exploited in the seventh and final section to identify $E^2(X)$ when $H_*(X; \mathbf{Z}_p)$ is coprimitive, p-odd.

The formulation that we present here is by no means final. It is merely an introduction to what may be obtained by way of applications of the results of **[8]**, **[9]**.

This paper is an outgrowth of my joint work with J. C. Moore and it is a pleasure to acknowledge my indebtedness to him for untold suggestions, help and improvements.

1. Some categories and constructions. We will begin by enlarging the category of Hopf spaces to contain more general objects. As we shall see this enlargement is forced upon us by the constructions that we wish to make.

The notation and terminology of **[8]** and **[9]** will be adhered to.

NOTATION AND CONVENTION. p will denote a fixed prime and $\mathbf{Z}_p = \mathbf{Z}/p\mathbf{Z}$. We will use the notation H_*X for $H_*(X; \mathbf{Z}_p)$ and similarly for cohomology. The mod p Steenrod algebra will be denoted by $\mathscr{A}^*$ and its dual by $\mathscr{A}_*$.

DEFINITION. A pseudo-Hopf space mod p consists of a pair, (X, ϕ), where X is a space and

$$\phi: H_*X \otimes H_*X \to H_*X$$

provides H_*X with the structure of a homology Hopf algebra compatible with the coaction of $\mathscr{A}_*$.

Examples abound.

(1) Any Hopf space (X, μ) together with the induced Hopf algebra structure μ_* on H_*X.

(2) If $p = 2$, then any sphere S^n is a pseudo-Hopf space mod 2 in the obvious way. Similarly if p is odd then the odd spheres, S^{2n+1}, are pseudo-Hopf spaces mod p in a unique way.

(3) Let X be a space with $H_*(X)$ of finite type and $H^*(X) = \mathscr{U}(M)$, for some unstable $\mathscr{A}^*$-module M **[5]**, **[16]**. Then there is an induced primitive Hopf algebra structure on H^*X and thus by duality H_*X inherits the structure of a coprimitive Hopf algebra under $\mathscr{A}_*$.

DEFINITION. If (X, ϕ) and (Y, ψ) are pseudo-Hopf mod p spaces, a map $f: X \to Y$ is a map of pseudo-Hopf spaces mod $p, f: (X, \phi) \to (Y, \psi)$ iff $f_*: H_*X \to H_*Y$ is a morphism of the given Hopf algebra structures.

One readily checks that compositions behave well and thus we have the category $P - \mathscr{H}Sp/p$ of pseudo-Hopf spaces mod p and their morphisms.

In a similar fashion we obtain the category $P - \mathscr{H}\mathscr{F}Sq/p$ of pseudo-Hopf fibre square mod p. An object of this category consists of a fibre square

$$\begin{array}{ccc} E & \longrightarrow & E_0 \\ {\scriptstyle \pi}\downarrow & & \downarrow{\scriptstyle \pi_0} \\ B & \xrightarrow{\ f\ } & B_0 \end{array}$$

where all the spaces and morphisms of $\mathscr{F}$ are in the category $\mathscr{H}Sp/p$ (i.e., choices of Hopf algebra structures on H_*E, H_*E_0, H_*B, H_*B_0 have been made, etc.) A morphism in $P - \mathscr{H}\mathscr{F}\ Sq/p$ is a morphism of fibre squares whose component morphisms lie in $P - \mathscr{H}Sp/p$.

The category $P - \mathscr{H}Sp/p$ has minimal structure for defining the coprimitive fibre square. We proceed to this now.

DEFINITION. If $(X, \mu) \in \operatorname{obj} P - \mathscr{H}Sp/p$ the coprimitive fibre square of X, $\mathscr{F}\langle -1\rangle(X)$, is the fibre square

$$\begin{array}{ccc} X\langle -1\rangle & \longrightarrow & LQH_*X \\ {\scriptstyle \pi\langle -1\rangle}\downarrow & & \downarrow \\ X & \xrightarrow{\ \phi\langle -1\rangle\ } & KQH_*X \end{array} \left.\vphantom{\begin{array}{c}a\\a\\a\end{array}}\right\} \begin{array}{l}\text{the pathspace}\\ \text{fibration over } KQH_*X\end{array}$$

where $\phi\langle -1\rangle$ is characterized by requiring that the diagram

$$\begin{array}{ccc} H_*X & \xrightarrow{\ \nu\ } & QH_*X \\ & {\scriptstyle \phi\langle -1\rangle_*}\searrow & \downarrow{\scriptstyle h} \quad = \text{Hurecwicz Map} \\ & & H_*KQH_*X \end{array}$$

be commutative.

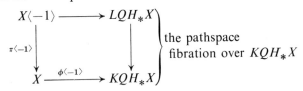

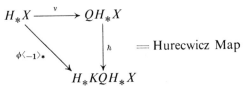

We are now presented with the following problem: If $(X, \mu) \in \text{obj } P - \mathscr{H}Sp/p$ does there exist a Hopf algebra structure $\mu\langle -1\rangle$ on $H_*X\langle -1\rangle$ such that $\mathscr{F}\langle -1\rangle(X)$ is a pseudo-Hopf fibre square mod p?

Unfortunately we are unable to answer this question in the affirmative for arbitrary X. For this reason we shall introduce two further categories of Hopf-like spaces.

DEFINITION. A Hopf space mod p is an H-space (X, μ) such that H_*X is a Hopf algebra. (Note that the only point at issue is the associativity of the multiplication $\mu_*: H_*X \otimes H_*X \to H_*X$.)

The category of Hopf spaces mod p has as objects the Hopf spaces mod p and as morphisms H-morphisms. We denote this category by $\mathscr{H}Sp/p$. There is a natural forgetful function $\mathscr{H}Sp/p \to P - \mathscr{H}Sp/p$.

WARNING. This is not the inclusion of a full subcategory. But note that $\mathscr{H}Sp/p$ is a full subcategory of the category of H-spaces and H-maps.

In the obvious manner we may also introduce the category of Hopf fibre squares mod p, which we denote by $\mathscr{H}\mathscr{F}Sq/p$.

Note that we do not demand that a Hopf space mod p be homotopy associative, but only that the induced Hopf algebra structure on H_*X be associative. For example S^7 is a Hopf space mod p for all p although it is not a Hopf space.

REMARK. Clearly the notion of Hopf space mod p is largely a technical convenience, its utility arising from our inability to verify that several rather natural constructions on Hopf spaces lead again to Hopf spaces. For a more detailed discussion of this point see the Appendix.

THEOREM 1.1. *If (X, μ) is a Hopf space* mod p, *X simply connected, then there exists a natural structure $\mu\langle -1\rangle$ of a Hopf space* mod p *on $X\langle -1\rangle$. Equipped with this structure $\mu\langle -1\rangle$, the fibre square $\mathscr{F}\langle -1\rangle(X)$ becomes a Hopf fibre square* mod p.

A proof of the preceding proposition may be found in the appendix. Note that by construction $X\langle -1\rangle$ is an H-space and the only point at issue is the associativity of H_*X.

The assertion of naturality is to be interpreted as follows: if $(X, \mu), (Y, \phi) \in \text{obj } P - \mathscr{H}Sp/p$ and $f:(X, \mu) \to (Y, \phi)$ is a morphism in $P - \mathscr{H}Sp/p$, then f induces a morphism of Hopf fibre squares mod p $f\langle -1\rangle: \mathscr{F}\langle -1\rangle(X) \to \mathscr{F}\langle -1\rangle(Y)$.

ACKNOWLEDGEMENT. Theorem 1.1 overcomes the difficulty that $X\langle -1\rangle$ need not be a Hopf space. This difficulty was pointed out to us by J. D. Stasheff (see the appendix for a more thorough discussion). Theorem 1.1 thus overcomes certain technical difficulties present in [**9**, Example 8.3].

DEFINITION. A pseudo-Hopf space mod p, (X, μ), is said to be abelian iff H_*X is abelian.

The full subcategory of $P - \mathscr{H}Sp/p$ generated by the abelian objects is denoted by $P - Ab\mathscr{H}Sp/p$. We may also form the obvious category $P - Ab\mathscr{H}\mathscr{F}Sq/p$ of pseudo-abelian-Hopf fibre squares mod p.

THEOREM 1.2. *If* $(X, \mu) \in \operatorname{obj} P - Ab\mathscr{H}Sp/p$, *X simply connected, then there is a natural structure*, $\mu\langle -1\rangle$, *of an abelian pseudo Hopf space* mod p *on* $X\langle -1\rangle$. *Equipped with this structure* $\mu\langle -1\rangle$, *the fibre square* $\mathscr{F}\langle -1\rangle(X)$ *becomes an object of* $P - Ab\mathscr{H}\mathscr{F}Sq/p$.

The proof of the preceding proposition may be found in the appendix.

It follows from Theorem 1.1 and Theorem 1.2 that we may, for certain (X, μ), iterate the construction $\langle -1\rangle$ to obtain a tower of fibrations over X. This tower is called the coprimitive tower of X, and seems to yield useful information on the mod p part of the homotopy of X. However, to avoid certain delicate convergence conditions we find it convenient to modify somewhat the coprimitive tower.

The following definition was suggested by R. E. Stong and replaces a more cumbersome one of the author.

DEFINITION. Let (X, μ) be a simply connected pseudo-Hopf space mod p and $k \in \mathbf{Z}$, $k \geq 0$. Define $Q(k)H_*X$ to be the following graded abelian group,

$$\begin{aligned}[Q(k)H_*X]_n &= 0 \quad \text{if} \quad n \leq k,\\ &= H_n(X; \mathbf{Z}) \quad \text{if} \quad n = k+1, k+2,\\ &= (QH_*X)_n \quad \text{if} \quad n > k+2.\end{aligned}$$

DEFINITION. If (X, μ) is a simply connected pseudo-Hopf space mod p let $\mathscr{F}_nX$ denote the fibre square

$$\begin{array}{ccc} X_n & \longrightarrow & LQ(n)H_*X \\ {\scriptstyle \pi(n)}\downarrow & & \downarrow \\ X & \xrightarrow{\phi(n)} & KQ(n)H_*X \end{array}$$

where $\phi(n)$ is characterized by the commutative diagrams

$$\phi(n)_* = h: H_j(X; \mathbf{Z}) \to H_j(KQ(n)H_*X; \mathbf{Z}) \qquad j = k+1, k+2,$$

$$\begin{array}{ccc} H_j(X; \mathbf{Z}_p) & \longrightarrow & QH_j(X; \mathbf{Z}_p) \\ & {\scriptstyle \phi(n)_*}\searrow & \downarrow \\ & & H_j(KQ(n)H_*X; \mathbf{Z}_p) \end{array} \qquad j > k+2.$$

Analogous to Theorem 1.1 and Theorem 1.2 we then have:

THEOREM 1.3. *If* $(X, \mu) \in \operatorname{obj} HSp/p$ *and X is n-connected then there exists a natural H-space structure* $\mu_n : X_n \times X_n \to X_n$ *such that* $(X_n, \mu_n) \in \operatorname{obj} \mathscr{H}Sp/p$ *and* $\mathscr{F}_n(X)$ *is a Hopf fibre square* mod p. *Moreover* X_n *is* $n+1$*-connected.*

PROOF. To prove that X_n is $n+1$-connected consider the homotopy exact sequence of the fibration $X_n \to X \to KQ(n)H_*X$. In low degrees we have

$$\begin{array}{ccccccccc} & & & & & & H_{n+2}(X;\mathbf{Z}) & & \\ & & & & & & \| & & \\ 0 \leftarrow \pi_{n+1}(KQ(n)H_*X) & \overset{1}{\longleftarrow} & \pi_{n+1}X & \leftarrow & \pi_{n+1}X_n & \leftarrow & \pi_{n+2}(KQ(n)H_*K) & \overset{2}{\leftarrow} & \pi_{n+2}X \\ \| & & & & & & & & \\ H_{n+1}(X;\mathbf{Z}) & & & & & & & & \end{array}$$

Now $\phi(n)_*$ may be identified with the Hurewicz map. Thus the map marked 1 is an isomorphism and the map marked 2 is an epimorphism. Thus $\pi_{n+1}X_n = 0$. The fact that $\pi_i X_n = 0$, $i < n+1$ is elementary.

The assertions about μ_n and $\mathscr{F}_n X$ may be obtained by a simple modification of the proof of Theorem 1.1. (See the appendix.) □

THEOREM 1.4. *If $(X, \mu) \in \operatorname{obj} P - Ab\mathscr{H}Sp/p$ and X is n-connected then there exists a natural abelian Hopf algebra structure $\mu_n: H_*X \otimes H_*X \to H_*X$ such that $(X_n, \mu_n) \in \operatorname{obj} P - Ab\mathscr{H}Sp/p$ and $\mathscr{F}_n(X)$ is a pseudo-abelian-Hopf fibre square* mod p.

PROOF. The details are similar to Theorem 1.3. □

The functors $\mathscr{F}_n$ will provide us with a suitable tower of fibrations over X. We turn to this next.

2. **The tower and spectral sequence.** The results of §1 show that it will be convenient to perform our constructions in one of the two categories $\mathscr{H}Sp/p$ or $P - Ab\mathscr{H}Sp/p$, and the categories of fibre squares over $\mathscr{H}Sp/p$ and $P - Ab\mathscr{H}Sp/p$. Most of our results will therefore have two parts to them.

DEFINITION. (1) If $X \in \operatorname{obj} \mathscr{H}Sp/p$ and X is simply connected then we may define inductively Hopf fibre squares mod p, $\mathscr{F}(-n)(X)$, by setting $X(0) = X$ and

$$\begin{array}{ccc} X(-n) = X(-n+1)_n & \longrightarrow & LQ(n)H_*(X(-n+1)) \\ \Big\downarrow{\scriptstyle \pi(-n)} & & \Big\downarrow \\ X(-n+1) & \xrightarrow{\phi(-n)} & KQ(n)H_*(X(-n+1)). \end{array}$$

Note that Theorem 1.3 is used to provide the inductive step.

(2) If $X \in \operatorname{obj} P - Ab\mathscr{H}Sp/p$ and X is simply connected then we may define inductively the pseudo-abelian Hopf fibre squares mod p, $\mathscr{F}(-n)(X)$, by setting $X(0) = X$ and

$$\begin{array}{ccc} X(-n) = X(-n+1)_n & \longrightarrow & LQ(n)H_*(X(-n+1)) \\ \Big\downarrow{\scriptstyle \pi(-n)} & & \Big\downarrow \\ X(-n+1) & \xrightarrow{\phi(-n)} & KQ(n)H_*(X(-n+1)). \end{array}$$

Note that Theorem 1.4 is used to provide the inductive step.

If $X \in \text{obj } \mathscr{H}Sp/p$ or $\text{obj } P - Ab\mathscr{H}Sp/p$ then the fibre squares $\{\mathscr{F}(-n)(X)\}$ may be strung together to form a tower of fibrations

$$
\begin{array}{ccccc}
 & & \vdots & & \\
F(-n) & \xrightarrow{e(-n)} & X(-n) & \xrightarrow{\phi(-n-1)} & K(-n) \\
 & & \Big\downarrow{\scriptstyle \pi(-n)} & & \\
F(-n+1) & \longrightarrow & X(-n+1) & \xrightarrow{\phi(-n)} & K(-n+1) \\
 & & \Big\downarrow & & \\
 & & \vdots & & \\
 & & \Big\downarrow & & \\
F(-1) & \longrightarrow & X(-1) & \longrightarrow & K(-1) \\
 & & \Big\downarrow{\scriptstyle \pi(-1)} & & \\
 & X = & X(0) & \xrightarrow{\phi(-1)} & K(0),
\end{array}
$$

where $\phi(-n)$ classifies $\pi(-n)$ which has fibre $F(-n)$. We refer to this tower of fibrations as the modified coprimitive tower of X. It is but a slight modification of the coprimitive tower, introduced in **[10]**, which is a geometric analog of the coprimitive series of a homology Hopf algebra (see **[8]**).

NOTATION. If X is a simply connected space denote by $\pi_*(X; \mathbf{Z}_p)$ the homotopy groups of X with coefficients in $\mathbf{Z}_p$. (See **[4]**, **[7]** for the basic facts about $\pi_*(\ ; \mathbf{Z}_p)$.) Recall that for $p > 2$ $\pi_*(X; \mathbf{Z}_p)$ is a $\mathbf{Z}_p$-module, but when $p = 2$ $\pi_*(X; \mathbf{Z}_2)$ may have 4-torsion. We will concentrate mostly on the case of p-odd.

Let X be fixed with $X \in \text{obj } \mathscr{H}Sp/p$ or $\text{obj } P - Ab\mathscr{H}Sp/p$, X simply connected. The homotopy exact triangles

$$
\begin{array}{ccc}
\pi_*(X(-n); \mathbf{Z}_p) & \longrightarrow & \pi_*(X(-n+1); \mathbf{Z}_p) \\
 \nwarrow & & \swarrow \\
 & \pi_*(F(-n); \mathbf{Z}_p) &
\end{array}
$$

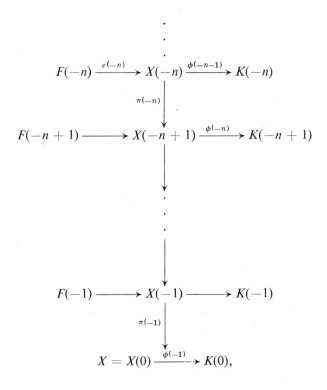

of the fibrations $\{\pi(-n)\}$ in the modified coprimitive tower of X may be spliced together to form an exact couple

$$\mathscr{C}(X) \qquad \begin{array}{ccc} A(X) & \longrightarrow & A(X) \\ & \nwarrow \qquad \swarrow & \\ & E(X). & \end{array}$$

More precisely we set

$$A(X)_{r,s} = \pi_{r+s}(F(r); \mathbf{Z}_p) \qquad r \leq 0$$
$$E_{r+1,s}(X) = \pi_{r+s}(X(r); \mathbf{Z}_p) \qquad r \leq 0.$$

Note the index shift in defining $E(X)_{*,*}$. The indexing looks as though we had used the homotopy exact triangles of the fibration $X(-n) \to X(-n+1) \to K(-n+1)$ instead of $F(-n) \to X(-n) \to X(-n+1)$.

Let $\{E^r(X), d^r(X)\}$ denote the spectral sequence of the exact couple $\mathscr{C}(X)$. Note that $\{E^r(X), d^r(X)\}$ is a second quadrant spectral sequence. Indeed, since $X(-n)$ is $n+1$-connected (Theorem 1.3 or 1.4) $E^1_{r,s}(X) = 0$ unless $s - 2n - 1 \geq 0$. The differential $d^r(X)$ has degree $(-r, r-1)$.

We refer to $\{E^r(X), d^r(X)\}$ as the modified coprimitive spectral sequence of X; the mod-C^p spectral sequence of X for short.

PROPOSITION 2.1. *$\{E^r(X), d^r(X)\}$ is a functor from the categories $\mathscr{H}Sp/p$, $P - Ab\mathscr{H}Sp/p$ to the category of second quadrant homology spectral sequences of $\mathbf{Z}_p$-modules, p odd and $\mathbf{Z}_4$-modules, $p = 2$.*

PROOF. The functoriality is a simple consequence of the naturality in Theorem 1.3 and Theorem 1.4 and a simple induction (i.e., one shows that $f: X \to Y$ induces a morphism of the coprimitive tower of X to that of Y). The different behavior for $p = 2$, $p > 2$ was noted previously. □

PROPOSITION 2.2. *If $X \in \text{obj}\, \mathscr{H}Sp/p$ or $P - Ab\mathscr{H}Sp/p$ then the* mod-C^p *spectral sequence $\{E^r(X), d^r(X)\}$ converges in the naive sense to $\pi_*(X; \mathbf{Z}_p)$.*

PROOF. This is an elementary consequence of the fact that $X(-n)$ is $n+1$-connected. □

On the basis of Proposition 2.2 we may regard $\{E^r(X), d^r(X)\}$ as an unstable Adams spectral sequence with $\mathbf{Z}_p$-coefficients.

In §7 we will identify $E^2(X)$ when H_*X is coprimitive and p is odd. The remainder of this paper is devoted to preparing the way for this identification.

3. **Elementary properties of the functor** $\langle -1 \rangle : \mathscr{H}_*\mathscr{H}/k \to \mathscr{H}_*\mathscr{H}/k$. Throughout this section k will denote a fixed field of characteristic $p \neq 0$. In §2 of **[8]** we introduced the functors $[-1]$, $\langle -1 \rangle : \mathscr{H}_*\mathscr{H}/k \to \mathscr{H}_*\mathscr{H}/k$. Our objective in this section is to establish the following property for these functors.

THEOREM 3.1. *The functors $[-1]$, $\langle -1 \rangle : \mathscr{H}_*\mathscr{H}/k \to \mathscr{H}_*\mathscr{H}/k$ preserve monomorphisms and epimorphisms.*

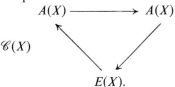

WARNING. In general the functors $\langle -1\rangle$ and $[-1]$ are not exact. For example suppose that $H \in \operatorname{obj} \mathscr{H}_*\mathscr{H}/k$ and $\operatorname{sol} H = 2$ (such gadgets certainly exist). Consider the exact sequence

$$k \to H[-1] \to H \to CH \to k.$$

Assuming that the functor $[-1]$ is exact we then obtain the exact sequence

$$k \to H[-2] \to H[-1] \to (CH)[-1] \to k.$$

By hypothesis $H[-2] = k$ and by definition $(CH)[-1] = k$. Thus we have an exact sequence

$$k \to H[-1] \to k$$

and hence $H[-1] = k$ contrary to the assumption that $\operatorname{sol} A = 2 > 1$. A similar example may be constructed to show $\langle -1\rangle$ is not exact; not even as a functor $\mathscr{H}^c_*\mathscr{H}/k \to \mathscr{H}^c_*\mathscr{H}/k$.

Later in this section (see Theorem 3.4) we will introduce a class of exact sequences in $\mathscr{H}_*\mathscr{H}/k$ whose exactness is preserved by $[-1]$ and $\langle -1\rangle$.

Familiarity with the material of [**8** §2] will be assumed throughout this section, and we will freely use the notation, definitions and results established there.

LEMMA 3.2. *If $A, B \in \operatorname{obj} \mathscr{H}_*\mathscr{H}/k$ are abelian and $f: A \to B$ is an epimorphism of homology Hopf algebras then $f\langle -1\rangle : A\langle -1\rangle \to B\langle -1\rangle$ is an epimorphism.*

PROOF. Recall that for $C \in \operatorname{obj} \mathscr{H}^c_*\mathscr{H}/k$ that $C\langle -1\rangle = k\xi C$. The rest is routine. □

NOTATION. Let $A \in \operatorname{obj} \mathscr{H}_*\mathscr{H}/k$, and let

$$\theta(A): A \amalg A \to A \pi A = A \otimes A$$

be the natural morphism in $\mathscr{H}_*\mathscr{H}/k$ from the coproduct to the product. Denote by $C(A, A) \subset A \amalg A$ the kernel in $\mathscr{H}_*\mathscr{H}/k$ of $\theta(A)$; i.e., $C(A, A) = A \amalg A \backslash\backslash \theta(A)$.

PROPOSITION 3.3. *If $A, B \in \operatorname{obj} \mathscr{H}_*\mathscr{H}/k$ and $f: A \to B$ is an epimorphism of homology Hopf Algebras, then $C(f, f): C(A, A) \to C(B, B)$ is an epimorphism.*

PROOF. Consider the commutative diagram

$$\begin{array}{ccccc} A \amalg A \backslash\backslash f \amalg f \backslash\backslash \phi & \longrightarrow & A \amalg A \backslash\backslash f \amalg f & \xrightarrow{\phi} & A \backslash\backslash f \otimes A \backslash\backslash f \\ \downarrow & & \downarrow & & \downarrow \\ C(A, A) & \longrightarrow & A \amalg A & \xrightarrow{\theta(A)} & A \otimes A \\ \downarrow & & \downarrow{\scriptstyle f \amalg f} & & \downarrow{\scriptstyle f \otimes f} \\ C(B, B) & \longrightarrow & B \amalg B & \xrightarrow{\theta(B)} & B \otimes B \end{array}$$

By definition the two right-hand columns and the bottom two rows are exact. We contend that the top row is also exact. To prove this it suffices to show that

$$\phi: A \amalg A \backslash\!\backslash f \amalg f \to A \backslash\!\backslash f \otimes A \backslash\!\backslash f$$

is an epimorphism. To this end consider the commutative diagram

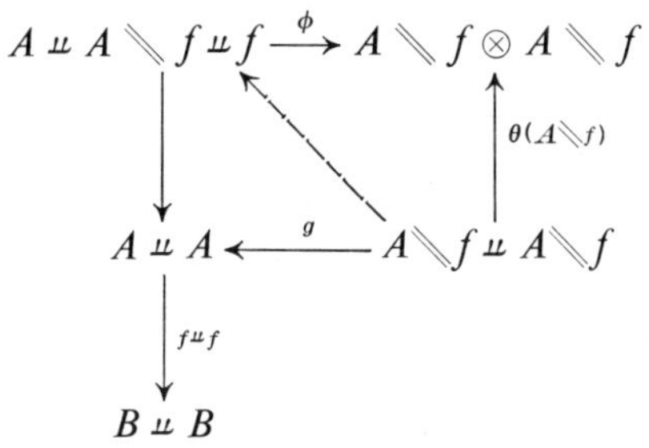

Since $(f \amalg f)g = *$ the existence of the dotted morphism is assured. Since $\theta(A \backslash\!\backslash f)$ is epic we may conclude that ϕ is also.

Thus in our original diagram the rows and two right-hand columns are exact. Therefore by a suitable version of the 3×3 lemma the left-hand column is exact and hence $C(A, A) \to C(B, B)$ is an epimorphism. □

PROOF OF THEOREM 3.1. It follows from Proposition 3.3 and the definition of the functor $[-1]$ that $[-1]$ preserves epimorphisms. To prove that $\langle -1 \rangle$ preserves epimorphisms consider the commutative diagram of exact sequences

$$\begin{array}{ccccccccc}
 & & & & & & k & & \\
 & & & & & & \downarrow & & \\
 & & k & & & & (CA)\langle -1 \rangle & & \\
 & & \downarrow & & & & \downarrow & & \\
k & \to & A[-1] & \to & A & \to & CA & \to & k \\
 & & \downarrow & & \| & & \downarrow & & \\
k & \to & A\langle -1 \rangle & \to & A & \to & C^pA & \to & k \\
 & & \downarrow & & & & \downarrow & & \\
 & & A\langle -1 \rangle /\!/ A[-1] & & & & k & & \\
 & & \downarrow & & & & & & \\
 & & k & & & & & &
\end{array}$$

An elementary diagram chase shows that $A\langle -1 \rangle /\!/ A[-1]$ is naturally isomorphic to $(CA)\langle -1 \rangle$. Therefore for any $H \in \operatorname{obj} \mathscr{H}_*\mathscr{H}/k$ we have a natural exact sequence of homology Hopf algebras

$$k \to H[-1] \to H\langle -1 \rangle \to (CH)\langle -1 \rangle \to k.$$

Now suppose that $f: A \to B$ is an epimorphism of homology Hopf algebras. We then obtain a diagram

$$
\begin{array}{ccccccccc}
k & \to & A[-1] & \to & A\langle -1\rangle & \to & (CA)\langle -1\rangle & \to & k \\
 & & \downarrow {\scriptstyle f[-1]} & & \downarrow {\scriptstyle f\langle -1\rangle} & & \downarrow {\scriptstyle [Cf]\langle -1\rangle} & & \\
k & \to & B[-1] & \to & B\langle -1\rangle & \to & (CB)\langle -1\rangle & \to & k
\end{array}
$$

with exact rows. Since f is an epimorphism, so is Cf. Thus by Lemma 3.2 $(Cf)\langle -1\rangle$ is an epimorphism. We have already seen that $f[-1]$ is an epimorphism and thus the above diagram shows that $f\langle -1\rangle$ is an epimorphism.

The proof that $[-1]$ and $\langle -1\rangle$ preserve monomorphisms is elementary. □

THEOREM 3.4. *If* $k \to H' \xrightarrow{f'} H \xrightarrow{f''} H'' \to k$ *is an exact sequence of homology Hopf algebras such that*

$$0 \to QH' \to QH \to QH'' \to 0$$

is an exact sequence of k-modules, then

$$k \to H'[-1] \to H[-1] \to H''[-1] \to k$$

and

$$k \to H'\langle -1\rangle \to H\langle -1\rangle \to H''\langle -1\rangle \to k$$

are exact sequences of homology Hopf algebras.

PROOF. The arguments are similar and we present the details for the second case only.

Consider the diagram

$$
\begin{array}{ccccc}
H'\langle -1\rangle & \to & H\langle -1\rangle & \to & H''\langle -1\rangle \\
\downarrow & & \downarrow & & \downarrow \\
H' & \longrightarrow & H & \longrightarrow & H''\langle -1\rangle \\
\downarrow & & \downarrow & & \downarrow \\
C^pH' & \longrightarrow & C^pH & \longrightarrow & C^pH''.
\end{array}
$$

By definition the columns are exact. By hypothesis the middle row is exact. To show that the bottom row is exact recall that for any $A \in \operatorname{obj} \mathscr{H}_*\mathscr{H}/k$ the natural map $QA \to QC^pA$ is an isomorphism. From this and the fact that C^pA is coprimitive the hypothesis implies the exactness of the bottom row.

Hence the top row is exact by a suitable version of the 3×3 lemma. □

COROLLARY 3.5. *Suppose that* $k \to H' \to H \to H'' \to k$ *is an exact sequence of homology Hopf algebras such that*

$$0 \to QH' \to QH \to QH'' \to 0$$

is an exact sequence of k-modules. Then if H' and H'' are abelian so is H and if H' and H'' are coprimitive so is H.

PROOF. By Theorem 3.4

$$k \to H'[-1] \to H[-1] \to H''[-1] \to k$$

and

$$k \to H'\langle -1\rangle \to H\langle -1\rangle \to H''\langle -1\rangle \to k$$

are exact and the result now follows from the definition of $[-1]$ and $\langle -1\rangle$. □

4. **Elementary properties of the tower.** Let X be either an object of $\mathscr{H}Sp/p$ or $P - Ab\mathscr{H}Sp/p$. Let $\mathscr{F}(-n)(X)$ be the nth-modified coprimitive fibre square of X; i.e., the fibre square

$$\begin{array}{ccc} X(-n) & \longrightarrow & L(-n+1) \\ {\scriptstyle \pi(-n)}\downarrow & & \downarrow \\ X(-n+1) & \xrightarrow{\phi(-n)} & K(-n+1) \end{array}$$

defined inductively as in §2. Let $F(-n)$ be the fibre of $\pi(-n)$ and $e(-n): F(-n) \to X(-n)$ the inclusion of the fibre.

THEOREM 4.1. (1) *If $X \in \operatorname{obj} \mathscr{H}Sp/p$ is simply connected, then $X(-n) \in$* obj *$\mathscr{H}Sp/p$ and is $n+1$-connected. Moreover $\mathscr{F}(-n)(X)$ is a Hopf fibre square* mod p.

(2) *If $X \in \operatorname{obj} P - Ab\mathscr{H}Sp/p$ is simply connected, then $X(-n) \in \operatorname{obj} P - Ab\mathscr{H}Sp/p$ and is $n+1$-connected. Moreover $\mathscr{F}(-n)(X)$ is a pseudo-abelian Hopf fibre square* mod p.

PROOF. This is a summary of results obtained in §§1 and 2. □

We will employ the above notation and results throughout the remainder of this section without explicit mention. We will assume that X is simply connected and equipped with enough additional structure to be an object of either of the categories $\mathscr{H}Sp/p$ or $P - Ab\mathscr{H}Sp/p$.

PROPOSITION 4.2. (1) *$\pi(-n)_* : H_*X(-n) \to H_*X(-n+1)$ is a morphism* of *Hopf algebras;*

(2) *the natural morphism of homology Hopf algebras $\pi(-n)_*: H_*X(-n) \to H_*X(-n+1) \backslash\backslash \phi(-n)_*$ is an epimorphism, and*

(3) $H_*X(-n+1) \backslash\backslash \phi(-n)_* = (H_*X(-n+1))\langle -1\rangle$.

PROOF. This follows as in [**9**, Example 8.1]. □

PROPOSITION 4.3. (1) *There is an exact sequence of homology Hopf algebras*

$$H_*F(-n) \xrightarrow{e(-n)*} H_*X(-n) \backslash\backslash \pi(-n)_* \to E_*(-n) \to \mathbf{Z}_p$$

where

$$E_*(-n) = E[s^{-1}P_{-2}C^pH_*X(-n+1)];$$

(2) *if* $p > 2$ *there is an isomorphism of homology coalgebras*

$$H_*X(-n) \cong (H_*X(-n+1))\langle -1\rangle \otimes E_*(-n) \otimes S_*(-n)$$

where

$$S_*(-n) = \operatorname{Im}\{e(-n)_* : H_*F(-n) \to H_*X(-n)\};$$

(3) *if* $p = 2$ *and* $P_{-2}C^pH_*X(-n+1) = 0$ *then there is an isomorphism of homology coalgebras*

$$H_*X(-n) \cong (H_*(X(-n+1))\langle -1\rangle \otimes S_*(-n)).$$

PROOF. The second and third assertions are consequences of the first, Theorem 4.1, Proposition 4.2 and the facts that $S_*(-n)$ is injective as a homology coalgebra (by **[1]**, **[3]** and **[9]**, Proposition 4.1]) and $E_*(-n)$ is if $p > 2$.

The exact sequence of the first assertion follows as in Example 8.1 of **[9]**. The identification of $E_*(-n)$ is a routine exercise in the use of the material developed in §§4–6 of **[9]** and we leave to the reader the proof of the following more general result. □

PROPOSITION 4.4. *Suppose that k is a field,*

$$\begin{array}{ccc} E & \longrightarrow & E_0 \\ {\scriptstyle \pi}\downarrow & & \downarrow{\scriptstyle \pi_0} \\ B & \xrightarrow{f} & B_0 \end{array}$$

is a pseudo-Hopf fibre square mod p *with F the fibre of the fibrations* π *and* π_0. *In addition suppose that*

(1) H_*B_0 *is injective as a homology coalgebra;*
(2) $H_*E_0 \cong k$;
(3) H_*F *is coprimitive.*

Then there is an exact sequence of homology Hopf algebras

$$H_*F \to H_*E \backslash\!\backslash \pi_* \to E_* \to k$$

where $E_* = E[s^{-1}P_{-2}\operatorname{im} f_*]$ *as Hopf algebras.* □

A similar result holds for pseudo-abelian Hopf fibre squares.

A slightly stronger result than Proposition 4.3 may be obtained when H_*X is coprimitive of finite type. We turn to this now.

PROPOSITION 4.5. *Suppose that H_*X is coprimitive of finite type. Then $H_*X(-n)$ is also coprimitive of finite type. Moreover if $\{F^{-t}H^*X(-n)\}$ is the filtration associated to the cohomology Eilenberg-Moore spectral sequence of $\mathscr{F}(-n)(X)$ and*

$$P(-n) = \overline{F^{-1}H^*X(-n)},$$

then $H^*X(-n) \cong \mathscr{U}(P(-n))$ as a Hopf algebra over $\mathscr{A}^*(p)$.

PROOF. This follows as in [**8**, Theorem 6.4] and [**14**, Theorem 5.9]. □

Consider now the mod-C^p tower of X;

$$\begin{array}{ccccc}
 & & \vdots & & \\
F(-n) & \to & X(-n) & \to & K(-n) \\
 & & \downarrow{\scriptstyle \pi(-n)} & & \\
 & & X(-n+1) & & \\
 & & \downarrow & & \\
 & & \vdots & & \\
 & & \downarrow & & \\
F(-1) & \to & X(-1) & \to & K(-1) \\
 & & \downarrow{\scriptstyle \pi(-1)} & & \\
 & X = & X(0) & \to & K(0).
\end{array}$$

NOTATION. If $m \leq n$ let $\pi(-m, \ldots, -n) = \pi(-m) \cdots \pi(-n)$.

PROPOSITION 4.6. *The image of the morphism*

$$\pi(-m, \ldots, -n)_* : H_*X(-n) \to H_*X(-m+1)$$

is exactly the sub-Hopf algebra $(H_*X(-m))\langle -n + m - 1\rangle$.

PROOF. Let $n = m + t$. For $t = 1$ this is just Proposition 4.2 (2) and so we may proceed by induction on t.

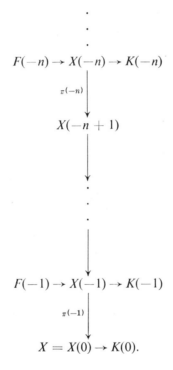

Consider the shortened tower

$$\begin{array}{c} X(-m-t) \\ \Big\downarrow{\scriptstyle \pi(-m-t)} \\ X(-m-t+1) \\ \Big\downarrow \\ \vdots \\ \Big\downarrow{\scriptstyle \pi(-m)} \\ X(-m+1). \end{array}$$

By the inductive hypothesis we have exact sequences

$$H_*X(-m-t) \to (H_*X(-m-t+1))\langle -1 \rangle \to \mathbf{Z}_p$$

and

$$H_*X(-m-t+1) \to (H_*X(-m+1))\langle -t \rangle \to \mathbf{Z}_p.$$

From Theorem 3.1 we obtain an epimorphism

$$(H_*X(-m-t+1))\langle -1 \rangle \to (H_*X(-m+1))\langle -t \rangle\langle -1 \rangle$$
$$= (H_*X(-m+1))\langle -t-1 \rangle,$$

and the result follows. □

5. **Some functors of unstable $\mathscr{A}^*$-modules.** In this section we will develop some properties of the category of unstable modules over the Steenrod algebra $\mathscr{A}^*(p)$. These will be of use in the identification of $E^2(X)$ in §7.

We assume familiarity with **[16]**.

NOTATION AND CONVENTION. Throughout this section p will be a prime and $\mathscr{A}^*$ the Steenrod algebra mod p. An $\mathscr{A}^*$-module will mean a positively graded left $\mathscr{A}^*$-module. We will use upper index notation for the grading. The dual of $\mathscr{A}^*$ will be denoted by $\mathscr{A}_*$. An $\mathscr{A}_*$-comodule will mean a positively graded left $\mathscr{A}_*$-comodule and lower index notation will be employed.

DEFINITION. An $\mathscr{A}^*$-module M is called unstable if

$$(p=2)\colon\ Sq^i x = 0 \quad \text{for all } x \in M^j, \quad j < i$$
$$(p>2)\colon\ P^i x = 0 \quad \text{for all } x \in M^j, \quad j < 2i$$
$$\beta P^i x = 0 \quad \text{for all } x \in M^{2i}.$$

If X is a space then H^*X is an unstable $\mathscr{A}^*$-module, which is of course the motivation for the definition. Note further that if X is a pseudo Hopf space mod p then PH^*X is an unstable $\mathscr{A}^*$-module.

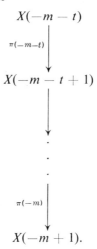

If M is an unstable $\mathscr{A}^*$-module and $N \subset M$ is an $\mathscr{A}^*$-module then N is unstable and so is M/N.

The category of unstable $\mathscr{A}^*$-modules will be denoted by $\mathscr{UM}^*/\mathscr{A}^*$. Note that it is a full subcategory of $\mathscr{M}/\mathscr{A}^*$.

Similarly we have the category of unstable comodules under $\mathscr{A}_*$ which we denote by $\mathscr{UC}o\mathscr{M}_*/\mathscr{A}_*$. For all practical purposes we may consider the categories $\mathscr{UM}^*/\mathscr{A}^*$ and $\mathscr{UC}o\mathscr{M}_*/\mathscr{A}_*$ as dual categories. More precisely the full subcategories generated by the objects of finite type are in duality.

PROPOSITION 5.1. *The categories $\mathscr{UM}^*/\mathscr{A}^*$ and $\mathscr{UC}o\mathscr{M}_*/\mathscr{A}_*$ are abelian.*

PROOF. This is elementary. □

DEFINITION. If $M \in \text{obj } \mathscr{UM}^*/\mathscr{A}^*$ we define a morphism $\zeta_M : M \to M$ by

$$(p = 2): \zeta_M(x) = Sq^n x \quad \text{for all } x \in M^n$$

$$(p > 2): \zeta_M(x) = P^n x \quad \text{for all } x \in M^{2n}$$

$$= \beta P^n x \quad \text{for all } x \in M^{2n+1}.$$

Note that ζ_M is not a morphism in the category. When $p = 2$, ζ_M doubles degrees and gives M the structure of an abelian restricted Lie algebra. If $p > 2$ then roughly speaking (M, ζ_M) is an abelian restricted Lie algebra with Bockstein.

Note that $\zeta_M : M \to M$ is $\mathbf{Z}_p$-linear and thus we may define ker ζ_M and coker ζ_M to be the kernel and cokernels in $\mathscr{M}^*/\mathbf{Z}_p$.

LEMMA 5.2. *If $M \in \text{obj } \mathscr{UM}^*/\mathscr{A}^*$ then* im ζ_M *and* ker ζ_M *are $\mathscr{A}^*$-submodules of M.*

PROOF. When $p = 2$ the Adem relations yield

$$Sq^i \zeta_M(x) = \zeta_M Sq^{i/2}(x) \quad i \text{ even}$$

$$= 0 \quad i \text{ odd}$$

from which the result follows for $p = 2$. A similar calculation applies when $p > 2$. □

DEFINITION. If $M \in \text{obj } \mathscr{UM}^*/\mathscr{A}^*$ then $\Omega^* M$ is defined by

$$(\Omega^* M)^q = (M/\text{Im } \zeta)^{q+1}.$$

If $f: M' \to M''$ is a morphism of unstable $\mathscr{A}^*$-modules then $\Omega^* f: \Omega^* M' \to \Omega^* M''$ is defined in the obvious fashion. It is a morphism in $\mathscr{UM}^*/\mathscr{A}^*$.

We denote by $\sigma_M : M \to \Omega^* M$ the natural morphism of degree -1.

PROPOSITION 5.3 (MASSEY-PETERSON). *The functor*

$$\Omega^* : \mathscr{UM}^*/\mathscr{A}^* \to \mathscr{UM}^*/\mathscr{A}^*$$

is an additive right exact functor and $\sigma : \text{Id} \to \Omega^*$ is a natural transformation of functors of degree -1.

PROOF. The proof is routine and left to the reader. □

ACKNOWLEDGEMENT. The definition of the functor Ω^* is due to Massey-Peterson [5] who developed its elementary properties.

EXAMPLES. (1) The functor Ω^* is not left exact. For consider the $\mathscr{A}^*$-modules defined by

$$M' = \{(M')^n \mid (M')^n = \begin{Bmatrix} 0 & \text{if } n \neq 2p^2 \\ x\mathbf{Z}_p & \text{if } n = 2p^2 \end{Bmatrix}$$

$$M'' = \{(M'')^n \mid (M'')^n = \begin{Bmatrix} 0 & \text{if } n \neq 2p \text{ or } 2p^2 \\ u\mathbf{Z}_p & \text{if } n = 2p \\ v\mathbf{Z}_p & \text{if } n = 2p^2 \end{Bmatrix}.$$

The $\mathscr{A}^*$-action on M' is trivial and on M'' is determined by requiring $P^p u = v$.

Let $f: M' \to M''$ be the monomorphism given by $f(x) = v$. One may readily check that $\Omega^* f: \Omega^* M' \to \Omega^* M''$ is the trivial morphism, although $\Omega^* M' \neq 0$.

(2) Let A be a simply connected graded abelian group. Then the results of Cartan and Serre [1], [13] show that $PH^*\Omega KA = \Omega^* PH^* KA$. This, in a large part, is the motivation for the definition of Ω^*.

DEFINITION. If $M \in \operatorname{obj} \mathscr{UM}^*/\mathscr{A}^*$, then $\Sigma^* M$ is defined by $(\Sigma^* M)^{q+1} = M^q$. One readily checks that Σ^* is a functor from $\mathscr{UM}^*/\mathscr{A}^*$ to itself.

PROPOSITION 5.4. *If $M \in \operatorname{obj} \mathscr{UM}^*/\mathscr{A}^*$ then*

(1) *there is a natural isomorphism*

$$\alpha_M : \Omega^*\Sigma^* M \to M$$

(2) *there is a natural morphism of $\mathscr{A}^*$-modules*

$$\beta_M : M \to \Sigma^*\Omega^* M.$$

PROOF. By definition $\Omega^*\Sigma^* M = M$. Thus α_M is really the identity. β_M is just the obvious quotient mapping. □

We now readily obtain,

PROPOSITION 5.5. $(\alpha, \beta): \Omega^* \dashv \Sigma^* : (\mathscr{UM}^*/\mathscr{A}^*, \mathscr{UM}^*/\mathscr{A}^*)$ *is an adjoint pair of functors.* □

DEFINITION. If $n \geq 0$ let $\mathbf{B}(n)$ denote the set of all elements of $\mathscr{A}^*$ such that for any $M \in \operatorname{obj} \mathscr{UM}^*/\mathscr{A}^*$ and $\alpha \in \mathbf{B}(n)$,

$$\alpha(x) = 0 \quad \text{for all } x \in M^j, \qquad j \leq n.$$

One readily checks that $\mathbf{B}(n)$ is a left ideal in $\mathscr{A}^*$ and hence $\mathscr{A}^*/\mathbf{B}(n)$ is an $\mathscr{A}^*$-module. The ideal $\mathbf{B}(n)$ occurs naturally in topology as the annihilator ideal (in $\mathscr{A}^*$) of the fundamental class $i \in H^n(\mathbf{Z}_p, n; \mathbf{Z}_p)$, see [1], [13], [14].

DEFINITION. $\mathbf{P}(n) = (\Sigma^*)^n(\mathscr{A}^*/\mathbf{B}(n))$.

PROPOSITION 5.6. $\mathbf{P}(n)$ *is an unstable* $\mathscr{A}^*$*-module, and indeed is a projective object in* $\mathscr{U}\mathscr{M}^*/\mathscr{A}^*$.

PROOF. Routine. □

REMARK. $\mathbf{P}(n) = PH^*(\mathbf{Z}_p, n; \mathbf{Z}_p)$.

Let V be a graded strictly positive $\mathbf{Z}_p$-module. A free unstable $\mathscr{A}^*$-module generated by V consists of an unstable $\mathscr{A}^*$-module $\mathbf{F}(V)$ together with a homomorphism $\theta: V \to \mathbf{F}(V)$ of graded $\mathbf{Z}_p$-modules such that, given any unstable $\mathscr{A}^*$-module M and a morphism $f: V \to M$ of $\mathbf{Z}_p$-modules there exists a unique morphism $\hat{f}: \mathbf{F}(V) \to M$ of unstable $\mathscr{A}^*$-modules such that the diagram of $\mathbf{Z}_p$-modules

$$\begin{array}{ccc} V & \xrightarrow{\theta} & \mathbf{F}(V) \\ {\scriptstyle f}\downarrow & \swarrow {\scriptstyle \hat{f}} & \\ M & & \end{array}$$

is commutative.

PROPOSITION 5.7. *If* V *is a strictly positive* $\mathbf{Z}_p$*-module then the free unstable* $\mathscr{A}^*$*-module generated by* V *exists and is unique up to a canonical isomorphism.*

PROOF. For the existence choose a basis $\{v_\alpha\}$ for V as a $\mathbf{Z}_p$-module. Let $\mathbf{P}_\alpha = \mathbf{P}\,(\deg v_\alpha)$ and set $\mathbf{F}(V) = {}_\alpha\oplus P_\alpha$. Let $\theta: V \to \mathbf{F}(V)$ be the obvious map. It is easy to verify that $(\mathbf{F}(V), \theta)$ solves the required universal mapping problem. Uniqueness is routine. □

Let M be an unstable $\mathscr{A}^*$-module. Then for the moment we may forget the $\mathscr{A}^*$-module structure and form the free unstable $\mathscr{A}^*$-module $\mathbf{F}\Sigma^*M$. For any $x \in \Sigma^*M$ and $\alpha \in \mathscr{A}^*$, $\alpha \circ x$ is now defined by remembering that Σ^*M is an $\mathscr{A}^*$-module. Let $N_M \subset \mathbf{F}\Sigma^*M$ denote the $\mathscr{A}^*$-submodule generated by the elements,

$$(p = 2): Sq^i \circ x - Sq^i x \quad \text{such that } i < \deg x$$

$$(p > 2): P_p^i \circ x - P_p^i x \quad \text{such that } i < 2 \deg x$$

$$\beta^\circ x - \beta x.$$

DEFINITION. With the notation employed above, let $\mathbf{B}^*M = \mathbf{F}\Sigma^*M/N_M$. $\mathbf{B}^*: \mathscr{U}\mathscr{M}^*/\mathscr{A}^* \circlearrowleft$ is called the classifying functor. (For $f \in \operatorname{morph} \mathscr{U}\mathscr{M}^*/\mathscr{A}^*$, $\mathbf{B}^*f$ is defined in the obvious way.) Let $\tau: M \to \mathbf{B}^*M$ be the natural morphism of degree $+1$.

PROPOSITION 5.8. *The functor* $\mathbf{B}^*: \mathscr{U}\mathscr{M}^*/\mathscr{A}^* \to \mathscr{U}\mathscr{M}^*/\mathscr{A}^*$ *is an additive right exact functor.*

PROOF. Suppose that

$$0 \to M' \xrightarrow{f'} M \xrightarrow{f''} M'' \to 0$$

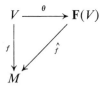

is an exact sequence of unstable $\mathscr{A}^*$-modules. Introduce the commutative diagram

$$\begin{array}{ccccccccc}
 & & 0 & & 0 & & 0 & & \\
 & & \downarrow & & \downarrow & & \downarrow & & \\
 & & N' & \xrightarrow{\varphi'} & N & \xrightarrow{\varphi''} & N'' & & \\
 & & \downarrow{\scriptstyle \eta'} & & \downarrow{\scriptstyle \eta} & & \downarrow{\scriptstyle \eta''} & & \\
0 & \longrightarrow & \mathbf{F}\Sigma^*M' & \xrightarrow{F'} & \mathbf{F}\Sigma^*M & \xrightarrow{F''} & \mathbf{F}\Sigma^*M'' & \longrightarrow & 0 \\
 & & \downarrow{\scriptstyle \pi'} & & \downarrow{\scriptstyle \pi} & & \downarrow{\scriptstyle \pi''} & & \\
 & & \mathbf{B}^*M' & \xrightarrow{B^*f'} & \mathbf{B}^*M & \xrightarrow{B^*f''} & \mathbf{B}^*M'' & & \\
 & & \downarrow & & \downarrow & & \downarrow & & \\
 & & 0 & & 0 & & 0 & &
\end{array}$$

where the columns are exact by definition. The middle row is easily seen to be exact also.

Next we note the φ'' is onto. For recall that N'' is generated as an $\mathscr{A}^*$-module by elements $\alpha \circ x - \alpha x$. Where $\alpha \in \mathscr{A}^*$, $x \in \Sigma^*M''$. Since f'' is surjective so is Σ^*f''. Thus there exists $y \in \Sigma^*M$ such that $\Sigma^*f''y = x$. Thus $\alpha \circ y - \alpha y \in N$ and $\varphi''(\alpha \circ y - \alpha y) = \alpha \circ x - \alpha x$. Since φ'' is a morphism of $\mathscr{A}^*$-modules it follows that φ'' is onto as claimed.

We are now ready to show that the bottom row is right exact.

(1) $\mathbf{B}^*f''$ is epic: For we have by commutativity $\mathbf{B}^*(f'')\pi = \pi''F''$. Now π'' and F'' are epimorphisms and hence so is $\mathbf{B}^*f''$.

(2) exactness at $\mathbf{B}^*M$: Note first that we have $\mathbf{B}^*f''\mathbf{B}^*f' = \mathbf{B}^*(f''f') = \mathbf{B}^*(0) = 0$.

Next suppose that $x \in \mathbf{B}^*M$ with $B^*f''(x) = 0$. Choose $y \in \mathbf{F}\Sigma^*M$ such that $\pi y = x$. Then $\pi''F''(y) = 0$ so $F''(y) = \eta''(u)$. Since φ'' is surjective there exists $v \in N$ such that $\varphi''(v) = u$. Then we have

(a): $$\begin{aligned} F''(y - \eta v) &= F''(y) - F''\eta(v) \\ &= F''(y) - \eta''\varphi''(v) \\ &= F''(y) - \eta''(u) = 0 \end{aligned}$$

(b): $$\pi(y - \eta v) = \pi y - \pi\eta(v) = \pi y = x.$$

From (a) it follows that there exists $w \in \mathbf{F}^*\Sigma^*M'$ such that $F'w = y - \eta v$. From (b) we obtain

$$\mathbf{B}^*f'\pi'w = \pi F'w = x.$$

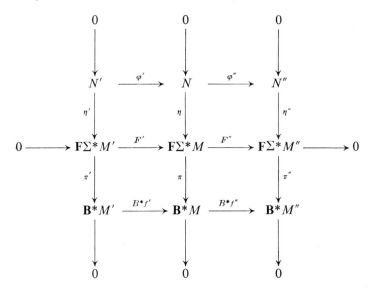

Thus $x \in \operatorname{im} \mathbf{B}^*f'$ and exactness at $\mathbf{B}^*M$ follows.

Additivity is routine. □

We shall have need of a more delicate exactness property of $\mathbf{B}^*$ in §7. We turn to this now.

Suppose that A is a graded simply connected abelian group of finite type. Let $\mathbf{P}_A = PH^*(KA)$. It follows easily from **[1]**, **[13]**, and the definition of $\mathbf{B}^*$ that $\mathbf{B}^*\mathbf{P}_A = \mathbf{P}_{sA}$, where s is the suspension functor.

We recall from **[1]**, **[13]**, that $\mathbf{P}_A$ has a basis $\{i_j, \beta^{r_j} i_j\}$ as an unstable $\mathscr{A}^*$-module. If $M \subset \mathbf{P}_A$ is an $\mathscr{A}^*$-submodule with generators $\{\mu_s = \Sigma \eta_{sj} i_j + \Sigma \zeta_{sj} \beta^{r_j} i_j\}$ then it may be verified directly from the definitions that $\mathbf{B}^*M$ is the $\mathscr{A}^*$-submodule of $\mathbf{B}^*\mathbf{P}_A = \mathbf{P}_{sA}$ generated by the elements $\{\tau\mu_s = \Sigma \eta_{sj} \tau i_j + \Sigma \zeta_{sj} \beta^{r_j} \tau i_j\}$ where $\tau : \mathbf{P}_A \to \mathbf{P}_{sA}$ is the transgression.

Combining this with Proposition 5.8 we obtain:

COROLLARY 5.9. *Suppose that A is a simply connected graded abelian group of finite type and*

$$0 \to M' \to \mathbf{P}_A \to M'' \to 0$$

is an exact sequence of unstable $\mathscr{A}^*$-modules. Then

$$0 \to \mathbf{B}^*M' \to \mathbf{B}^*P_A \to \mathbf{B}^*M'' \to 0$$

is an exact sequence of unstable $\mathscr{A}^*$-modules. □

PROPOSITION 5.10. *There are natural transformations of functors*

$$\mathrm{Id} \to \Omega^*\mathbf{B}^* \quad \textit{and} \quad \mathbf{B}^* \to \Sigma^*.$$

PROOF. This is routine. We define the second one only. Let

$$M \in \operatorname{obj} \mathscr{U}\mathscr{M}^*/\mathscr{A}^*.$$

Then $\widehat{\mathrm{id}} : \Sigma^*M \to \Sigma^*M$ induces a morphism of $\mathscr{A}^*$-modules $\widehat{\mathrm{id}} : \mathbf{F}\Sigma^*M \to \Sigma^*M$. It is immediate that $\widehat{\mathrm{id}} \mid N = 0$. Thus $\widehat{\mathrm{id}}$ induces a morphism of $\mathscr{A}^*$-modules $s_M : \mathbf{B}^*M = \mathbf{F}\Sigma^*M/N_M \to \Sigma^*M$. It is easy to extend the definition of s_M to a natural transformation $\mathbf{B}^* \to \Sigma^*$ as required. □

The functor $\mathbf{B}^*$ is constructed by analogy with the work of Cartan **[1]** and Serre **[13]**. Indeed their work shows:

PROPOSITION 5.11. (1) *If A is a connected abelian group then*

$$PH^*KsA = \mathbf{B}^*PH^*KA.$$

(2) *If A is a simply connected abelian group then*

$$PH^*Ks^{-1}A = \Omega^*PH^*KA. \quad □$$

PROPOSITION 5.12. *If A is a simply connected $\mathbf{Z}_p$-module then QH_*KA is an injective object of $\mathscr{U}\mathscr{C}o\mathscr{M}_*/\mathscr{A}_*$. Dually, if A is of finite type PH^*KA is a projective object of $\mathscr{U}\mathscr{M}^*/\mathscr{A}^*$.* □

We note that for any strictly positive $\mathbf{Z}_p$-module V, $\mathbf{F}(V)$ is a projective object in the category $\mathcal{UM}^*/\mathcal{A}^*$ and thus we readily obtain,

PROPOSITION 5.13. (1) *The category* $\mathcal{UM}^*/\mathcal{A}^*$ *has enough projectives, and dually,*

(2) *the category* $\mathcal{UC}o\mathcal{M}_*/\mathcal{A}_*$ *has enough injectives.* □

Since the category $\mathcal{UC}o\mathcal{M}_*/\mathcal{A}*$ has enough injectives we may proceed in the usual fashion [3] to construct the derived functors of $M \,\square_{\mathcal{A}_*} N$. These will be denoted by $\operatorname{UnCotor}^{\mathcal{A}*}_{*,*}(M, N)$.

REMARK. Roughly speaking the functor $\operatorname{UnCotor}^{\mathcal{A}*}(,)$ is "dual" to the functor $\operatorname{Unext}_{\mathcal{A}^*}(\ ,\ )$ of [5]. Indeed under suitable finite type hypotheses there is an isomorphism $\operatorname{UnCotor}^{\mathcal{A}_*}(M, N) \simeq \operatorname{Unext}_{\mathcal{A}^*}(M^*, N)$. (Compare [2] page 87.)

PROPOSITION 5.14. *If* $M \in \operatorname{obj} \mathcal{UM}^*/\mathcal{A}^*$ *and* $\zeta : M \to M$ *is a monomorphism, i.e.* $\ker \zeta = 0$, *then there is a natural isomorphism* $\mathbf{B}^*\Omega^*M \simeq M$ *of* $\mathcal{A}*$*-modules.*

PROOF. A routine exercise in the definitions. □

REMARK. If A is a graded simply connected abelian group and $M \subset PH^*KA$ is an $\mathcal{A}^*$-submodule, then $\ker \zeta_M = 0$. This follows easily from [1], [13].

PROPOSITION 5.15. *If* $M \in \operatorname{obj} \mathcal{UM}^*/\mathcal{A}^*$ *is a projective object then so are* $\mathbf{B}^*M$ *and* Ω^*M.

PROOF. One readily checks that

$$\mathbf{B}^*\mathbf{P}(n-1) \simeq \mathbf{P}(n) \simeq \Omega^*\mathbf{P}(n+1),$$

and the result follows. □

REMARKS ON DUALITY. For the sake of convenience we have worked almost exclusively in the category $\mathcal{UM}^*/\mathcal{A}^*$. We could dually have defined the functors $\mathbf{B}_*$, Ω_*, Σ_* on $\mathcal{UC}o\mathcal{M}_*/\mathcal{A}_*$ dual to the functors $\mathbf{B}^*$, Ω^*, Σ^*. The functors $\mathbf{B}_*$, Ω_*, and Σ_* would then enjoy the properties dual to $\mathbf{B}^*$, Ω^*, Σ^*. In dealing with homotopy groups the category $\mathcal{UC}o\mathcal{M}_*/\mathcal{A}_*$ is more natural to work in. We will adopt this course in §7.

6. **The algebra** $\mathcal{H}^*$. One of the useful properties enjoyed by the $K(\mathbf{Z}_p, n)$-spaces is that the Hurewicz map provides an isomorphism

$$h : \pi_*(K(\mathbf{Z}_p, n)) \to \mathbf{Z}_p \,\square_{\mathcal{A}_*} QH_*(\mathbf{Z}_p, n; \mathbf{Z}_p).$$

Relative to the homotopy theory $\pi_*(\ ; \mathbf{Z}_p)$, p an odd prime, all the $K(\pi, n)$ spaces (π-abelian) exhibit analogous behavior. We turn to this first.

ACKNOWLEDGEMENT. The results of this section are due in large measure to J. C. Moore.

CONVENTION. Throughout this section p is an odd prime.

Let $f\colon \mathscr{A}^* \to E[x]$, where $\deg x = 1$, be defined by

$$f(\beta) = x \qquad f(P^k) = 0,\ k > 0.$$

One readily checks that f is a morphism of homology Hopf algebras.

DEFINITION. $\mathscr{H}^* = \mathscr{A}^* \setminus\!\!\setminus f^*$.

Note that $\mathscr{H}^*$ is a normal sub-Hopf algebra of $\mathscr{A}^*$ and that there is an exact sequence of homology Hopf algebras

$$\mathbf{Z}_p \to \mathscr{H}^* \to \mathscr{A}^* \to E[\beta] \to \mathbf{Z}_p.$$

NOTATION. If $x, y \in \mathscr{A}^*$ then we set $[x, y] = xy - (-1)^{\deg x \deg y} yx$. Observe that $\mathscr{H}^*$ is generated as an algebra by the elements P^k and $[\beta, P^k]$, $k > 0$, and that a minimal set of generators is provided by P^{p^k}, $[\beta, P^{p^k}]$, $k > 0$. Thus $\mathscr{H}^*$ is $2p - 3$ connected.

$\mathscr{H}^*$ may be described quite simply in terms of Milnor's characterization of the dual $\mathscr{A}_*$ of the Steenrod algebra. We recall this now **[16]**.

If p is an odd prime then

$$\mathscr{A}_* = E[x_0, x_1, \ldots] \otimes P[y_1, \ldots]$$

where

$$\deg x_i = 2p^i - 1 \qquad \deg y_i = 2p^i - 2$$

and $\nabla\colon \mathscr{A}_* \to \mathscr{A}_* \otimes \mathscr{A}_*$ is given by

$$\nabla y_k = \Sigma y_{k-i}^{p^i} \otimes y_i$$

$$\nabla x_k = x_k \otimes 1 + \Sigma y_{k-i}^{p^i} \otimes x_i.$$

Since x_0 is primitive $E[x_0]$ is a sub-Hopf algebra of $\mathscr{A}_*$. Thus $\mathscr{A}_* /\!\!/ E[x_0]$ is again a Hopf algebra. If we denote by $\mathscr{H}_*$ the dual of $\mathscr{H}^*$ then it is easy to see that

$$\mathscr{H}_* \simeq \mathscr{A}_* /\!\!/ E[x_0]$$

as Hopf algebras.

The algebra $\mathscr{H}^*$ has the following two nice properties:

(1) the natural map $\mathscr{H}^* \to H^*(\mathbf{Z}, \infty; \mathbf{Z}_p)$ given by action on the counit is an isomorphism of coalgebras;

(2) $\mathscr{H}^*$ acts naturally on the cohomology mod p of any space.

In addition, a careful examination of the results of [**1**] shows:

THEOREM 6.1. *If A is a connected graded abelian group and p is an odd prime then the Hurewicz map induces an isomorphism*

$$h:\pi_*(KA;\mathbf{Z}_p)\to \mathbf{Z}_p \square_{\mathscr{H}_*} Q_{\mathscr{H}_*} KA.$$

DEFINITION. Let $\mathscr{H}^*(n) \subset \mathscr{H}^*$ be defined by $\mathscr{H}^*(n) = \mathbf{B}(n) \cap \mathscr{H}^*$.

If we think of $\mathscr{H}^*$ as acting on $H^*(\mathbf{Z}, n; \mathbf{Z}_p)$ then $\mathscr{H}^*(n)$ is just the annihilator ideal of the fundamental class (reduced mod p).

DEFINITION. An $\mathscr{H}^*$-module is unstable iff for all $x \in M^j$ and $\alpha \in \mathscr{H}^*(n)$, $j \leq n \Rightarrow \alpha \cdot x = 0$.

We thus obtain the full subcategory of $\mathscr{M}^*/\mathscr{H}^*$ generated by the unstable $\mathscr{H}^*$-modules. We denote this category by $\mathscr{UM}^*/\mathscr{H}^*$. Note that there is a natural forgetful functor

$$\Phi^*: \mathscr{UM}^*/\mathscr{A}^* \to \mathscr{UM}^*/\mathscr{H}^*$$

which roughly speaking forgets the action of β.

Dually we also have the category $\mathscr{UC}o\mathscr{M}_*/\mathscr{H}_*$.

From [**1**] again we obtain (compare Proposition 5.11):

THEOREM 6.2. *If A is a connected graded abelian group then QH_*KA is an injective object of $\mathscr{UC}o\mathscr{M}_*/\mathscr{H}_*$. Dually if A is of finite type then PH^*KA is a projective object of $\mathscr{UM}^*/\mathscr{H}^*$.* □

PROPOSITION 6.3. *The category $\mathscr{UM}^*/\mathscr{H}^*$ has enough projectives and dually the category $\mathscr{UM}_*/\mathscr{H}_*$ has enough injectives.*

PROOF. Routine. □

Thus we may form the derived functors of the functor $M \square_{\mathscr{H}_*} N$ in the standard fashion [**3**]. These will be denoted by $\text{UnCotor}^{\mathscr{H}_*}_{*,*}(M, N)$.

PROPOSITION 6.4. (1) *The forgetful functor $\Phi^*: \mathscr{UM}^*/\mathscr{A}^* \to \mathscr{UM}^*/\mathscr{H}^*$ preserves projectives;*

(2) *the forgetful functor $\Phi_*: \mathscr{UC}o\mathscr{M}_*/\mathscr{A}_* \to \mathscr{UC}o\mathscr{M}_*/\mathscr{H}_*$ preserves injectives;*

(3) *if $M \in \text{obj}\, \mathscr{UM}^*/\mathscr{A}^*$ and Φ^*M is a projective in $\mathscr{UM}^*/\mathscr{H}^*$ then $\Phi^*\mathbf{B}^*M$ is a projective in $\mathscr{UM}^*/\mathscr{H}^*$;*

(4) *if $N \in \text{obj}\, \mathscr{UC}o\mathscr{M}_*/\mathscr{A}_*$ and Φ_*N is an injective in $\mathscr{UC}o\mathscr{M}_*/\mathscr{H}_*$ then $\Phi_*\mathbf{B}_*N$ is an injective in $\mathscr{UC}o\mathscr{M}_*/\mathscr{H}_*$.*

PROOF. Routine. □

7. $E^2(X)$: **A special case.** Throughout this section as in the previous, p is an odd prime. X will denote either a Hopf space mod p or a pseudo-abelian Hopf

space mod p. The mod C^p-tower of X will be denoted by

$$\begin{array}{ccccc}
 & & \vdots & & \\
 & & \downarrow & & \\
F(-n) & \to & X(-n) & \to & K(-n) \\
 & & \downarrow{\scriptstyle \pi(-n)} & & \\
F(-n+1) & \to & X(-n+1) & \to & K(-n+1) \\
 & & \downarrow & & \\
 & & \vdots & & \\
 & & \downarrow & & \\
F(-1) & \to & X(-1) & \to & K(-1) \\
 & & \downarrow{\scriptstyle \pi(-1)} & & \\
 & X = & X(0) & \to & K(0)
\end{array}$$

The mod C^p-spectral sequence of X will be denoted by $\{E^r(X), d^r(X)\}$. Our objective in this section is to prove:

THEOREM 7.1. *Let p be an odd prime. Let X be a simply connected pseudo-abelian Hopf space* mod p *with H_*X coprimitive of finite type. Then there is a natural isomorphism*

$$E^2(X) = \mathrm{UnCotor}_{\mathscr{H}_*}(\mathbf{Z}_p, QH_*X).$$

PROOF. For each integer n we have exact sequences

$$0 \to QH_*X(-n) \xrightarrow{\varphi(-n-1)_*} QH_*K(-n) \to Q(H_*K(-n) /\!/ \varphi(-n-1)_*) \to 0.$$

In addition it follows inductively from Proposition 4.5 that $H_*X(-n)$ is coprimitive. Thus we have exact sequences of homology Hopf algebras

$$\mathbf{Z}_p \to H_*X(-n) \to H_*K(-n) \to H_*K(-n) /\!/ \varphi(-n-1)_* \to \mathbf{Z}_p.$$

Thus we obtain from Proposition 4.3 an isomorphism

$$QH_*X(-n-1) \simeq \Omega_*Q(H_*K(-n) /\!/ \varphi(-n-1)_*).$$

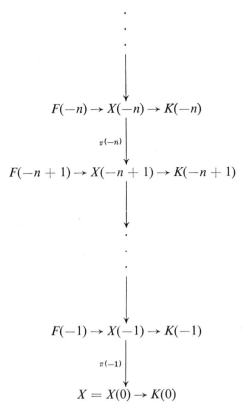

Hence we have exact sequences

$$0 \to QH_*X \to QH_*K(0) \to QH_*K(0) /\!/ \varphi(-1)_* \to 0$$

$$0 \to \Omega_* QH_*K(0) /\!/ \varphi(-1)_* \to QH_*K(-1) \to QH_*K(-1) /\!/ \varphi(-2)_* \to 0$$

$$\vdots \qquad\qquad \vdots \qquad\qquad \vdots$$

$$0 \to \Omega_* QH_*K(-n) /\!/ \varphi(-n-1)_* \to QH_*K(-n-1) \to QH_*K(-n-1) /\!/ \varphi(-n-2)_* \to 0$$

$$\vdots \qquad\qquad \vdots \qquad\qquad \vdots$$

It follows from the dual of Proposition 5.13 and the succeeding remarks that

$$\mathbf{B}_*\Omega_* QH_*K(-n) /\!/ \varphi(-n-1)_* \cong QH_*K(-n) /\!/ \varphi(-n-1)_*.$$

Thus applying the functor $\mathbf{B}_*$ n-times to the nth sequence, we obtain in view of Corollary 5.9 exact sequences

$$0 \to QH_*X \to QH_*K(0) \to QH_*K(0) /\!/ \varphi(-1)_* \to 0$$

$$0 \to QH_*K(0) /\!/ \varphi(-1)_* \to \mathbf{B}_*QH_*K(-1) \to \mathbf{B}_*Q_*H_*K(-1) /\!/ \varphi(-2)_* \to 0$$

$$\vdots \qquad\qquad \vdots \qquad\qquad \vdots$$

$$0 \to \mathbf{B}_*^n QH_*K(-n) /\!/ \varphi(-n-1)_* \to \mathbf{B}_*^{n+1} QH_*K(-n-1) \to \mathbf{B}_*^{n+1} QH_*K(-n-1) /\!/ \varphi(-n-2)_* \to 0$$

Pasting these short exact sequences together we obtain a long exact sequence of $\mathscr{A}_*$-comodules

$$\mathscr{X}: \qquad 0 \to QH_*X \to QH_*K(0) \to \mathbf{B}_*QH_*K(-1) \to \cdots.$$

It follows from Proposition 6.2 and Proposition 6.4 that $\mathscr{X}$ is an injective resolution of QH_*X in the category $\mathscr{UCoM}_*/\mathscr{H}_*$. From Theorem 6.1 and naturality it follows that

$$E^1(X) \cong \mathbf{Z}_p \,\square_{\mathscr{H}}\, \mathscr{X}$$

as differential modules. Thus

$$E^2(X) \cong \operatorname{UnCotor}^{\mathscr{H}*}(\mathbf{Z}_p, QH_*X).$$

The naturality of this isomorphism is elementary. □

APPENDIX

Let X be a simply connected pseudo-Hopf space mod p with coprimitive fibre square [9, Example 8.1]

$$\mathscr{F}\langle -1\rangle(X) \qquad \begin{array}{ccc} X\langle -1\rangle & \longrightarrow & LQH_*X \\ \Big\downarrow{\scriptstyle \pi\langle -1\rangle} & & \Big\downarrow \\ X & \xrightarrow{\varphi\langle -1\rangle} & KQH_*X \end{array}$$

We will study conditions that assure a "natural" pseudo-Hopf fibre square mod p or Hopf fibre square mod p structure on $\mathscr{F}\langle -1\rangle(X)$.

NOTATION. X, $\mathscr{F}\langle -1\rangle(X)$ are as above;

$K\langle X\rangle = KQH_*X,$

$\varphi\langle X\rangle = \varphi\langle -1\rangle,$

$\{E^r\langle X\rangle, d^r\langle X\rangle\}$ denotes the homology spectral sequence of the fibre square $\mathscr{F}\langle -1\rangle(X)$, with $\mathbf{Z}_p$-coefficients [3], [8].

LEMMA A.1.

$$E^2\langle X\rangle = (H_*X)\langle -1\rangle \otimes \mathrm{Cotor}^{H_*K\langle X\rangle /\!/ \varphi\langle X\rangle_*}(\mathbf{Z}_p, \mathbf{Z}_p).$$

PROOF. See [9; Example 8.1]. □

NOTATION. $\{F_{-n}H_*X\langle -1\rangle\}$ denotes the filtration on $H_*X\langle -1\rangle$ that is associated to the spectral sequence $\{E^r\langle X\rangle, d^r\langle X\rangle\}$.

$E^0H_*X\langle -1\rangle$ is the associated graded object. Note that $E^0H_*X\langle -1\rangle = E^\infty\langle X\rangle$.

PROPOSITION A.2. *$E^0H_*X\langle -1\rangle$ admits a natural homology Hopf algebra structure.*

PROOF. From [8, Theorem A.1]. □

REMARK. In the context of Proposition A.2 natural means that for any morphism $f: X \to Y$ of pseudo-Hopf spaces mod p, the induced map

$$f\langle -1\rangle : X\langle -1\rangle \to Y\langle -1\rangle$$

yields a morphism of homology Hopf algebras $E^0f\langle -1\rangle_* : E^0H_*X\langle -1\rangle \to E^0H_*Y\langle -1\rangle$.

DEFINITION. A triple of pseudo-Hopf spaces mod p and their morphism

$$X' \xrightarrow{f'} X \xrightarrow{f''} X''$$

is said to be Q-left exact mod p if

$$\mathbf{Z}_p \to H_*X' \to H_*X \to H_*X''$$

is an exact sequence of homology Hopf algebras and

$$0 \to QH_*X' \to QH_*X \to QH_*X''$$

is an exact sequence of $\mathbf{Z}_p$-modules.

LEMMA A.3. *If* $0 \to V' \to V \to V''$ *is an exact sequence of graded* $\mathbf{Z}_p$*-vector spaces then* $KV' \to KV \to KV''$ *is a Q-left exact triple of Hopf spaces* mod p.

PROOF. Routine. □

PROPOSITION A.4. *Suppose that*

$$\mathbf{Z}_p \to H' \to H \to H''$$

is an exact sequence of homology Hopf algebras with the property that

$$0 \to QH' \to QH \to QH''$$

is in exact sequence of $\mathbf{Z}_p$*-modules. Then*

$$\mathbf{Z}_p \to H'\langle -1\rangle \to H\langle -1\rangle \to H''\langle -1\rangle$$

is an exact sequence of homology Hopf algebras.

PROOF. An easy consequence of Theorem 3.4. □

PROPOSITION A.5. *Suppose that* $X' \xrightarrow{f'} X \xrightarrow{f''} X''$ *is a Q-left exact triple of pseudo-Hopf spaces* mod p. *Then*

$$\mathbf{Z}_p \to H_*K\langle X'\rangle /\!\!/ \varphi\langle X'\rangle_* \to H_*K\langle X\rangle /\!\!/ \varphi\langle X\rangle_* \to H_*K\langle X''\rangle /\!\!/ \varphi\langle X''\rangle_*$$

is an exact sequence of homology Hopf algebras.

PROOF. Consider the diagram

$$\begin{array}{ccccc} C^pH_*X' & \to & H_*K\langle X'\rangle & \to & H_*K\langle X'\rangle /\!\!/ \varphi\langle X'\rangle_* \\ \downarrow & & \downarrow & & \downarrow \\ C^pH_*X & \to & H_*K\langle X\rangle & \to & H_*K\langle X\rangle /\!\!/ \varphi\langle X\rangle_* \\ \downarrow & & \downarrow & & \downarrow \\ C^pH_*X'' & \to & H_*K\langle X''\rangle & \to & H_*K\langle X''\rangle /\!\!/ \varphi\langle X''\rangle_*. \end{array}$$

By replacing H_*X'' by $\operatorname{im} f''_*$ we may assume that the two left-hand columns are exact. Since the rows are exact by definition it follows from the 3×3 lemma that the right-hand column is exact. □

PROPOSITION A.6. *Suppose that*

$$\mathbf{Z}_p \to H' \to H \to H''$$

is an exact sequence of homology Hopf algebras whose underlying homology coalgebras are injective. Then

$$\mathbf{Z}_p \to \operatorname{Cotor}^{H'}(\mathbf{Z}_p, \mathbf{Z}_p) \to \operatorname{Cotor}^{H}(\mathbf{Z}_p, \mathbf{Z}_p) \to \operatorname{Cotor}^{H''}(\mathbf{Z}_p, \mathbf{Z}_p)$$

is an exact sequence of abelian Hopf algebras.

PROOF. This follows from the structure of $\mathrm{Cotor}^{U(V)}(k, k)$ discussed in §3 of **[9]**. □

THEOREM A.7. *If $X' \to X \to X''$ is a Q-left exact triple of pseudo-Hopf spaces* mod *p then*

$$\mathbf{Z}_p \to E^0H_*X'\langle -1\rangle \to E^0H_*X\langle -1\rangle \to E^0H_*X''\langle -1\rangle$$

is an exact sequence of homology Hopf algebras.

PROOF. If follows from **[1]**, **[13]** and **[9**, Proposition 4.1] that for any $Y \in \mathrm{obj}\, P - \mathscr{H}Sp/p$ that $H_*K\langle Y\rangle /\!/ \varphi\langle Y\rangle_*$ is injective as a homology coalgebra. Therefore from Lemma A.1 and Proposition A.4, Proposition A.5 and Proposition A.6 it follows that

$$\mathbf{Z}_p \to E^2\langle X'\rangle \to E^2\langle X\rangle \to E^2\langle X''\rangle$$

is an exact sequence of homology Hopf algebras. From **[8**, Theorem A.1] it now follows that

$$\mathbf{Z}_p \to E^\infty\langle X'\rangle \to E^\infty\langle X\rangle \to E^\infty\langle X''\rangle$$

is an exact sequence of homology Hopf algebras, yielding the result. □

RECOLLECTIONS. If A is a connected algebra over $\mathbf{Z}_p$ then as in **[9**, Example 8.4] we may give KIA the structure of a Hopf space. If A is commutative then KIA is homotopy commutative. If $f: A \to B$ is a morphism of graded connected algebras, then f induces a morphism of Hopf spaces $KIf: KIA \to KIB$.

We will often use the notation KA for the Hopf space KIA.

LEMMA A.8. *Let A be a connected $\mathbf{Z}_p$-algebra. The natural map $\varphi\langle KA\rangle$: $KA \to K\langle KA\rangle$ is the unique map determined by the morphism of $\mathbf{Z}_p$-algebras $I\alpha: IA \to QH_*KA$, i.e., $\varphi\langle KA\rangle = K\alpha$.*

PROOF. Direct from the definitions. □

LEMMA A.9. *The monomorphism of homology Hopf algebras $\lambda: A \to H_*KA$ induces a monomorphism $Q\lambda: QA \to QH_*KA$.*

PROOF. Direct from the definition of the Hopf space structure **[9**, Example 8.4] on KA. □

LEMMA A.10. *Suppose that $f: A \to B$ is a morphism of connected $\mathbf{Z}_p$-algebras where B has trivial multiplication. Consider the fibre square*

$$\begin{array}{ccc} E & \longrightarrow & LB \\ \downarrow & & \downarrow \\ KA & \xrightarrow{Kf} & KB. \end{array}$$

Let $C \subset A$ be the ideal ker *f together with the identity in degree* 0 *(i.e., C is the kernel of f in the category of supplemented $\mathbf{Z}_p$-algebras). Let $D = B/fA$. Then E is a Hopf space and there is an isomorphism of Hopf spaces $E \simeq KC \times K(s^{-1}D)$.*

PROOF. Routine. □

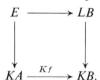

REMARKS. The only use that we shall make of Lemma A.10 is that E is a Hopf space, which a priori we do not know; we only know that E is an H-space. In the terminology of **[15]**, KA is an A_3-space. Is KA an A_n-space for $n > 3$? Probably careful attention to a construction of KA via simplicial sets settles this in the affirmative. Suppose that $f: A' \to A''$ is a morphism of connected $\mathbf{Z}_p$-algebras. Is $Kf: KA' \to KA''$ an A_3-map? Probably careful attention to a construction of KA via simplicial sets settles this in the affirmative. If so then Theorem 6.1 of **[15]** would neatly dispose of the need for Lemma A.10.

NOTATION. If A is a connected $\mathbf{Z}_p$-algebra let $Q^{\perp}A = QH_*KA/QA$.

PROPOSITION A.11. *If A is a connected $\mathbf{Z}_p$-algebra then $K\langle A\rangle\langle -1\rangle \simeq KA^2 \times K(s^{-1}Q^{\perp}A)$ as a Hopf space.*

PROOF. From Lemma A.8 it follows that $\varphi\langle KA\rangle = K\alpha$ and the result now follows from Lemma A.10. □

THEOREM A.12. *If X is a Hopf space* mod p *then $X\langle -1\rangle$ is a Hopf space* mod p *and $\mathscr{F}\langle X\rangle$ is a Hopf fibre square* mod p.

PROOF. Note that $X\langle -1\rangle$ is easily seen to be an H-space and thus the only issue at stake is the associativity of $H_*X\langle -1\rangle$.

Consider the natural morphism of Hopf spaces $\theta: X \to KH_*X$. The argument in Theorem A.7 shows that

$$E^0\theta\langle -1\rangle_*: E^0H_*X\langle -1\rangle \to E^0H_*[(KH_*X)\langle -1\rangle]$$

is a monomorphism of homology Hopf algebras. Therefore

$$\theta\langle -1\rangle_*: H_*X\langle -1\rangle \to H_*[(KH_*X)\langle -1\rangle]$$

is a monomorphism of quasi-Hopf algebras. Since the right-hand side is actually associative by Proposition A.11 it follows that $H_*X\langle -1\rangle$ is a Hopf algebra. □

PROBLEM. If X is a Hopf space, is the natural map $X \to K\langle X\rangle$ an A_3-map?

Suppose that $(X, \mu) \in \text{obj } P - Ab\mathscr{H}Sp/p$. Then H_*KH_*X is abelian and hence the natural morphism $\theta_*: H_*X \to H_*KH_*X$ is normal. Thus $H_*KH_*X /\!\!/ \theta_*$ is defined and is an abelian Hopf algebra over $\mathbf{Z}_p$. There is also a natural morphism of Hopf spaces $\zeta: KH_*X \to K(H_*KH_*X /\!\!/ \theta_*)$ and we readily obtain:

PROPOSITION A.13. *If $(X, \mu) \in \text{obj } P - Ab\mathscr{H}Sp/p$ and X is simply connected then*

$$X \xrightarrow{\theta} KH_*X \xrightarrow{\zeta} K(H_*KH_*X /\!\!/ \theta_*)$$

is a Q-left exact triple of Hopf spaces mod p. □

THEOREM A.14. *If $(X, \mu) \in$* obj $P - Ab\mathscr{H}Sp/p$ *then $X\langle -1\rangle$ admits a natural pseudo-abelian Hopf space* mod *p-structure and $\mathscr{F}\langle -1\rangle(X)$ is a pseudo-Hopf fibre square* mod p.

PROOF. From Proposition A.13 we obtain a Q-left exact triple of pseudo-Hopf spaces mod p

$$X \xrightarrow{\theta} KH_*X \xrightarrow{\zeta} K(H_*KH_*X /\!\!/ \theta_*)$$

and hence by Theorem A.7 we have an exact sequence of homology Hopf algebras

$$\mathbf{Z}_p \to E^0H_*X\langle -1\rangle \to E^0H_*((KH_*X)\langle -1\rangle) \to E^0H_*(K(H_*KH_*X /\!\!/ \theta_*)\langle -1\rangle).$$

By Theorem A.12 $(KH_*X)\langle -1\rangle$ and $(K(H_*KH_*X /\!\!/ \theta_*))\langle -1\rangle$ are Hopf spaces and

$$\zeta\langle -1\rangle_* : H_*[(KH_*X)\langle -1\rangle] \to H_*[(K(H_*KH_*X /\!\!/ \theta_*))\langle -1\rangle]$$

is a morphism of Hopf algebras. Hence

$$H_*[(KH_*X)\langle -1\rangle] \backslash\!\!\backslash \zeta\langle -1\rangle_* = \mathbf{Z}_p \,\square_{H_*[(KH_*X)\langle -1\rangle]}\, H_*[(K(H_*KH_*X /\!\!/ \theta_*))\langle -1\rangle]$$

is again a Hopf algebra under $\mathscr{A}_*$.

Since $\zeta \circ \theta$ is null homotopic it readily follows that the composite morphism of homology coalgebras

$$H_*X\langle -1\rangle \to H_*[(KH_*X)\langle -1\rangle] \to H_*[K(H_*KH_*X /\!\!/ \theta_*))\langle -1\rangle]$$

is trivial there is a natural morphism of homology coalgebras

$$H_*X\langle -1\rangle \to H_*[(KH_*X)\langle -1\rangle] \backslash\!\!\backslash \zeta\langle -1\rangle_*.$$

Since

$$E^0H_*X\langle -1\rangle \to E^0H_*[(KH_*X)\langle -1\rangle] \backslash\!\!\backslash E^0\zeta\langle -1\rangle_*$$

is an isomorphism of coalgebras it readily follows that

$$H_*X\langle -1\rangle \to H_*[(KH_*X)\langle -1\rangle] \backslash\!\!\backslash \zeta\langle -1\rangle_*$$

is an isomorphism of coalgebras under $\mathscr{A}_*$. But the right-hand side is an abelian Hopf algebra under $\mathscr{A}_*$ by Proposition A.11. Transferring this structure to $H_*X\langle -1\rangle$ via the above isomorphism of coalgebras gives $X\langle -1\rangle$ the structure of an object of $P - \mathrm{Ab}\mathscr{H}Sp/p$ as required. The rest is routine. □

PROBLEM. If $X \in \operatorname{obj} P - \mathscr{H}Sp/p$ is $\mathscr{F}\langle -1\rangle(X)$ a pseudo-Hopf fibre square mod p? Theorem A.14 gives an affirmative answer when H_*X is abelian.

There is a closely related and completely algebraic problem. Suppose that $H \in \operatorname{obj} \mathscr{H}_*\mathscr{H}/K$. Let $\rho : H \to \mathbf{U}H$ be the morphism of homology Hopf algebras constructed in [**9**, Proposition 3.6]. Is ρ a *normal* monomorphism?

Theorem 1.1 and Theorem 1.2 follow from essentially trivial modifications of the arguments used to prove Theorem A.12 and Theorem A.14, respectively. Details are left to the reader.

BIBLIOGRAPHY

1. H. Cartan, *Algebres d'Eilenberg-MacLane et homotopie*, Seminar Cartan ENS 1954/1955.

2. S. Eilenberg and J. C. Moore, *Homological algebra and fibrations*, Colloque de Topologie Bruxelles, 1964, pp. 81–90.

3. ———, Homology and fibrations I, Comment. Math. Helv. **40** (1966), 199–236.

4. P. J. Hilton, *Homotopy and duality*, Gordon and Breach, New York, 1966.

5. W. S. Massey and F. P. Peterson, *The mod 2 cohomology structure of certain fibre spaces*, Mem. Amer. Math. Soc. no. 74, 1967.

6. J. Milnor and J. C. Moore, *The structure of Hopf algebras*, Ann. of Math. (2) **81** (1965), 211–264.

7. J. C. Moore, *Some applications of homology theory to homotopy problems*, Ann. of Math. (2) **58** (1953), 325–350.
8. J. C. Moore and L. Smith, *Hopf algebras and multiplicative fibrations*. I, Amer. J. Math. **90** (1968), 1113–1149.
9. ———, *Hopf algebras and multiplicative fibrations*. II, Amer. J. Math. (to appear).
10. ———, (to appear).
11. F. P. Peterson, *Generalized cohomotopy groups*, Amer. J. Math, **78** (1956), 259–281.
12. D. L. Rector, *An unstable Adams spectral sequence*, Topology **5** (1966), 343–346.
13. J. P. Serre, *Cohomologie modulo 2 déspaces d'Eilenberg-MacLane*, Comment. Math. Helv. **27** (1953), 198–232.
14. L. Smith, *Cohomology of stable two stage Postnikov systems*, Illinois J. Math. **11** (1967), 310–329.
15. J. D. Stasheff, *Homotopy associative H-spaces*. I, II, Trans. Amer. Math. Soc., **108** (1963), 275–292, 293–312.
16. N. E. Steenrod and D. B. A. Epstein, *Cohomology operations*, Ann. of Math. Study No. 50, Princeton Univ. Press, Princeton, N.J., 1962.

PRINCETON UNIVERSITY

Stable Homotopy II

Peter Freyd

We shall use the language of our paper *Stable homotopy* [**1**] freely in the proofs. Those propositions marked with *, however, will be stated in more conventional language. A few new definitions will be necessary in the statement of some of the *-theorems; the definitions will, however, be given in conventional language. All spaces are finite complexes.

1. **Cancellation**

* PROPOSITION 1.1. *Given spaces X, Y, Z all in a stable range and a homotopy equivalence $X \vee Z \simeq Y \vee Z$ it is not necessarily the case that $X \simeq Y$ even if Z* is *a sphere.*

PROOF. We shall need the Schanuel Lemma here and throughout the paper. In an arbitrary abelian category, the Schanuel Lemma says that if

$$0 \to X_n \to X_{n-1} \to \cdots \to X_1 \to A \to 0$$

and

$$0 \to X_{-n} \to X_{-(n-1)} \to \cdots \to X_{-1} \to A \to 0$$

are exact, X_j projective for $-n < j < n$ then

$$X_n \oplus X_{-(n-1)} \oplus X_{n-2} \oplus \cdots \oplus X_{-(-1)^n} \simeq X_{-n} \oplus \cdots \oplus X_{(-1)^n}.$$

We shall use the Lemma in $\mathfrak{F}$, always in the case that $X_n \simeq X_{-n}$. Furthermore, either X_n or A will be torsion. Thus

SPECIAL SCHANUEL LEMMA 1.12. *Given exact sequences*

$$0 \to B \to X_n \to X_{n-1} \to \cdots \to X_1 \to A \to 0$$
$$0 \to B \to X_{-n} \to X_{-(n-1)} \to \cdots \to X_{-1} \to A \to 0$$

in $\mathfrak{F}$, $X_j \in \mathcal{S}$ for all j, and either A or B torsion, then

$$X_n \vee X_{-(n-1)} \vee \cdots \vee X_{-(-1)^n} \simeq X_{-n} \vee X_{n-1} \vee \cdots \vee X_{(-1)^n}.$$

The copy of B has been cancelled assuming that it is B that is torsion. If it is A that's torsion then we use co-Schanuel and cancel A. ▮LEMMA▮

Now let $\nu: S^3 \to S^0$ be a map of order 8 [2]. Let $S^3 \xrightarrow{\nu} S^0 \to C_\nu \to S^4$ and $S^3 \xrightarrow{3\nu} S^0 \to C_{3\nu} \to S^4$ be mapping cone sequences. Note that Im $(\nu) \simeq$ Im (3ν) and Ker $(\nu) \simeq$ Ker (3ν). Hence we may apply Schanuel's Lemma to the exact sequences

$$0 \to \mathrm{Im}\,(\nu) \to S^0 \to C_\nu \to S\,\mathrm{Ker}\,(\nu) \to 0$$

$$0 \to \mathrm{Im}\,(3\nu) \to S^0 \to C_{3\nu} \to S\,\mathrm{Ker}\,(3\nu) \to 0$$

to obtain

$$S^0 \vee C_{3\nu} \simeq S^0 \vee C_\nu.$$

(We could similarly obtain $C_\nu \vee S^4 \simeq C_{3\nu} \vee S^4$.) In general, given $\gamma: S^n \to S^0$, and an integer j prime to the order of γ we would obtain $S^0 \vee C_\gamma \simeq S^0 \vee C_{j\gamma}$. We may, of course, suspend the example into a stable range. It is not known to be a counter-example however, until we know that $C_\nu \not\simeq C_{3\nu}$. We need

PROPOSITION 1.2. *Given $\alpha: S^n \to S^0$, $\beta: S^n \to S^0$, if $C_\alpha \simeq C_\beta$ then $\alpha = \pm\beta$.*

PROOF. Let $f: C_\alpha \to C_\beta$ be an isomorphism. We obtain

$$\begin{array}{ccccccc}
S^n & \xrightarrow{\alpha} & S^0 & \longrightarrow & C_\alpha & \longrightarrow & S^{n+1} \\
\downarrow b & & \downarrow a & & \downarrow f & & \downarrow b \\
S^n & \xrightarrow{\beta} & S^0 & \longrightarrow & C_\beta & \longrightarrow & S^{n+1} \\
\downarrow b' & & \downarrow a' & & \downarrow f^{-1} & & \downarrow b' \\
S^n & \xrightarrow[\alpha]{} & S^0 & \longrightarrow & C_\alpha & \longrightarrow & S^{n+1}
\end{array}$$

where a, a', b, b' are integers. Note that

$$S^0 \xrightarrow{aa'-1} S^0 \to C_\alpha = 0,$$

hence $(aa' - 1) \cdot 1_{S^0}$ factors through α, hence $aa' - 1 = 0$. Similarly (dually $bb' = 1$. ▮ Cancellation fails.

* THEOREM 1.3. *Given X, X', Z all in a stable range, if $X \vee Z$ and $X' \vee Z$ are homotopically equivalent, then for the bouquet of spheres B with the same betti numbers as X there exists a homotopy equivalence $X \vee B \simeq X' \vee B$.*

PROOF. Given $X \in \mathscr{S}$, $\pi_*(X)$ is the graded stable homotopy group $\{\mathscr{S}(S^n, X)\}_{-\infty}^{+\infty}$. By $\pi_*/T(X)$ we mean the torsion-free part of $\pi_*(X)$, i.e.

$$\{\mathscr{S}(S^n, X)/\mathrm{Tor}\ \mathfrak{S}(S^n, X)\}_{-\infty}^{+\infty}.$$

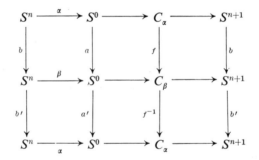

The values of π_*/T are graded free abelian groups, each component of which is of finite rank, almost all of which are trivial. Needing a name for the range of π_*/T we settle for $\sum \mathfrak{Mat}\,(Z)$, where $\mathfrak{Mat}\,(—)$ is intended to denote the category of matrices over a ring (—), and $\sum$ means the weak product over the integers.

We note that we may imbed $\sum \mathfrak{Mat}\,(Z)$ into $\mathcal{S}$ by assigning bouquets of spheres, the betti numbers of which are determined by the ranks of the free groups, and that

$$\textstyle\sum \mathfrak{Mat}\,(Z) \to \mathcal{S} \xrightarrow{\pi_*/T} \sum \mathfrak{Mat}\,(Z) = 1.$$

Furthermore note that

$$\mathcal{S} \xrightarrow{\pi_*/T} \textstyle\sum \mathfrak{Mat}\,(Z) \xrightarrow{-\otimes Q} \sum \mathfrak{Mat}\,(Q) \simeq H_*(—, Q)$$

because $\mathcal{S}(S^n, X) \oplus Q \simeq H_n(X, Q)$ [**2**(4.10)].

We shall define $B:\mathcal{S} \to \mathcal{S}$ to be the functor which assigns bouquets of spheres of the same betti numbers

$$\mathcal{S} \xrightarrow{\pi_*/T} \textstyle\sum \mathfrak{Mat}\,(Z) \to \mathcal{S} = B.$$

$B(X)$ will be written B_X.[1]

For any space X we can find a map $\varphi: B_X \to X$ such that $\pi_*/T(\varphi)$ is an isomorphism.

LEMMA 1.31. *Suppose X and X' are isomorphic in $\mathcal{S}$, $\varphi: B_X \to X$, $\varphi': B_{X'} \to X'$, π_*/T-isomorphisms. Then there is an integer $t > 0$ such that* Ker $(t\varphi) \simeq$ Ker $(t\varphi')$ *and* Cok $(t\varphi) \simeq$ Cok $(t\varphi')$.

PROOF. Let $f: X \to X'$ be an isomorphism. Note that there exists

$$\begin{array}{ccc} B_X & \xrightarrow{B\varphi} & B_X \\ {\scriptstyle g}\downarrow & & \downarrow{\scriptstyle B_f} \\ B_{X'} & \xrightarrow[B\varphi]{} & B_{X'} \end{array}$$

and that g is an isomorphism. Hence the diagram

$$\begin{array}{ccc} B_X & \xrightarrow{\varphi} & X \\ {\scriptstyle g}\downarrow & & \downarrow{\scriptstyle f} \\ B_{X'} & \xrightarrow[\varphi']{} & X' \end{array}$$

commutes when π_*/T is applied. $f\varphi - \varphi' g$ is killed by π_*/T, therefore it is

[1] Philosophers and people at Penn will understand why I think of this as the Betty-Flower Functor.

torsion. Let t be the exponent of the torsion part of $\mathscr{S}(B_X, X')$. Then

$$\begin{array}{ccc} B_X & \xrightarrow{t\varphi} & X \\ {\scriptstyle g}\downarrow & & \downarrow{\scriptstyle f} \\ B_{X'} & \xrightarrow[t\varphi']{} & X' \end{array}$$

commutes and the isomorphisms announced in the statement of the lemma are obtained. ▌LEMMA▌

Now suppose that $X \vee Z \simeq X' \vee Z$. Let $\varphi'': B_z \to Z$ be a π_*/T-isomorphism. By the above lemma we can find $t > 0$ such that

$$\text{Ker}\,(t(\varphi \oplus \varphi'')) \simeq \text{Ker}\,(t(\varphi' \oplus \varphi'')).$$

Noting that

$$\text{Ker}\,(t(\varphi \oplus \varphi'')) = \text{Ker}\,(t\varphi \oplus t\varphi'') = \text{Ker}\,(t\varphi) \oplus \text{Ker}\,(t\varphi'')$$

and similarly for φ' instead of φ we obtain

$$\text{Ker}\,(t\varphi) \oplus \text{Ker}\,(t\varphi'') \simeq \text{Ker}\,(t\varphi') \oplus \text{Ker}\,(t\varphi'').$$

Observe that an π_*/T-isomorphism is a rational homology isomorphism and therefore $t\varphi''$ is a rational homology isomorphism, and Ker $(t\varphi'')$ is torsion. It cancels Ker $(t\varphi) \simeq$ Ker $(t\varphi')$. Similarly Cok $(t\varphi) \simeq$ Cok $(t\varphi')$. We have exact sequences

$$0 \to \text{Ker}\,(t\varphi) \to B_X \to X \to \text{Cok}\,(t\varphi) \to 0$$
$$0 \to \text{Ker}\,(t\varphi') \to B_{X'} \to X' \to \text{Cok}\,(t\varphi') \to 0.$$

The Special Schanuel Lemma 1.12 gives $B_X \vee X' \simeq B_{X'} \vee X$. Finally note that the isomorphism $X \vee Z \simeq X' \vee Z$ implies that X and X' have the same betti numbers, hence $B_X \simeq B_{X'}$. ▌

The semigroup $\langle |\mathscr{S}|, \vee \rangle$ is not a cancellation semigroup by the first proposition. We define the congruence $X \equiv Y$ as usual by: $X \equiv Y$ iff there is Z such that $X \vee Z \simeq Y \vee Z$. Then $\langle |\mathscr{S}|, \vee \rangle / \equiv$ is a cancellation semigroup. By the above theorem $X \equiv Y$ iff $X \vee B_X \simeq Y \vee B_X$.

Let $\mathbf{H}_c^d$ be the family of homotopy types of finite complexes of connectivity c, dimension d. We shall always assume that $d \leq 2c + 1$, i.e. that we are in a meta-stable range. The example in the proof of Proposition 1 says that $\langle \mathbf{H}_4^9, \vee \rangle$ is not a cancellation semigroup.

The Freudenthal Theorem gives an embedding $\langle \mathbf{H}_c^d, \vee \rangle \to \langle |\mathscr{S}|, \vee \rangle$ for all $d \leqq 2c + 1$. The above theorem tells us that when we pass to the associated cancellation semigroups the map remains an embedding, i.e. that if X, $Y \in \mathbf{H}_c^d$ and $X \equiv V$ in $\mathscr{S}$, then there exists $B \in \mathbf{H}_c^d$ such that $X \vee B \simeq Y \vee B$.

2. **Spherical retracts.** We define the *TOTAL n-DIMENSIONAL SPHERICAL RETRACT* of a space X to be the largest bouquet of n-spheres B_n which retracts from X. There is a largest such because the nth betti number of X

is clearly an upper bound on its rank. For each n let B_n be the total n-dimensional retract of X, and let $f_n: B_n \to X, g_n: X \to B_n$ be such that $g_n f_n = 1$. Define $B_T = \bigvee B_n$ and $f: B_T \to X$ by $B_n \to B_T \to X = f_n$, $g: X \to B_T$ by $X \to B_T \to B_n = g_n$. Note that $\pi_*/T(gf) = 1$, hence that $gf = 1 - t$ where t is a torsion map. Clearly t is nilpotent. Therefore gf is an isomorphism and B_T retracts from X. Thus there is a largest bouquet of spheres that retracts from X and we may define it to be the *TOTAL SPHERICAL RETRACT* of X.

* PROPOSITION 2.1. *Total spherical retract is a $\equiv$ invariant. That is, if X, Y are in a stable range and if $X \vee Z \simeq Y \vee Z$ and a bouquet B retracts from X then it retracts from Y.*

PROOF. First a lemma, stated in $\mathscr{S}$.

If S^0 retracts from X and if $X \equiv Y$, then S^0 retracts from Y.

Let $B_X = B_0 \vee B_+ \vee B_-$ where B_0 is a bouquet of 0-dimensional spheres, B_+ positive dimensional, B_- negative dimensional. Let $X = X' \vee S^0$. We know that $S^0 \vee B_0$ retracts from $X \vee B_X$, hence from $Y \vee B_X$ by the last Theorem (1.3). It is apparent that the retraction $S^0 \vee B_0 \to Y \vee B_0 \vee B_+ \vee B_- \to S^0 \vee B_0 = 1$ is not upset if we remove B_+ since $(S^0 \vee B_0, B_+) = 0$, and then if we remove B_- since $(B_-, S^0 \vee B_0) = 0$. Hence we obtain $S^0 \vee B_0 \to Y \vee B_0 \to S^0 \vee B_0 = 1$. Let

$$f = S^0 \vee B_0 \to Y \to S^0 \vee B_0$$
$$g = S^0 \vee B_0 \to B_0 \to S^0 \vee B_0.$$

Then

$$f + g = 1.$$

The full subcategory of bouquets of 0-dimensional spheres is equivalent to the category of integral matrices $\mathfrak{Mat}\ (Z)$. With that equivalence in mind we observe that g is a $(b_0 + 1) \times (b_0 + 1)$ matrix, where b_0 is the 0th betti number of X and Y. Moreover g is of rank b_0 because it factors through 1_{B_0}, a $b_0 \times b_0$ matrix. We may diagonalize g: there exist automorphisms θ_1, θ_2 such that $\theta_1 g \theta_2$ is diagonal. Necessarily, one of its rows must be 0. We'll assume it's the top row. Therefore, the top row of $\theta_1 f \theta_2$ is unimodular, and we obtain

$$S^0 \to S^0 \vee B_0 \xrightarrow{\theta_1 f \theta_2} S^0 \vee B_0 \to S^0 = 1.$$

Recalling that f factors through Y we have that S^0 retracts from Y. Thus the lemma.

Now let $B_{T,0}$, $B'_{T,0}$ be the total 0-dimensional retracts of X and Y. And suppose $B_{T,0}$ is bigger than $B'_{T,0}$. In particular, then, $B_{T,0} = B'_{T,0} \vee S^0 \vee B''$ where B'' could be trivial. Let $X = X' \vee B_{T,0}$, $Y = Y'' \vee B'_{T,0}$. Note that $X' \vee S^0 \vee B'' \equiv Y'$. By the lemma, S^0 would retract from Y', a contradiction. Hence the total 0-dimensional retracts of X and Y are equal, similarly for all dimensions, hence their total spherical retracts are equal. ▮

* COROLLARY 2.2. *Given X and Y in a stable range. If $X \vee B \equiv Y$ where B is a bouquet of spheres whose betti numbers dominate those of X, then $X \vee B \simeq Y$.*

PROOF. By the above proposition $Y \simeq Y' \vee B$, and $X \equiv Y'$. By Theorem 1.3, $X \vee B_X \simeq Y' \vee B_X$. By the condition on the betti numbers, we obtain $X \vee B \simeq Y' \vee B$. ∎

PROPOSITION 2.3. *Given* $\gamma: S^n \to S^0$, $\gamma \neq 0$, *then the cone*, C_γ *is indecomposable in* $\langle |\mathscr{S}|, \vee \rangle / \equiv$.

PROOF. Suppose not. $C_\gamma \equiv X_1 \vee X_2$ where neither X_1, X_2 is trivial. By the stable Whitehead theorem, neither X_1 nor X_2 has trivial homology. But the homology of C_γ is that of $S^0 \vee S^{n+1}$ and hence X_1 and X_2 are homology spheres, therefore spheres. By the last Proposition 2.2 $C_\gamma \simeq S^0 \vee S^{n+1}$, i.e. $C_\gamma \simeq C_0$. By Proposition 1.2, $\gamma = \pm 0$. ∎

* PROPOSITION 2.4. $\langle \mathbf{H}_4^9, \vee \rangle / \equiv$ *is not a unique factorization semigroup, a fortiori, neither is* $\langle |\mathscr{S}|, \vee \rangle / \equiv$.

PROOF. Let $\alpha: S^3 \to S^0$ be a map of order 3, and $S^3 \xrightarrow{\alpha} S^0 \to C_\alpha \to S^4$, $S^3 \xrightarrow{\alpha+\nu} S^0 \to C_{\alpha+\nu} \to S^4$ mapping cone sequences. Note that the torsion object $\operatorname{Im}(\alpha + \nu)$ is isomorphic to $\operatorname{Im}(\alpha) \oplus \operatorname{Im}(\nu)$, because α and ν are of coprime order. We have exact sequences

$$0 \to \operatorname{Im}(\alpha + \nu) \to S^0 \to C_{\alpha+\nu} \to S^4 \to S \operatorname{Im}(\alpha + \nu) \to 0$$

$$0 \to \operatorname{Im}(\alpha) \oplus \operatorname{Im}(\nu) \to S^0 \vee S^0 \to C_\alpha \vee C_\nu \to S^4 \vee S^4 \to S \operatorname{Im}(\alpha) \oplus S \operatorname{Im}(\nu) \to 0.$$

The ends are isomorphic and torsion. Hence special Schanuel 1.12 yields

$$S^0 \vee C_\alpha \vee C_\nu \vee S^4 \simeq S^0 \vee S^0 \vee C_{\alpha+\nu} \vee S^4 \vee S^4.$$

We may cancel to obtain

$$C_\alpha \vee C_\nu \equiv C_{\alpha+\nu} \vee S^0 \vee S^4.$$

We may use Proposition 2.2 to obtain $C_\alpha \vee C_\nu \simeq C_{\alpha+\nu} \vee S^0 \vee S^4$. But even without the $\simeq$, just with the $\equiv$, we know that $\langle |\mathscr{S}|, \vee \rangle / \equiv$ fails to obey unique factorization, since C_α, C_ν, $C_{\alpha+\nu}$, S^0 and S^4 are all indecomposible therein. By suspending five times we may move the example into the pre-stable range.

Note that not even the number of indecomposable factors is invariant. Also note that total spherical retracts do not add up on wedges. The example can be duplicated for any $\gamma, \gamma': S^n \to S^0$ of coprime order. We wish to explicate the dependence of the failure of unique factorization upon the mixing of primes.

3. **Primary spaces.** Given $X \in \mathscr{S}$ consider the integers n such that there is a bouquet of spheres B and maps such that $X \to B \to X = n \cdot 1_X$. The set of such integers is clearly an ideal. Define $T(X)$ to be the nonnegative generator of that ideal. $T(X) = 1$ iff X is a bouquet of spheres (retracts of bouquets are bouquets [**2**(6.4)]). It may be observed that $T(X)$ is the exponent of X in the quotient category $\mathscr{S}/B$ obtained by killing maps which factor through bouquets. (The exponent of an object is the order of its identity map.) $T(X)$ is never zero, since we can choose a rational homology isomorphism $\varphi: B \to X$ and note that the exponent

of Cok (φ) is in the ideal ($T(X)$). Note that $T(X \vee Y) = \text{l.c.m.}\,(T(X), T(Y))$. Of more interest is the set $\mathbf{P}_X$ of prime divisors of $T(X)$:

$\mathbf{P}_X = \varnothing$ *iff* X *is a bouquet of spheres.*

$\mathbf{P}_{X \vee Y} = \mathbf{P}_X \cup \mathbf{P}_Y$.

If X *is torsion,* $\mathbf{P}_X$ *is the set of prime divisors of its exponent.*

For the last statement notice first that if $n \cdot 1_X = 0$ then clearly $n \in (T(X))$, hence $\mathbf{P}_X$ is contained in the set of prime divisors of n. For the reverse containment, let p be an arbitrary prime divisor of the exponent of X, and X_p the *p-primary* component of X. Suppose $T(X)$ is prime to p. Then $T(X_p) \mid T(X)$ and $T(X_p)$ is prime to p. But we have just observed that the prime divisors of $T(X_p)$ must be divisors of the exponent of X_p which is a power of p. Hence $X_p = 0$ and $X = 0$.

PROPOSITION 3.1. *Let* $\varphi : B \to X$ *be a* π_*/T*-isomorphism.* $\mathbf{P}_X$ *is equal to the set of prime divisors of the exponent of* Cok (φ).

PROOF. One containment is easy. If $n \cdot 1_{\text{Cok}(\varphi)} = 0$, then there exist $X \to B \xrightarrow{\varphi} X = n$. Hence $n \in T(X)$ and $\mathbf{P}_X$ is contained in the prime divisors of n.

For the other direction, let $X \xrightarrow{g} B' \xrightarrow{f} X = T(X) \cdot 1_X$. Note that there exists

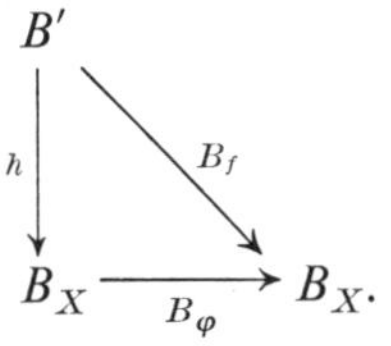

Hence,

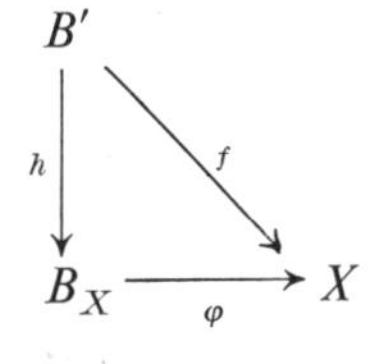

commutes after the applications of π_*/T. Therefore $f - \varphi h$ is torsion. Define $g' = X \xrightarrow{g} B' \xrightarrow{h} B_X$. Then $\varphi g' = T(X) - t'$ where t' is a torsion map. We may decompose t' as a sum of primary torsion elements $t' = \sum_p t_p$. Define $t'' = \sum_{p \in \mathbf{P}} t_p$. For some exponent $a \geq 0$, $(T(X))^a \cdot t'' = 0$. Hence $(T(X))^a \cdot t'$ is of order prime to $T(X)$ therefore divisible by $T(X)$. Let t be such that $(T(X))^{a+1} t = (T(X))^a \cdot t'$. Define $g'' = (T(X))^a \cdot g'$. Then $\varphi g'' - (T(X))^{a+1}(1 - t)$. Now, t is an element of a finite multiplicative semigroup, Tor (End (X)), and some power t^b is idempotent. Define $g''' = g'' \cdot (1 + t + t^2 + \cdots + t^{b-1})$. Then $\varphi g''' = (T(X))^{a+1}(1 - t^b)$.

Let $X' = \text{Im}\,(t^b)$. X' is a retract of X since t^b is idempotent. The exponent of X' is equal to the order of t^b, and hence not divisible by any prime in $\mathbf{P}_X$. But $\mathbf{P}_{X'} \subset \mathbf{P}_X$. It follows that $\mathbf{P}_{X'} = \varnothing$, and X' is a bouquet of spheres. But it is torsion. Hence $X' = 0$. Hence $t^b = 0$. Hence $\varphi g''' = (T(X))^{a+1}$. Now

$$X \xrightarrow{g'''} B_X \xrightarrow{\varphi} X \to \text{Cok}\,(\varphi) = 0.$$

Therefore $(T(X))^{a+1} \cdot 1_{\mathrm{Cok}(\varphi)} = 0$, and the prime divisors of the exponent of Cok (φ) divide $(T(X))^{a+1}$, hence are contained in $\mathbf{P}_X$. ∎

We will say that X is *p-primary* if $\mathbf{P}_X \subset \{p\}$, equivalently if there exist maps $X \to B \to X = p^n$ for some bouquet B of spheres, or equivalently for B the bouquet of spheres with the same betti numbers as X (equivalently, if $T(X)$ is a power of p).

The only spaces both p-primary and q-primary, $p \neq q$, are bouquets of spheres.

The family of p-primary spaces is closed under wedge composition and decomposition.

* THEOREM 3.2. *Let X be in a stable range. There is a set of primes $\mathbf{P}$, spaces $\{X_p\}_{p\in\mathbf{P}}$ in the same stable range, each X_p p-primary, and each with the same betti numbers as X such that*

$$X \vee \bigvee_{|\mathbf{P}|-1} B \simeq \bigvee_{\mathbf{P}} X_p$$

where B is the bouquet of spheres with the same betti numbers as X.

PROOF. $\mathbf{P} = \mathbf{P}_X$. Let $\varphi: B \to X$ be a π_*/T-isomorphism, $B \to X \to Y \to SB$ a mapping cone sequence. Because φ is a rational homology isomorphism, Y is torsion, a fortiori, Cok (φ) and S Ker (φ) are torsion. We split each into its primary components, and obtain for each prime p,

$$0 \to [\mathrm{Cok}\,(\varphi)]_p \to Y_p \to [\mathrm{Ker}\,(\varphi)]_p \to 0.$$

For $p \notin \mathbf{P}_X$ we know by the last proposition that $[\mathrm{Cok}\,(\varphi)]_p = 0$ hence that $Y_p \simeq [S\,\mathrm{Ker}\,(\varphi)]_p$, but $[S\,\mathrm{Ker}\,(\varphi)]_p$ being injective would retract from SB. But it is torsion. Hence trivial. Therefore $Y = \bigvee_{\mathbf{P}} Y_p$.

For each p, let $S^{-1}Y_p \to B \to X_p \to Y_p$ be a mapping cone sequence. We obtain, for each $p \in \mathbf{P}$ an exact sequence

$$0 \to [\mathrm{Ker}\,(\varphi)]_p \to B \to X_p \to [\mathrm{Cok}\,(\varphi)]_p \to 0,$$

where X_p is p-primary and has betti numbers equal to those of B. If we take the direct sum of these sequences we obtain

$$0 \to \bigoplus_{p\varepsilon\mathbf{P}} [\mathrm{Ker}\,(\varphi)]_p \to \bigvee_{|\mathbf{P}|} B \to \bigvee_{p\varepsilon\mathbf{P}} X_p \to \bigoplus_{p\varepsilon\mathbf{P}} [\mathrm{Cok}\,(\varphi)]_p \to 0$$

the ends of which are torsion and isomorphic to those of

$$0 \to \mathrm{Ker}\,(\varphi) \to B \to X \to \mathrm{Cok}\,(\varphi) \to 0.$$

The Special Schanuel Lemma yields

$$X \vee \bigvee_{|\mathbf{P}|} B \simeq B \vee \bigvee_{p\varepsilon\mathbf{P}} X_p.$$

The cancellation of the extra B follows from Proposition 2.2 (noting that if $|\mathbf{P}_X| \leq 1$ the theorem is immediate). ∎

If we make $\langle|\mathscr{S}|, \vee\rangle$ into a group, $\mathbf{G}$, the Grothendieck group, K_0, of $\mathscr{S}$, we have

* COROLLARY 3.3. $\mathbf{G}$ *is generated by the primary spaces.* ∎

We shall show far more. In particular that the X_p's of the theorem are unique up to $\equiv$. Note, incidentally, that the image of $\langle|\mathscr{S}|, \vee\rangle \to \mathbf{G}$ is the cancellation semigroup $\langle|\mathfrak{S}|, \vee\rangle/\equiv$. Also, that if we define $\mathbf{G}^d_c$ to be $\langle \mathbf{H}^d_c, \vee\rangle$ made into a group, then for $d \leq 2c + 1$, $\mathbf{G}^d_c$ is embedded in $\mathbf{G}$, and $\mathbf{G}^d_c$ is generated by the primary spaces therein. We observe that the one-to-oneness of $\mathbf{G}^d_c \to \mathbf{G}$ depends on Theorem 1.3.

4. **The localized categories.** We need to concentrate on each prime, individually. When studying rings we do so by localizing. In additive categories we do the same. Given an $+$'ive category $\mathfrak{A}$ and a prime integer p, we define $\mathfrak{A}_p$ to be the category with the same objects as $\mathfrak{A}$ and with maps represented by pairs $\langle f, m\rangle, f \in \mathfrak{A}$, m an integer prime to p, subject to the equivalence $\langle f, m\rangle \equiv \langle f', m'\rangle$ if for some m'' prime to p, $m'm''f = mm''f'$. We denote the equivalence class of $\langle f, m\rangle$ by f/m. The rules for addition and composition were learned in grade school. Note that if $\mathfrak{A}$ has one object, in other words $\mathfrak{A}$ is a ring, then $\mathfrak{A}_p$ is the localization, as defined for rings, at the prime p. In general, $\mathfrak{A}_p(A, B) = \mathfrak{A}(A, B) \otimes Z_{(p)}$ where $Z_{(p)}$ is the ring of p-adic rationals, $\{a/m \in Q \mid (m, p) = 1\}$. $\mathfrak{A} \to \mathfrak{A}_p$ kills precisely those maps which are torsion of order prime to p. Moreover, if $\mathfrak{Tor}_p(\mathfrak{A})$ is the full subcategory of p-torsion objects in $\mathfrak{A}$, then $\mathfrak{Tor}_p(\mathfrak{A}) \to \mathfrak{A} \to \mathfrak{A}_p$ is a full embedding and it is an equivalence from $\mathfrak{Tor}_p(\mathfrak{A})$ to $(\mathfrak{Tor}(\mathfrak{A}_p))$.

In $\mathfrak{A}_p$ every map of the form $m \cdot 1_A$, $(m, p) = 1$, is an automorphism. $\mathfrak{A}_p$ is universal in this respect. Given a category $\mathfrak{B}$ with this property and functor $\mathfrak{A} \to \mathfrak{B}$ there is a unique factorization through $\mathfrak{A} \to \mathfrak{A}_p$. Note that $H_*(—, Q)$, $H_*(—, Z_{(p)})$ and $H_*(—, Z/pZ)$ each are definable on $\mathscr{S}_p$.

In general, if $\mathfrak{A}$ is abelian then so is $\mathfrak{A}_p$, and $\mathfrak{A} \to \mathfrak{A}_p$ is exact. Moreover, if $\mathfrak{B}$ is abelian and $\mathfrak{A} \to \mathfrak{A}_p \to \mathfrak{B}$ is exact, then so is $\mathfrak{A}_p \to \mathfrak{B}$. It follows that $\mathfrak{A} \to \mathfrak{A}_p$ preserves projectives and injectives. We may consider the commutative

$$\begin{array}{ccc} \mathscr{S} & \longleftrightarrow & \mathfrak{F} \\ \downarrow & & \downarrow \\ \mathscr{S}_p & \longleftrightarrow & \mathfrak{F}_p \end{array}$$

It is clear that $\mathscr{S}_p \to \mathfrak{F}$ is a full embedding, and from the last paragraph, that the $\mathscr{S}_p$-objects form a resolving set of projectives in $\mathfrak{F}_p$ and a coresolving set of injectives. $\mathfrak{F}_p$ is a frobenius category. We do not know, a priori, that idempotents split in $\mathscr{S}_p$, i.e. that all $\mathfrak{F}_p$ projectives are isomorphic to $\mathscr{S}_p$ objects. Wait.

THEOREM 4.1. *Given X, Y in a stable range, $X \equiv Y$ in $\mathscr{S}$ iff $X \simeq Y$ in $\mathscr{S}_p$, each p.*

PROOF. If $X \equiv Y$ then $X \vee B \simeq Y \vee B$ for B a bouquet of spheres by Theorem 1.3. Hence in each $\mathscr{S}_p$, $X \vee B \simeq Y \vee B$. But note that the ring of endomorphisms of a sphere in $\mathscr{S}_p$ is a local ring, namely $Z_{(p)}$. Objects with local rings of endomorphisms may be cancelled in any $\oplus$'ive category [**1**(6.11)]. Hence $X \simeq Y$ in each $\mathscr{S}_p$.

Because rational homology factors through $\mathcal{S}_p$, we know from $X \simeq Y$ (in $\mathcal{S}_p$) that the betti numbers of X and Y are equal. Let $\varphi: B \to X$ and $\varphi': B \to Y$ be π_*/T-isomorphisms in $\mathcal{S}$, and let t be the exponent of the torsion part of $\mathcal{S}(B, X \vee Y)$. For given p,l et $f/n: X \to Y$ be an $\mathcal{S}_p$-isomorphism, $f \in \mathcal{S}(X, Y)$. We may assume that $n = 1$. (Multiply by n if it is not.) Let $f'/m: Y \to X$ be its inverse, $f' \in \mathcal{S}(Y, X)$. We may assume that $f'f = m \cdot 1_X$ in $\mathcal{S}$. Now let

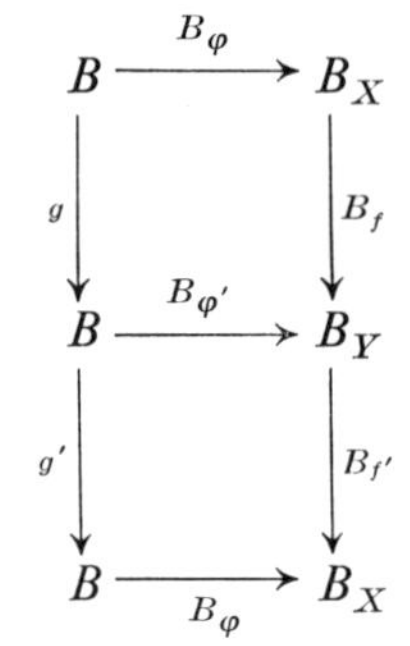

commute, $g'g = m \cdot 1_B$ g is an $\mathcal{S}_p$-isomorphism.

Note that the square

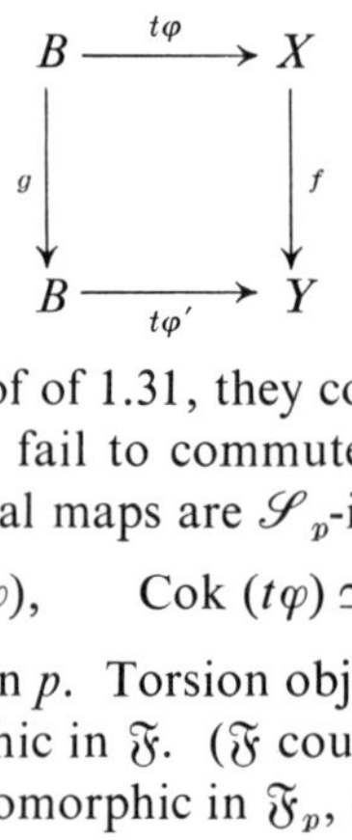

commutes in $\mathcal{S}$. (As in the proof of 1.31, they commute even without the t's after application of π_*/T, hence they fail to commute only because of torsion. That failure is killed by t.) The vertical maps are $\mathcal{S}_p$-isomorphisms. Thus

$$\text{Ker}\,(t\varphi) \simeq \text{Ker}\,(t\varphi), \qquad \text{Cok}\,(t\varphi) \simeq \text{Cok}\,(t\varphi') \text{ in } \mathcal{S}_p.$$

The choice of t did not depend on p. Torsion objects in $\mathfrak{F}$ are isomorphic in $\mathfrak{F}_p$ iff their p-components are isomorphic in $\mathfrak{F}$. ($\mathfrak{F}$ could be any $\oplus$'ive category.) Thus if two torsion objects in $\mathfrak{F}$ are isomorphic in $\mathfrak{F}_p$, each p, they are isomorphic in $\mathfrak{F}$ (still good for any $\oplus$'ive category). Schanuel again. We have exact sequences

$$0 \to \text{Ker}\,(t\varphi) \to B \to X \to \text{Cok}\,(t\varphi) \to 0$$
$$0 \to \text{Ker}\,(t\varphi') \to B \to Y \to \text{Cok}\,(t\varphi') \to 0$$

in $\mathfrak{F}$, with isomorphic torsion ends. (1.12) $B \vee X \simeq B \vee Y$. ∎

COROLLARY 4.2. *Given p-primary spaces X and Y in a stable range, $X \equiv Y$ iff $X \simeq Y$ in $\mathcal{S}_p$.*

PROOF. If $X \simeq Y$ in $\mathcal{S}_p$, they have the same betti numbers. Let $\varphi: B \to X$, $\varphi': B \to Y$ be π_*/T-isomorphisms. Because Ker (φ), Ker (φ'), Cok (φ), Cok (φ') are all p-torsion, 3.1 they are all 0 in $\mathcal{S}_q$ each $q \neq p$. Hence $X \simeq Y$ in $\mathcal{S}_q$ each $q \neq p$. The theorem above now suffices. ∎

We shall find repeated need for the observation

PROPOSITION 4.3. *X is p-primary iff it is isomorphic to a bouquet of spheres in $\mathscr{S}_q$ all $q \neq p$.*

PROOF. $\Rightarrow$ As in the proof of the above corollary. $\Leftarrow$ If $X \simeq B$ in $\mathscr{S}_q$ there exist maps in $\mathscr{S}$ $X \to B \to X = m \cdot 1_X$ where $q \neq m$. Noting that $m \in (T(X))$ we know that $q \nmid T(X)$. ∎

5. **Uniqueness theorems.** The complement of 3.2:

* THEOREM 5.1. *Let $\mathbf{P}$ be a finite set of primes, $\{X_p\}_{p\in\mathbf{P}}$, $\{X'_p\}_{p\in\mathbf{P}}$ collections such that X_p and X'_p are p-primary with the same betti numbers, all in a stable range. Then $\bigvee_{\mathbf{P}} X_p \equiv \bigvee_{\mathbf{P}} X'_p$ implies $X_p \equiv X'_p$ each p.*

PROOF. From $\bigvee_{\mathbf{P}} X_p \equiv \bigvee_{\mathbf{P}} X'_p$ we have $\bigvee_{\mathbf{P}} X_p \simeq \bigvee_{\mathbf{P}} X'_p$ in $\mathscr{S}_q$. But note that X_p and X'_p are isomorphic in $\mathscr{S}_q$, $q \neq p$, to bouquets of spheres according to the above Proposition 4.3, and the betti numbers force them to be the same bouquets. Hence we may cancel them, just as in the first half (the easy half) to Theorem 4.1, leaving us with $X_q \simeq X'_q$ in $\mathscr{S}_q$. Proposition 4.2 finishes the proof. ∎

This last theorem does not translate directly into a nice theorem about $\mathbf{G}$. We need a bit more.

PROPOSITION 5.2. *Given X there is a p-primary space X' such that $X \simeq X'$ in $\mathscr{S}_p$. We can assume, therefore that $\mathscr{S}_p$-objects are p-primary in $\mathscr{S}$.*

PROOF. Decompose X as in 3.2.

$$X \vee \bigvee_{|\mathbf{P}|-1} B \simeq X_p \vee \bigvee_{q\in\mathbf{P}-\{p\}} X_q.$$

In $\mathscr{S}_p$, $X_q \simeq B$ each $q \neq p$. We cancel the spheres to obtain $X \simeq X_p$ in $\mathscr{S}_p$. ∎

PROPOSITION 5.3. *If X is a p-primary space and if a bouquet B retracts from X in $\mathscr{S}_p$, then it does so in $\mathscr{S}$.*

PROOF. Suppose $X \simeq X' \vee B$ in $\mathscr{S}_p$. We may assume by the last proposition that X' is p-primary. Hence by 4.2 $X \equiv X' \vee B$. Theorem 2.1 said that spherical retracts are $\equiv$ invariants. ∎

* PROPOSITION 5.4. *The family of p-primary spaces in a given stable range, without spherical retracts is closed under wedges.*

PROOF. Suppose S retracts from $X_1 \vee X_2$ where X_1 and X_2 are p-primary. We recall that in any $\oplus$'ive category, an object with a local ring of endomorphisms not only cancels but is such that if it retracts from $X_1 \oplus X_2$ it retracts from one of them. Hence in $\mathscr{S}_p$, S retracts from either X_1 or X_2 and by the last proposition, it does so in $\mathscr{S}$. ∎

* THEOREM 5.5. *Let $\mathbf{G}_s$ be the subgroup of $\mathbf{G}$ freely generated by spheres. For each p, let $\mathbf{G}_{\$p}$ be the subgroup of $\mathbf{G}$ generated by p-primary spaces without spherical retracts. Then $\mathbf{G}$ is internally isomorphic to $\mathbf{G}_s \oplus \sum \mathbf{G}_{\$p}$.*

PROOF. We have shown that $\mathbf{G}_s \cup \bigcup \mathbf{G}_{sp}$ generates $\mathbf{G}$ in 3.2. To show that $\mathbf{G}_s \oplus \sum \mathbf{G}_{\natural p} \to \mathbf{G}$ is one-to-one we let $B \vee \bigvee_p X_p \equiv B' \vee \bigvee_p X'_p$ where B, B' are bouquets of spheres, X_p, X'_p are p-primary without spherical retracts, each p. In $\mathscr{S}_p$ we have $X_p \vee B \vee \bigvee_{q \neq p} X_q \simeq X'_p \vee B' \vee \bigvee_{q \neq p} X'_q$ by 4.1, X_q and X'_p are isomorphic to bouquets of spheres in $\mathscr{S}_p$ each $q \neq p$ by 4.3. We may cancel as many spheres as we can. If any spheres are left we would obtain either that X_p or X'_p has a spherical retract in $\mathscr{S}_p$, hence in $\mathscr{S}$, (5.3), a contradiction. Thus $X_p \simeq X'_p$ in $\mathscr{S}_p$ and by 4.2, $X_p \equiv X'_p$. Finally $B = B'$ just by a betti number argument. ▮[2]

6. **The Grothendieck ring.** $\mathbf{G}$ has a ring structure, smash product as multiplication, S^0 the unit. $B:\mathbf{G} \to \mathbf{G}_s$ is a ring homomorphism, where $\mathbf{G}_s$ can be considered to be the ring of Poincaré polynomials. The augmentation ideal will be denoted $\mathbf{G}^*$. It consists of elements of the form $[X] - [Y]$ where X and Y have the same betti numbers. We define $\mathbf{G}^*_p$ to be the subgroup of $\mathbf{G}^*$ of elements of the form $[X] - [Y]$ where X and Y are p-primary. Note that $\mathbf{G}^*$ is generated by $\bigcup \mathbf{G}^*_p$ by Theorem 3.2. Indeed that theorem can be restated nicely as

$$[X] - [B_X] = \sum([X_p] - [B_X])$$

where $[X_p] - [B_X] \in \mathbf{G}^*_p$ each p. It follows easily from 5.1, that

* THEOREM 6.1. *$\mathbf{G}^*$ is internally isomorphic to $\sum_p \mathbf{G}_p$.* ▮

The theorem is a statement about an abelian group. But it is also a statement about ideals, since

* THEOREM 6.2. *$\mathbf{G}^*_p$ is an ideal.*

PROOF. Let $[X] - [Y] \in \mathbf{G}^*_p$. X and Y have the same betti numbers by definition of $\mathbf{G}^*_p \subset \mathbf{G}^*$. Let $\varphi: B \to X$, $\varphi': B \to Y$ be π_*/T-isomorphisms, and let $\psi: X \to B$ be such that $\varphi\psi = p^n \cdot 1_X$ (3.1). ψ and φ' are isomorphisms in $\mathscr{S}_q$ each $q \neq p$, hence $f: X \to Y, f = \varphi'\psi$ is an isomorphism in $\mathscr{S}_q$.

---

[2] We may prove here the freeness of $\mathbf{G}_{\natural p}$, (hence of $\mathbf{G}$). The category $\mathscr{S}/B$ has the property that every map has a finite ring of endomorphisms. (Recall that $\mathscr{S}/B$ is obtained from $\mathscr{S}$ by killing maps which factor through spheres.) Let $\mathfrak{A}$ be the category obtained from $\mathscr{S}/B$ by splitting idempotents. Every indecomposable object in $\mathfrak{A}$ has a local ring of endomorphisms, hence $\langle |\mathfrak{A}|, \oplus \rangle$ is free. The Grothendieck group, $k_0(\mathscr{S}/B)$ is embedded in $k_0(\mathfrak{A})$ (because $\mathscr{S}/B$ is a full subcategory of $\mathfrak{A}$), hence $k_0(\mathscr{S}/B)$ is free. Now we need only show that $\mathbf{G}_{sp}$ is embedded in $K_0(\mathscr{S}/B)$ For that is suffices to know

PROPOSITION 5.6. *If X and X' are p-primary, without spherical retracts with the same betti numbers, and if they are isomorphic in $\mathscr{S}/B$, then $X \equiv X'$.*

PROOF. If X and X' are isomorphic in $\mathscr{S}/B$ three exists a retraction $X \to X' \vee B \to X = 1$, B a bouquet of spheres. Let C be such that $X \vee C \simeq X' \vee B$. In $\mathscr{S}_p$ we can easily show that C is a bouquet of spheres. The equality of betti numbers yields $C \simeq B$ in $\mathscr{S}_p$. But note that C is p-primary. Hence C is a bouquet of spheres 4.3. Thus $X \equiv X'$. ▮

We shall want more than the freeness of $\mathbf{G}$. We shall want, for example, a basis. Note that at this point we do not know if indecomposable p-primary spaces remain indecomposable in $\mathscr{S}/B$.

For any Z consider $Z \wedge ([X] - [Y]) = [Z \wedge X] - [Z \wedge Y]$. The map $1_Z \wedge f$ remains an isomorphism in $\mathscr{S}_q$, each $q \neq p$. If we decompose

$$[Z \wedge X] - [B] = \sum [X_n] - [B]$$
$$[Z \wedge Y] - [B] = \sum [Y_n] - [B]$$

where $B = B_{Z\wedge X}$, we note that the $\mathscr{S}_q$-isomorphisms $Z \wedge X \simeq Z \wedge Y$ yield $\mathscr{S}_q$-isomorphisms $\bigvee X_n \simeq \bigvee Y_n$ from which we may cancel to obtain $X_q \equiv Y_q$ all $q \neq p$. Because $[Z \wedge X] - [Z - Y] = \sum([X_n] - [Y_n])$ we have $[Z \wedge X] - [Z \wedge Y] \in \mathbf{G}_p^*$. ▮

* Corollary 6.3. *If X and Y are primary for different primes, in a stable range, then $(X \wedge Y) \vee (B_{X\wedge Y}) \simeq (B_X \wedge Y) \vee (B_Y \wedge X)$, where $B_{(-)}$ is the bouquet of spheres with the betti numbers of (—). (The right-hand side of $\simeq$ is a wedge of suspensions of copies of X and Y.)*

Proof. $([X] - [B_X]) \wedge ([Y] - [B_Y]) \equiv 0$ by the last theorem. Hence $(X \wedge Y) \vee (B_X \wedge B_Y) \equiv (B_X \wedge Y) \vee (B_Y \wedge X)$. The $\equiv$ becomes a $\simeq$ because of 2.2. ▮

Note that we do not need X and Y to be primary, as long as $\mathbf{P}_X \cap \mathbf{P}_Y = \varnothing$. ("Coprime spaces"?)

Corollary 6.4. *Let $\alpha: S^n \to S^0$, $\beta: S^n \to S^0$ be of coprime orders, C_α, C_β their mapping cones. Then $C_\alpha \wedge C_\beta \equiv C_{\alpha+\beta} \vee S^{n+1}C_{\alpha+\beta}$.*

Proof. By the last corollary and the subsequent remark we have

$$(C_\alpha \wedge C_\beta) \vee (S^0 \vee S^{n+1} \vee S^{n+1} \vee S^{2n+2}) \simeq C_\alpha \vee S^{n+1}C_\alpha \vee C_\beta \vee S^{n+1}C_\beta.$$

By the proof of 2.4, the right-hand side is isomorphic to

$$C_{\alpha+\beta} \vee S^{n+1}C_{\alpha+\beta} \vee S^0 \vee S^{n+1} \vee S^{n+1} \vee S^{2n+2}. \quad ▮$$

7. **The basic theorem of local stable homotopy.** Before we prove it, we wish to observe a few consequences of the theorem we choose to call the Basic Theorem of Local Stable Homotopy:

If X is an indecomposable p-primary space in $\mathscr{S}$ then $\operatorname{End}_{\mathscr{S}_p}(X)$ *is a local ring.*

First, it follows that every indecomposable in $\mathscr{S}_p$ has a local ring of endomorphism, because if X is indecomposable in $\mathscr{S}_p$ and p-primary, it is, a fortiori, indecomposable in $\mathscr{S}$. Hence the Krull-Schmidt theorem holds in $\mathscr{S}_p$, as in any category in which every indecomposable object has a local ring of endomorphisms. That is, $\langle |\mathscr{S}_p|, \vee\rangle$ is free. Let $\langle |\mathscr{S}|, \vee\rangle_p$ be the sub-semi-group of $\langle |\mathscr{S}|, \vee\rangle$ consisting of p-primary spaces. The image of $\langle |\mathscr{S}|, \vee\rangle_p \to \langle |\mathscr{S}_p|, \vee\rangle$ is $\langle |\mathscr{S}|, \vee\rangle_p/\equiv$ according to 4.2. But (5.2) says it is an onto map, i.e. $\langle |\mathscr{S}|, \vee\rangle_p/\equiv$ is isomorphic to the free semigroup $\langle |\mathscr{S}_p|, \vee\rangle$. Hence making the cancellation semigroups into groups:

Theorem 7.1. $\mathbf{G}_s \otimes \mathbf{G}_{\$p}$ *is freely generated by the indecomposables.* ▮

This we may restate as

* THEOREM 7.2. *If* $X_1 \vee \cdots \vee X_n \equiv X_1' \vee \cdots \vee X_n'$ *and each* X_i, X_j' *is an indecomposable p-primary space, then* $n = m$ *and for some permutation* π, $X_i \equiv X_{\pi(i)}$ *all i.* ∎

We may combine this theorem with 5.5 to obtain

* THEOREM 7.3. **G** *is freely generated by primary spaces.* ∎

And for the ring-version:

* THEOREM 7.4. $\mathbf{G}_p^*$ *is freely generated by* $\{[X] - [B_X]\ X$ *an indecomposable p-primary space*$\}$.[3]

In preparation for the basic theorem:

THE LOCAL STABLE WHITEHEAD THEOREM. 7.6. $H_*(-, Z_p)\colon \mathscr{S}_p \to (Z_p\text{-modules})$ *reflect isomorphisms, where* Z_p *may be interpreted as either the prime field or the p-adic rationals.*

PROOF. Suppose $f/m\colon X \to Y \in \mathscr{S}$ is an $H_*(-, Z_p)$ isomorphism. It suffices to consider the case that $m = 1, f \in \mathscr{S}(X, Y)$. Let $X \to Y \to C \to SX \to SY$ be a mapping cone sequence $H_*(C, Z_p) = 0$. Hence $H_*(C, Z)$ is torsion prime to p. Hence $p \cdot 1_C$ is an $H_*(-, Z)$ isomorphism. By the nonlocal stable Whitehead theorem $p \cdot 1_C$ is an isomorphism. Thus $\mathrm{End}_{\mathscr{S}}(C) \xrightarrow{p} \mathrm{End}_{\mathscr{S}}(C)$ is onto. But $\mathrm{End}_{\mathscr{S}}(C)$ is finitely generated, thus torsion prime to p. Hence $C = 0$ is $\mathscr{S}_p$ and f is an isomorphism in $\mathscr{S}_p$. ∎

COROLLARY 7.7. *If* $g \in \mathrm{End}_{\mathscr{S}_p}(X)$, *then* $1 - pg$ *is an automorphism.*

* THE BASIC THEOREM OF LOCAL STABLE HOMOTOPY. 7.8. *Given a p-primary indecomposable space X in a stable range, then* $\mathrm{End}_{\mathscr{S}}(X) \otimes Z_{(p)}$ *is a local ring.*

PROOF. (Algebraists please note that $Z_{(p)}$ is the ring of p-adic rationals, not the completion of that ring.)

Let $\varphi\colon B \to X$ be a π_*/T-isomorphism, let p^n be large enough to kill the torsion part of $\mathscr{S}_p(B, X)$ and large enough so that $p^n \geq p$. Given $f/m \in \mathrm{End}_{\mathscr{S}_p}(X)$,

---

[3] Of more technical interest:

PROPOSITION 7.5. *Idempotents split in* $\mathscr{S}/B$.

PROOF. Suppose $X \in \mathscr{S}$ and $X \simeq \oplus X_p$ in $\mathfrak{A}$, (following the notation of remarks proceeding 5.6) where X_p is p-torsion in $\mathfrak{A}$. The decomposition suggested in Theorem 3.2 $X \vee B \simeq B' \vee VX_p'$ where each X_p' is p-primary and without spherical retract, yields $X_p' \simeq X_p$ in $\mathfrak{A}$. Hence X_p is in $\mathscr{S}/B$.

Now note that we have an epimorphism $\mathrm{End}_{\mathscr{S}_p}(X) \to \mathrm{End}_{\mathscr{S}/B}(X)$ for p-primary X. Clearly $(1/m) \cdot 1_X$, $(p, m) = 1$ exists in $\mathscr{S}/B$ because X is there p-torsion). The basic theorem tells us that if X is indecomposable p-primary then $\mathrm{End}_{\mathscr{S}/p}(X)$ is local, hence $\mathrm{End}_{\mathscr{S}/B}(X)$ is local. In particular, X does not have idempotents in $\mathscr{S}/B$. Now given arbitrary p-torsion $X \in \mathscr{S}/B$ suppose that $X \simeq X_1 \oplus \cdots \oplus X_n$ in $\mathfrak{A}$ where each X_i is indecomposable. We may assume that X is without spherical retracts in $\mathscr{S}$. We may decompose it in $\mathscr{S}$, $X \simeq X_1' \oplus \cdots \oplus X_m'$. Now each X_i' is, by the above remarks, indecomposable in $\mathscr{S}/B$ and hence $n = m$ and for each i, $X_i \simeq X_j'$ in $\mathfrak{A}$ some j. Thus $X_i \in \mathscr{S}/B$, and idempotents split in $\mathscr{S}/B$ because every object of $\mathfrak{A}$ is in $\mathscr{S}/B$. ∎

$f \in \mathrm{End}_{\mathscr{S}}(X)$, $(m, p) = 1$ let $g \in \mathrm{End}_{\mathscr{S}}(B)$ be a map such that

$$\begin{array}{ccc} B & \xrightarrow{B_\varphi} & B_X \\ {\scriptstyle g}\downarrow & & \downarrow{\scriptstyle B_f} \\ B & \xrightarrow[B_\varphi]{} & B_X \end{array}$$

commutes. Then $f\varphi - \varphi g$ is torsion. Thus

$$\begin{array}{ccc} B & \xrightarrow{p^n\varphi} & X \\ {\scriptstyle g/m}\downarrow & & \downarrow{\scriptstyle f/m} \\ B & \xrightarrow[p^n\varphi]{} & X \end{array}$$

commutes in $\mathscr{S}_p$ as in the proof of (1.3).

We obtain a ring homomorphism $\Phi : \mathrm{End}_{\mathscr{S}_p}(X) \to \mathrm{End}_{\mathscr{S}_p}(\mathrm{Cok}\,(p^n\varphi))$, the kernel of which consists of those maps which factor through $p^n\varphi$. By the corollary above, Φ reflects units because if $\Phi(f)$ is a unit then $\Phi(f^m) = 1$ some $m > 0$ (the image of Φ is finite). Hence $\Phi(f^m - 1) = 0$ and $f^m = 1 + p^n\varphi h$. By the corollary f^m is a unit, therefore f is a unit.

We shall show that Cok $(p^n\varphi)$ is indecomposable. This will suffice for the theorem because if Cok $(p^n\varphi)$ is indecomposable then its ring of endomorphisms being finite is local. Hence if $a, b, n \in \mathrm{End}_{\mathscr{S}_p}(X)$, $a + b = u$, a, b nonunits, u a unit we could obtain a contradiction via Φ.

Because Cok $(p^n\varphi)$ is p-torsion (3.1) its $\mathscr{S}$-endomorphisms are the same as its $\mathscr{S}_p$-endomorphisms. We shall work entirely in $\mathscr{S}$ and $\mathfrak{F}$ from now on, the situation being:

$0 \to K \to B \to X \to F \to 0$ exact in $\mathfrak{F}$

K torsion

F p-torsion

B a bouquet of spheres.

To show: X indecomposable $\Rightarrow$ F indecomposable. We shall, in fact, show the contrapositive: If F has a nontrivial idempotent then so does X. We could finish the theorem by lifting the idempotent. But we don't know how. Rather, we will, please, *extend* the idempotent up through the mapping cone sequence to SX.

First note that we may assume that F is reduced, i.e. no projective subobjects. If it were not, then X would have a torsion retract.

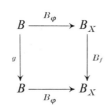

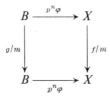

Second, note that K is reduced, (if not, then B would have a torsion retract, which it can not). Let $B \to X \to Y \to SB$ be a mapping cone sequence. Let I be the image of $B \to X$. We have exact sequences

$$0 \to F \to Y \to SK \to 0$$
$$0 \to SK \to SB \to SI \to 0$$
$$0 \to SI \to SX \to SF \to 0.$$

We shall extend idempotents from F to Y, reduce to SK, extend to SB reduce to SI, extend to SX. Note that nontriviality is preserved by extension. For the two reductions,

LEMMA 7.81. *If $0 \to A' \to E \to A'' \to 0$ is exact in an abelian category, E injective, A' reduced, then any idempotent on E which extends a nontrivial idempotent on A', reduces to a nontrivial idempotent on A''.*

PROOF. Let

$$\begin{array}{ccccc} A' & \longrightarrow & E & \longrightarrow & A'' \\ \downarrow{\scriptstyle e'} & & \downarrow{\scriptstyle e} & & \downarrow{\scriptstyle e''} \\ A' & \longrightarrow & E & \longrightarrow & A'' \end{array}$$

commute. If e'' were trivial we may without loss of generality assume that $e'' = 0$ (otherwise consider $1 - e$). Then we would obtain a map $E \to A'$ such that $E \to A' \to E = e$. Hence Im (e) is an injective subobject of A. ▮LEMMA▮

Two of the extensions are covered by

LEMMA 7.82. *Let $0 \to A \to E \to A'' \to 0$ be exact, E injective,* End (A'') *finite. Then any idempotent on A may be extended to E.*

PROOF. Let e be an idempotent on A. Let

$$\begin{array}{ccccc} A & \longrightarrow & E & \longrightarrow & A'' \\ \downarrow{\scriptstyle e} & & \downarrow{\scriptstyle f} & & \downarrow{\scriptstyle g} \\ A & \longrightarrow & E & \longrightarrow & A'' \end{array}$$

commute. Let g^n be idempotent. Noting that f^n extends $e^n = e$ we may assume that f and g were chosen to be f^n and g^n, i.e. that g is already idempotent. Then

$$\begin{array}{ccccc} A & \longrightarrow & E & \longrightarrow & A'' \\ \downarrow{\scriptstyle 0} & & \downarrow{\scriptstyle f^2-f} & & \downarrow{\scriptstyle 0} \\ A & \longrightarrow & E & \longrightarrow & A'' \end{array}$$

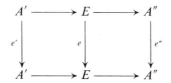

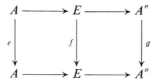

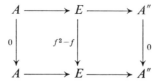

commutes, from which we obtain two equations: $(f^2 - f)^2 = 0$; $a(f^2 - f) = 0$ where a is the exponent of A''. We may assume that $a \geq 2$.

$$(1 - f)^a = 1 - af + f^2 P_1(f)$$

where $P_1(f)$ is a polynomial in f.

$$(1 - f)^a = 1 - af + a(f^2 - f) + f^2 P_1(f) = 1 - f^2 P(f)$$

where $P(f) = a - P_1(f)$.

$$(1 - f)^a (1 - f)^a = (1 - f)^a - (1 - f)^a f^2 P(f) = (1 - f)^a.$$

That is, $(1 - f)^a$ is an idempotent. It is an extension of the idempotent $1 - e$. Such is the perversity of computation. On the other hand $1 - (1 - f)^a$ is idempotent and it extends e. ▌LEMMA▌

Only one extension is left: from K to B. We note that K is p-torsion (any non-p-torsion must be isomorphic to the non-p-torsion of Y, hence injective.) We thus have the Basic Theorem once we have

LEMMA 7.83. *If K is a p-torsion subobject of a bouquet of spheres B, then every idempotent on K extends to an idempotent on B.*

PROOF. We need a functor. Let G' be the subfunctor of $(—, B)$ defined by $f \in G'(A)$ if there is a bouquet B', a torsion map $B' \to B$ and a factorization $A \to B' \to B = f$. Note that $G'(B)$ is a two sided ideal of End (B). Let R be the ring End $(B)/G'(B)$. Define $G(A)$ to be the R-module $(A, B)/G'(A)$. (Whether it is a left or right module depends on too many conventions. It doesn't matter. Whatever the conventions, note that (A, B) is an End (B) module, and that the submodule generated by $G'(B)$ is in $G'(A)$.)

Now for the geometry: $G'(B)$ is the torsion ideal of End (B), and R is a product of integral matrix rings $M_1 \times M_2 \times \cdots \times M_n$ (determined by the betti numbers of B). We have a map $R \to$ End (B) which retracts End $(B) \to R$. End of geometry.

G is a functor from $\mathfrak{F}$ to the category of R-modules, $\mathscr{G}^R$. But $\mathscr{G}^R \simeq \mathscr{G}^{M_1} \times \mathscr{G}^{M_2} \times \cdots \times \mathscr{G}^{M_n}$. For integral matrix rings M we have $\mathscr{G}^M \simeq \mathscr{G}^Z$, hence $\mathscr{G}^R \simeq \mathscr{G} \times \mathscr{G} \times \cdots \times \mathscr{G}$.

G carries monomorphisms into epimorphisms. It carries B into a small projective. It carries K into a p-torsion object. $G(B) \to G(K) \to 0$.

In $\mathscr{G}^R$ idempotents on p-torsion quotients of small projectives lift to idempotents on the projectives.

This statement is true for $\mathscr{G}^R$ because it is true for $\mathscr{G} \times \cdots \times \mathscr{G}$. It is true for $\mathscr{G} \times \cdots \times \mathscr{G}$ because it is true for $\mathscr{G}$. It is true for $\mathscr{G}$ because small projectives are free and because of classical integral linear algebra.

Given an idempotent e on K we have an idempotent f such that

$$\begin{array}{ccc} G(B) & \longrightarrow & G(K) \\ \downarrow{\scriptstyle f} & & \downarrow{\scriptstyle e} \\ G(B) & \longrightarrow & G(K) \end{array}$$

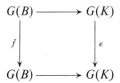

The retraction $R \to \text{End}(B) \to R$ gives us an idempotent $f \in \text{End}(B)$ such that $G(f) = \bar{f}$. Thus we have a G-commutative diagram

$$\begin{array}{ccc} K & \longrightarrow & B \\ {\scriptstyle e}\downarrow & & \downarrow{\scriptstyle f} \\ K & \longrightarrow & B \end{array}$$

where e and f are idempotent. Giving $K \to B$ the name u, we have $G(fu - ue) = 0$. Hence there exists a bouquet B', a torsion map $B' \to B$ and a factorization $K \to B' \to K = fu - ue$. B' is injective, u monomorphic, hence there exists a factorization $K \to B' = K \to B \to B'$. Define $t_1 = B \to B' \to B$. Then $fu - ue = K \to B' \to B = K \to B \to B' \to B = t_1 u$ and

$$\begin{array}{ccc} K & \longrightarrow & B \\ {\scriptstyle e}\downarrow & & \downarrow{\scriptstyle f+t_1} \\ K & \longrightarrow & B \end{array}$$

commutes. Note that t_1 is torsion. Define $t_{n+1} = t_n f + f t_1 + t_n t_1$, and note inductively that $(f + t_1)^n = f + t_n$. The t's are torsion. There are only a finite number of torsion elements in End (B). Hence for some positive n and j, $(f + t_1)^n = (f + t_1)^{n+j}$. Now the familiar argument: since $2j, 3j, \ldots$ work as well as j we may assume that j is at least as large as $2n$, that is, $(f + t_1)^n = (f + t_1)^{2n+k}$, $k \geq 0$. Then $(f + t_1)^{n+k}$ is idempotent. But this idempotent is an extension of e^{n+k} and e^{n+k} is e. ∎

COROLLARY 7.9. *Idempotents split in $\mathcal{S}_p$. That is, all projectives in $\mathfrak{F}_p$ are isomorphic to objects in $\mathcal{S}_p$.*

PROOF. Let P be a projective object in $\mathfrak{F}_p$. Because we may cover P with an object in $\mathcal{S}_p$ we have $P \oplus P' \in \mathcal{S}_p$. Let $X_1, X_2, \ldots, X_n$ be indecomposable $\mathcal{S}_p$-objects such that $P \oplus P' \simeq X_1 \oplus X_2 \oplus \cdots \oplus X_n$. Because each X_i has a local ring of endomorphisms we may find $i_1, i_2, \ldots, i_j$ such that $P \simeq X_{i_1} \oplus \cdots \oplus X_{i_j}$. ∎

8. **The infinite localization.** The Dold Lemma [**2** (4.9)] says that the natural maps $\mathfrak{F}(A, B) \otimes Q \to (H_*(A, Q), H_*(B, Q))$ where the values of $H_*(—, Q)$ are understood to be graded vector spaces, is an isomorphism. The most amusing consequence of this observation is the statement that the rank of $\mathcal{S}(X, Y)$ is equal to the inner product of the betti-vectors of X and Y.

If we understand $\mathfrak{A} \otimes Q$ to be the category whose objects are those of $\mathfrak{A}$ and whose maps are equivalence classes of pairs $\langle f, m\rangle$, $f \in \mathfrak{A}$, $m \neq 0$ subject to $f/m = f'/m' : f \exists\, m'' \neq 0$, $m'm''f = mm''f'$ then the above isomorphism tells us that $\mathcal{S} \otimes Q$ is isomorphic to the category of graded vector spaces (0-degree maps) globally finite in dimension — in our previous notation $\mathcal{S} \otimes Q \simeq \sum \mathfrak{Mat}\,(Q)$.

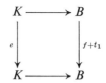

Exactly the same is true for $\mathfrak{F} \otimes Q$, i.e. $\mathscr{S} \otimes Q \to \mathfrak{F} \otimes Q$ is an equivalence of categories. ($\mathfrak{F} \otimes Q$ is the Gabriel quotient obtained by killing torsion objects.) Everything in $\mathfrak{F} \otimes Q$ is a bouquet of spheres.

We shall find it useful to use

PROPOSITION 8.1. *$\mathfrak{F} \otimes Q$ has global dimension 0, i.e. every epimorphism and monomorphism split.* ∎

While considering the isomorphism $\mathscr{S}(X, Y) \otimes Q \to (H_*(X, Q), H_*(Y, Q))$ it is convenient to record a piece of the folklore:

PROPOSITION 8.2. *There are only a finite number of homotopy types in a given stable range of the same homology type.*

PROOF. Given a graded abelian group $\{\ldots, 0, G_0, G_1, \ldots, G_n, 0, \ldots\}$, let $\mathbf{K}$ be the set of spaces X such that

$$\begin{aligned} H_j(X) &= G_j \qquad 0 \le j \le n \\ &= 0 \qquad \text{otherwise,} \end{aligned}$$

let $\mathbf{K}'$ be the spaces X' such that

$$\begin{aligned} H_j(X') &= G_j \qquad 1 \le j \le n \\ &= 0 \qquad \text{otherwise.} \end{aligned}$$

We know inductively that $\mathbf{K}'$ has only a finite number of isomorphism types. We shall assume notationally, that $\mathscr{S}$ is a skeletal category, i.e. isomorphic objects are equal.

Given $X \in \mathbf{K}$, let M_0 be the Moore space of G_0 let $M \to X$ be an H_0-isomorphism, let

$$M_0 \to X \to X' \xrightarrow{f} SM_0$$

be a mapping cone sequence. $X' \in \mathbf{K}$. Moreover $H_*(f) = 0$, hence from

$$\mathscr{S}(X', SM_0) \to \mathscr{S}(X', SM_0) \otimes Q \simeq (H_*(X', Q), H_*(M_0, Q))$$

we know that $f \in \operatorname{Tor}[\mathfrak{S}(X', SM_0)]$. Hence

$$X \in \bigcup_{X' \in \mathbf{K}'} \{\operatorname{Cone}(f) \,|\, f \in \operatorname{Tor}(S^{-1}X', M_0)\},$$

a finite set. ∎

* COROLLARY 8.3. *$\equiv$ classes are finite.* ∎

9. **Injective envelopes in $\mathfrak{F}$.**

PROPOSITION 9.1. *If $0 \to A \to X \to A'' \to 0$ is exact in $\mathfrak{F}$, $X \in \mathscr{S}$, then X is the injective envelope of A iff A'' is torsion and reduced.*

PROOF. $\Leftarrow$ Suppose $A' \subset X$, $A \cap A' = 0$. Then we have an exact sequence $0 \to A \oplus A' \to X \to A''/A' \to 0$, the right-hand end clearly torsion. Lemma 7.82 applies. We may extend the idempotent $n_1 p_1$ to X, the kernel of the extension being injective and containing A' and not meeting A. Hence if $A' \neq 0$ we would obtain an injective subobject of $X/A = A''$.

$\Rightarrow$ If A'' has an injective subobject, it would retract from X (since it is projective). We would have an injective subobject of X not meeting A. If A'' were not torsion consider the splitting in $\mathfrak{F} \otimes Q$,

(8.1)
$$\begin{array}{ccc} A'' & & \\ {\scriptstyle f/m}\downarrow & \searrow^{1} & \\ A & \longrightarrow & A'' \end{array}$$

Hence for some $m > 0$ we have in $\mathfrak{F}$,

$$\begin{array}{ccc} A'' & & \\ \downarrow & \searrow^{m} & \\ X & \longrightarrow & A'' \end{array}$$

Note that

$$\begin{array}{ccc} \operatorname{Ker}(m \cdot 1_A) & \longrightarrow & A'' \\ \downarrow & & \downarrow \\ A & \longrightarrow & X \end{array}$$

is a pullback. Let p be a prime not dividing m. $\operatorname{Ker}(p \cdot 1_{A''}) \cap \operatorname{Ker}(m \cdot 1_{A''}) = 0$. Moreover $\operatorname{Ker}(p \cdot 1_{A''}) \neq 0$ because

$$(A'', Y) \xrightarrow{(p, Y)} (A'', Y)$$

being onto implies that (A'', Y) is torsion all spaces $Y \in \mathscr{S}$. Finally, then

$$\operatorname{Ker}(p \cdot 1_{A''}) \to A'' \to X$$

is monomorphic, and its image does not meet A. ▮

We know that some objects in $\mathfrak{F}$ do not have injective envelopes (e.g. $\operatorname{Ker}(2 \cdot 1_{S^0})$). But

PROPOSITION 9.2. *Direct summands of $\mathfrak{F}$ objects with injective envelopes have injective envelopes.*

PROOF. Suppose $A_1 \oplus A_2$ has an injective envelope X. Consider

$$0 \to A_1 \oplus A_2 \to X \to A'' \to 0.$$

By the above proposition, A'' is torsion. By 7.82, $u_1 p_1$ extends to an idempotent on X. The image of the extension is the injective envelope of A_1, again by the above proposition. ▮

We note that Lemma 7.82 said that idempotents extend to injective envelopes, and this in turn implies the last proposition (by a slightly longer argument). Also note, that the converse is true. To be sure that we have said something about the category $\mathfrak{F}$ consider the category of modules over $Z_{(6)} = \{a/b \in Q \mid (6, b) = 1\}$. $Z_{(6)}$ is projective and indecomposable and a co-envelope of its quotient $Z_{(6)/(6)}$, that is, if $X \xrightarrow{f} Z_{(6)} \to Z_{(6)/(6)}$ is onto, then f is already onto (because (6) is the intersection of all the maximal proper subobjects of $Z_{(6)}$). But $Z_{(6)/(6)} = Z/6Z = Z_2 \oplus Z_3$. In the dual category, therefore, we have an example of an idempotent failing to extend to an injective envelope.

10. **Injective envelopes in $\mathfrak{F}_p$.**

THEOREM 10.1. *Every object in $\mathfrak{F}_p$ has an injective envelope.*

PROOF. Every injective in $\mathfrak{F}_p$ is a finite sum of indecomposables. By the Basic Theorem 7.7, every indecomposable injective has a local ring of endomorphisms. Hence we need only show

LEMMA 10.11. *Let $\mathfrak{A}$ be an abelian category with enough injectives, in which every injective is a finite sum of indecomposables.*

If indecomposable injectives have local rings of endomorphisms then every object has an injective envelope.

And conversely.

PROOF. The "and conversely" is well known (e.g. [**1**, 6D]).

First, suppose E is an indecomposable injective, $A \subset E$ nontrivial. We wish to show that E is the injective envelope of A. Suppose $B \subset E$, $A \cap B = 0$. Let

$$\begin{array}{ccc} A \oplus B & \longrightarrow & E \\ {\scriptstyle u_1 p_1}\downarrow & & \downarrow {\scriptstyle f} \\ A \oplus B & \longrightarrow & E \end{array}$$

commute. Note that $A \subset \operatorname{Ker}(1 - f)$ hence $1 - f$ is a nonunit. Because $\operatorname{End}(E)$ is local, f is a unit. But $f \mid B = 0$. Hence $B = 0$.

Given A, let n be an integer such that A may be embedded in an n-fold sum of indecomposable injectives, $E_1 \oplus E_2 \oplus \cdots \oplus E_n$ but not in any subsum. We shall show such an embedding is an essential extension.

First, given a monomorphism $A \to E_1 \oplus E_2 \oplus \cdots \oplus E_n$ we observe that the pullback

$$\begin{array}{ccc} P & \longrightarrow & E_1 \\ \downarrow & & \downarrow {\scriptstyle u_1} \\ A & \longrightarrow & E_1 \oplus E_2 \oplus \cdots \oplus E_n \end{array}$$

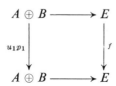

is nontrivial because $P = \operatorname{Ker}(A \to E_1 \oplus \cdots \oplus E_n \to \operatorname{Cok}(u_1))$, and A can not be embedded in $\operatorname{Cok}(u_1) = E_2 \oplus \cdots \oplus E_n$.

Now let $B \to E_1 \oplus \cdots \oplus E_n$ be a monomorphism. Let I be the smallest set of indices such that B may be embedded in $\bigoplus_{i \in I} E$. We may assume that $I = \{1, \ldots, j\}$.

If $j < n$, then we know inductively that $E_1 \oplus \cdots \oplus E_j$ is the injective envelope of B. Hence

$$B \to E_1 \oplus \cdots \oplus E_n = B \to E_1 \oplus \cdots \oplus E_j \xrightarrow{f} E_1 \oplus \cdots \oplus E_n,$$

f a monomorphism. If $A \cap \operatorname{Im}(f) = 0$ then A is embeddable in $\operatorname{Cok}(f)$. But $E_1 \oplus \cdots \oplus E_j \oplus \operatorname{Cok}(f) \simeq E_1 \oplus \cdots \oplus E_n$ and by the cancellation of objects with local rings of endomorphism, $\operatorname{Cok}(f) = E_{j+1} \oplus \cdots \oplus E_n$ which contradicts the definition of n. Hence $A \cap \operatorname{Im}(f) \neq 0$, that is, the pullback

$$\begin{array}{ccc} P_1 & \longrightarrow & E_1 \oplus \cdots \oplus E_j \\ \downarrow & & \downarrow \\ A & \longrightarrow & E_1 \oplus \cdots \oplus E_n \end{array}$$

is nontrivial. Hence the pullback

$$\begin{array}{ccc} P_2 & \longrightarrow & B \\ \downarrow & & \downarrow \\ P_1 & \longrightarrow & E_1 \oplus \cdots \oplus E_j \end{array}$$

is nontrivial. But

$$\begin{array}{ccc} P_2 & \longrightarrow & B \\ \downarrow & & \downarrow \\ A & \longrightarrow & E_1 \oplus \cdots \oplus E_n \end{array}$$

is a pullback, and A meets B.

If $j = n$, then both A and B meet E_1 nontrivially, and by the first part of the proof, they must therefore meet nontrivially therein. ∎

COROLLARY 10.2. *If* $X \to Y \xrightarrow{f} Z \to SX \xrightarrow{Sf} SY$ *in* $\mathcal{S}_p$ *is exact in* $\mathfrak{F}_p$, *then* $Z \simeq \operatorname{Cone}(f)$ *in* $\mathcal{S}_p$.

PROOF. Let E be the injective envelope of $\operatorname{Cok}(f)$. Note that $E/\operatorname{Cok}(f)$ is reduced. $Z = E \oplus E'$ and E' embeds in $S \operatorname{Ker}(f)$. Then $S \operatorname{Ker}(f) = E' \oplus E/\operatorname{Cok}(f)$. Hence Z is the sum of the injective envelope of $\operatorname{Cok}(f)$ and the injective part of $S \operatorname{Ker}(f)$. Injective parts are uniquely definable because of the Basic Theorem 7.7.

Exactly the same must be true for $\operatorname{Cone}(f)$. Hence $Z \simeq \operatorname{Cone}(f)$. ∎

*COROLLARY 10.3. *Let* $X \xrightarrow{f} Y \to Z \to SX \xrightarrow{Sf} SY$ *be in a stable range and suppose that an exact sequence results upon the application of any cohomology functor. Then* $Z \vee B_Z \simeq \text{Cone}(f) \vee B_Z$.

PROOF. The condition is equivalent to the fact that $X \to Y \to Z \to SX \to SY$ is exact in $\mathfrak{F}$. By the last corollary, therefore, $Z \simeq \text{Cone}(f)$ in $\mathcal{S}_p$ each p. By 4.1, then, $Z \equiv \text{Cone}(f)$. Finally use 1.3. ∎

We should note that for $\gamma: S^n \to S^0$ and j an integer prime to the order of γ, that $\text{Cok}(\gamma) \simeq \text{Cok}(j\gamma)$, $S\,\text{Ker}(\gamma) \simeq S\,\text{Ker}(j\gamma)$. Hence we obtain an exact sequence

$$S^n \xrightarrow{\gamma} S^0 \to \text{Cone}(j\gamma) \to S^{n+1} \xrightarrow{S\gamma} S^1.$$

Thus the condition of the corollary does not imply isomorphism 1.2 just $\equiv$.

Note that for cohomology theories K with finitely generated values $Z \equiv Z'$ implies $K(Z) \simeq K(Z')$. Hence the condition of the corollary implies that $K(Z) \simeq K(\text{Cone}(f))$. Whereas, mapping cones are not characterized by exactness properties, the cohomology (and homology, homotopy, cohomotopy, etc.) of mapping cones is characterized.

BIBLIOGRAPHY

1. P. J. Freyd, *Stable homotopy*, Proceedings of Conference on Categorical Algebra (La Jolla, 1965), Springer-Verlag, New York, 1966.

2. H. Toda, *Composition methods in homotopy groups of spheres*, Ann. of Math. Studies no. 49, Princeton Univ. Press, Princeton, N.J., 1962.

UNIVERSITY OF PENNSYLVANIA

On a theorem of Wilder

J.-L. Verdier

1. **Wilder's theorem.** In this section, we state a theorem due to M. Zisman and the author. Proofs can be found in [**1**, Exposé 9].

1.1. Let X be a locally compact space of finite cohomological dimension [1], A a unitary ring which is a flat algebra over some commutative ring and let $F^\cdot$ be a complex of sheaves of right A-modules on X which has only finitely many nonzero cohomology sheaves. Let $F^\cdot \to \Omega(F^\cdot)$ be a resolution (i.e. a morphism of complexes which induces isomorphisms on the cohomology) of $F^\cdot$ by injective sheaves of A-modules. For any paracompact open subset $U \subset X$, the nth cohomology module of the complex $\Gamma(U, \Omega(F^\cdot))$ (resp. $\Gamma_c(U, \Omega(F^\cdot))$) is denoted by $H^n(U, F^\cdot)$ (resp. $H_c^n(U, F^\cdot)$) and is called the nth hypercohomology module (resp. the nth hypercohomology module with compact supports) of U with values in $F^\cdot$.

Since the complex $\Gamma(U, \Omega(F^\cdot))$ (resp. $\Gamma_c(U, \Omega(F^\cdot))$) has only finitely many nonzero cohomology modules, there exists a complex $L(U, F^\cdot)$ (resp. $L_c(U, F^\cdot)$) whose components are projective modules and a map of complexes ϕ_U: $L(U, F^\cdot) \to \Gamma(U, \Omega(F^\cdot))$ (resp. $\phi_{c,U}: L_c(U, F^\cdot) \to \Gamma_c(U, \Omega(F^\cdot))$) such that

(1) the component $L^n(U, F^\cdot)$ (resp. $L_c^n(U, F^\cdot)$) is zero for n big enough; and

(2) the map ϕ_U (resp. $\phi_{c,U}$) induces isomorphisms on the cohomology. Moreover if $U \subset V$ are two paracompact open subsets of X, the restriction map $\rho_{V,U}: \Gamma(V, \Omega(F^\cdot)) \to \Gamma(U, \Omega(F^\cdot))$ (resp. the extension map $\varepsilon_{U,V}: \Gamma_c(U, \Omega(F^\cdot)) \to \Gamma_c(V, \Omega(F^\cdot))$) can be lifted to the projective resolutions. The lifting is unique up to homotopy, is called once again the restriction map (resp. the extension map) and is denoted once again by $\rho_{V,U}$ (resp. $\varepsilon_{U,V}$).

1.2 Definition. Let $(L_i,\ i \in I,\ i < j \mapsto \psi_{i,j}: L_i \mapsto L_j)$ be a directed (resp. inverse) system of complexes of right A-modules (we require only that the transition morphisms agree up to homotopy), and let $[m, n] \subset \mathbf{Z}$ be an interval. The system $(L_i, i \in I, \psi_{i,j})$ is said to be essentially of finite type of amplitude contained in $[m, n]$ if, for any $i \in I$, there exists a $j > i$ (resp. $j < i$), a complex $L_{i,j}$ whose components $L_{i,j}^k$ are projective modules of finite type, zero whenever $k \notin [m, n]$, and a diagram commutative up to homotopy:

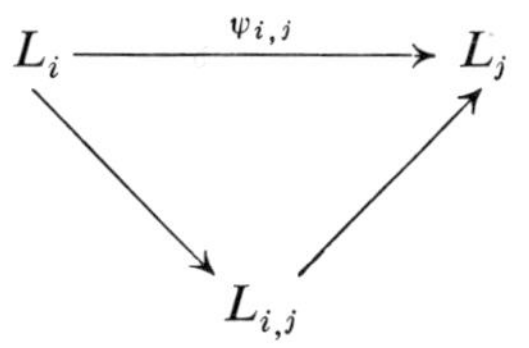

resp.

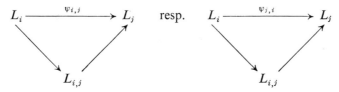

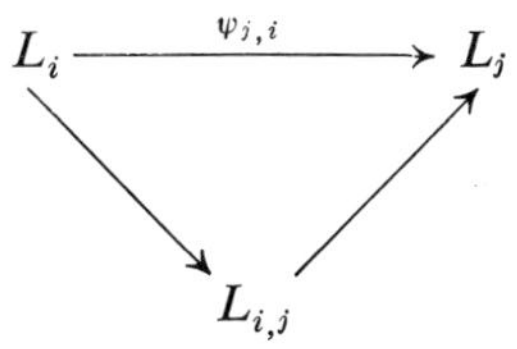

1.3 In the following theorem we use the notation of 1.1.

THEOREM. *The following conditions are equivalent:*

(i) *There exists an interval* $[m, n] \subset \mathbf{Z}$ *such that for any compact* K *in* X, *the directed system indexed by the paracompact open neighborhood of* K:

$$(U \to L(U, F^{\cdot}),\ V \subset U \mapsto {}_{\rho_{U,V}}: L(U, F^{\cdot}) \to L(V, F^{\cdot}))$$

is essentially of finite rank of amplitude contained in $[m, n]$.

(ii) *There exists an interval* $[m', n'] \subset \mathbf{Z}$ *such that for any compact* K *in* X, *the inverse system indexed by the paracompact open neighborhood of* K:

$$(U \to L(U, F^{\cdot}),\ V \subset U \to {}_{\rho_{U,V}}: L(U, F^{\cdot}) \to L_c(U, F^{\cdot}))$$

is essentially of finite rank of amplitude contained in $[m, n]$.

(iii) *Same as* (i) *but consider only the compact subsets reduced to one point.*

(iv) *Same as* (ii) *but consider only the compact subsets reduced to one point.*

1.4 DEFINITION. A complex of sheaves $F^{\cdot}$ on X is said to be perfect if it has the equivalent properties of Theorem 1.3.

It is clear that these properties depend only on $F^{\cdot}$ and not on the different resolutions.

1.5 COROLLARY. *Let* $f: X \to Y$ *be a proper map between two locally compact spaces of finite cohomological dimension, and* $F^{\cdot}$ *a perfect complex of injective sheaves*[1] *of* A*-modules on* X. *Then the complex* $f_*(F^{\cdot})$ *is a perfect complex on* Y. *In particular, when* Y *is a point and* X *compact, the complex* $L(X, F^{\cdot})$ *is homotopic to a bounded complex of projective* A*-modules of finite rank.*

1.6 PROPOSITION. *Let* $0 \to F'^{\cdot} \to F^{\cdot} \to F''^{\cdot} \to 0$ *be an exact sequence of complexes of sheaves on* X. *If two of the complexes of the sequence are perfect, so is the third one.*

1.7 PROPOSITION. *Assume that* A *is right noetherian and that* $F^{\cdot}$ *has a finite flat amplitude (i.e. that* $F^{\cdot}$ *has a resolution by a bounded complex whose components are sheaves with flat stalks). Then* $F^{\cdot}$ *is perfect if and only if it has one of the two following equivalent properties:*

(a) *For any* $x \in X$, *and for any open neighborhood* U *of* x, *there exists an open neighborhood* V *of* x, $V \subset U$, *such that for any* $n \in \mathbf{Z}$, *the restriction map* $H^n(U, F^{\cdot}) \to H^n(V, F^{\cdot})$ *has a finitely generated image.*

[1] Or more generally of sheaves acyclic for the direct image functor.

(b) *For any $x \in X$ and for any open neighborhood U of x, there exists an open neighborhood V of x, $U \subset V$, such that for any $n \in \mathbf{Z}$, the extension map*

$$H_c^n(U, F^\cdot) \to H_c^n(V, F^\cdot)$$

has a finitely generated image.

Applying Theorem 1.3 and Proposition 1.7 to the case $A = \mathbf{Z}$ and $F =$ constant sheaf free of rank one, we get Wilder's theorem [**2**].

2. **The Wall invariant.**

2.1 Let $f: X \to Y$ be a continuous map between locally compact spaces and let G be a sheaf of A-modules on X (say on the right). For any open set U of Y denote by $\psi(U)$ the family of closed subsets S of the space $f^{-1}(U)$ such that the map f induces a proper map from S to U. We denote by $f_! G$ the sheaf

$$U \to \Gamma_{\psi(U)}(f^{-1}(U), G).$$

The sheaf $f_! G$ is called the direct image of G with proper supports.

2.2 Let X be a connected, locally simply connected, finite dimensional, compact space and $X^\sim \xrightarrow{p} X$ its universal covering. Choose a base point in $X^\sim$. Then for any complex of sheaves of right A-modules $F^\cdot$ on X, the complex $p_! p * F^\cdot$ is canonically equipped with a right $A[\Pi_1(X)]$-structure. Furthermore, when $F^\cdot$ is perfect one checks immediately by local inspection that $p_! p^* F^\cdot$ is a perfect $A[\Pi_1(X)]$-complex. Therefore, in that case, the complex $L(X, p_! p^* F^\cdot)$ (1.1) (the resolutions being taken in the category of right $A[\Pi_1(X)]$-modules) is homotopic to a bounded complex $C^\cdot(X, F^\cdot)$ of projective $A[\Pi_1(X)]$-modules of finite rank.

2.3 Definition. Let X be a connected, locally simply connected, finite dimensional, compact space and let $F^\cdot$ be a perfect complex of sheaves of right A-modules on X. The element $\sum_{i \in \mathbf{Z}} (-1)^i \operatorname{cl}(C^i(X, F^\cdot))$, in the Grothendieck group $K(A[\Pi_1(X)])$ of projective right $A[\Pi_1(X)]$-modules of finite type, is called the *Wall Invariant* of $F^\cdot$ and denoted by $W_A(X, F^\cdot)$.

It is easy to check that the element $W_A(X, F^\cdot)$ does not depend on the choice of the point in $X^\sim$ used to define the action of $\Pi_1(X)$ and does not depend on the choice of the different resolutions.

As an immediate consequence of the definitions we have the following proposition:

2.4 Proposition. (1) *Let $u: F^\cdot \to F'^\cdot$ be a map between two perfect complexes which induces an isomorphism on the sheaves of cohomology. Then $W_A(X, F^\cdot) = W_A(X, F'^\cdot)$.*

(2) *Let $0 \to F'^\cdot \to F^\cdot \to F''^\cdot \to 0$ be an exact sequence of perfect complexes. Then $W_A(X, F^\cdot) = W_A(X, F'^\cdot) + W_A(X, F''^\cdot)$.*

In particular if $F^\cdot[1]$ denotes the complex $F^\cdot$ shifted one degree to the left, we have $W_A(X, F^\cdot[1]) = -W_A(X, F^\cdot)$.

2.5 Let $f: Y \to X$ be a continuous map of connected, locally simply connected, finite dimensional compact spaces. The map f induces a homomorphism

$$\Pi_1(f): \Pi_1(Y) \to \Pi_1(X)$$

so that the ring $A[\Pi_1(X)]$ becomes an $A[\Pi_1(Y)]$-algebra. The tensor product— $\otimes_{A[\Pi_1(Y)]} A[\Pi_1(X)]$ defines a map denoted $f_*: K(A[\Pi_1(Y)]) \to K(A[\Pi_1(X)])$ on the corresponding Grothendieck groups. Let now $F^\cdot$ be a perfect complex of right A-modules on Y. Assume that the components of $F^\cdot$ are acyclic for the functor direct image by f. The direct image $f_*F^\cdot$ is then a perfect complex on X (1.5).

2.6 PROPOSITION. *With the notation of* (2.5), *we have*

$$W(X, f_*F^\cdot) = f_* W(Y, F^\cdot).$$

PROOF. Let $p: X^\sim \to X$ the universal covering of X and $X^\sim \times_X Y$ the fiber product. We have a cartesian square:

$$\begin{array}{ccc} X^\sim \times_Y X & \xrightarrow{f^\sim} & X^\sim \\ \downarrow{\scriptstyle pr_2} & & \downarrow{\scriptstyle p} \\ Y & \xrightarrow{f} & X \end{array}$$

and the map $\mathrm{pr}_2: X^\sim \times_X Y \to Y$ is a principal covering with group $\Pi_1(X)$. Furthermore, the base change theorem for direct images by proper morphism yields a canonical isomorphism, compatible with the action of $\Pi_1(X)$; $f_*\mathrm{pr}_{2!}\,\mathrm{pr}_2^* F^\cdot \simeq p_! p^* f_* F^\cdot$.

Let $\mathrm{pr}_{2!}\,\mathrm{pr}_2^* F^\cdot \to I^\cdot$ be a resolution by injective sheaves of right $A[\Pi_1(X)]$-modules. The complex $f_*I^\cdot$ is an injective resolution of the complex $p_! p^* f_* F^\cdot$ and we have $\Gamma(X, f_*I^\cdot) \simeq \Gamma(Y, I^\cdot)$. By Corollary 1.5, there exists a bounded complex whose components are finitely generated projective $A[\Pi_1(X)]$-modules $P^\cdot$ and a resolution $P^\cdot \to \Gamma(Y, I^\cdot)$. We have therefore the equality $W_A(X, f_*F^\cdot) = \sum_{i \in \mathbf{Z}} (-1)^i \mathrm{cl}(P^i)$. Let $q: Y^\sim \to Y$ be the universal covering of Y. There exists a canonical isomorphism $q_! q^* F^\cdot \otimes_{A[\Pi_1(Y)]} A[\Pi_1(X)] \simeq \mathrm{pr}_{2!}\,\mathrm{pr}_2^* F^\cdot$. Hence we have a homotopy equivalence $C^\cdot(Y, F^\cdot) \otimes_{A[\Pi_1(Y)]} A[\Pi_1(X)] \to P^\cdot$ and the proposition follows.

2.7 COROLLARY. *Let X be a finite polyhedron (geometric realization of a finite semisimplicial complex), and let $F^\cdot$ be a complex of sheaves on X which induces on each cell a perfect complex. Then $F^\cdot$ is perfect. Let $\chi(F^\cdot) \in K(A)$ be the Euler-Poincaré characteristic of $F^\cdot$ in the Grothendieck group of finitely generated projective right A-modules. Then*

$$W_A(X, F^\cdot) = \chi(F^\cdot) \cdot \mathrm{cl}(A[(\Pi_1 X)]).$$

PROOF. Follows, by induction on the number of cells, from Propositions 2.4 and 2.6.

2.8 Let X be a connected, locally simply connected, finite dimensional compact space on which the constant sheaf $\mathbf{Z}$ is perfect. Let $T_X^{\cdot}$ be the orientation complex introduced by Borel and Moore [3]. Then the complex $T_X^{\cdot}$ is perfect [1]. The cohomology groups $H_i(X, \mathbf{Z}) = H^{-i}(X, T_X^{\cdot})$ are the homology groups of X. The results quoted above (1.6) show that the homology groups of $X^{\sim}$ equipped with their natural action of $\Pi_1(X)$ are the homology modules of a finite $\mathbf{Z}[\Pi_1(X)]$-complex whose components are finitely generated projective $\mathbf{Z}[\Pi_1(X)]$-modules, this complex being well defined up to homotopy, so that our Wall invariants are natural translations in the theory of sheaves of the invariant introduced by Wall.

In particular, when the Borel-Moore homology of X is isomorphic to the singular homology, the element $W_{\mathbf{Z}}(X, T_X^{\cdot})$ is the invariant introduced by Wall [4].

3. **Properties of the Wall invariant.** In this section we want to show that some properties of the Wall invariant are immediate consequences of the definitions and of the Poincaré duality theorem (see [5] for analogous results in singular homology).

3.1 Let $F^{\cdot}$ be a complex of sheaves with finitely many nonzero cohomology sheaves on a locally compact, finite dimensional space X. Let $T_{X,A}^{\cdot}$ be the orientation complex of X with respect to the ring A [3]. (This is a complex of sheaves of A-bimodules injective on the right.) We denote by $D_{X,A}F^{\cdot}$ the complex $\mathscr{H}om_A^{\cdot}(F^{\cdot}, T_{X,A}^{\cdot})$ ($\mathscr{H}om^{\cdot}$ is the complex of sheaf homomorphisms). The complex $D_{X,A}F^{\cdot}$ is a complex of sheaves of right A^0-modules and is called the dual complex of $F^{\cdot}$. When $F^{\cdot}$ is perfect, $D_{X,A}F^{\cdot}$ is perfect [1, Exposé 9].

Let $F^{\cdot} \mapsto S(F^{\cdot})$ be a bounded c-soft resolution of $F^{\cdot}$ and $A \to I^{\cdot}(A)$ be a resolution by A-bimodules injective on the right. The complex of presheaves on X:

$$U \mapsto \mathrm{Hom}_A^{\cdot}(\Gamma_c(U, S(F^{\cdot}), I^{\cdot}(A)), \tag{3.1.1}$$

is a complex of flabby sheaves of right A^0-modules [3], and is denoted by $F^{\cdot *}$. It follows from the Poincaré duality theorem [1, Exposé 4] that the complexes $D_{X,A}F^{\cdot}$ and $F^{\cdot *}$ have isomorphic injective resolutions.

3.2 Let Π be a group and let P be a projective right $A[\Pi]$-module of finite type. Set

$$P^* = \mathrm{Hom}_{A[\Pi]}(P, A[\Pi]). \tag{3.2.1}$$

The module P^* is a right projective $A^0[\Pi]$-module of finite type, (Π acting on the right via its left action on $A[\Pi]$). The map $P \mapsto P^*$ induces an isomorphism $K(A[\Pi]) \to K(A^0[\Pi])$ on the corresponding Grothendieck groups.

3.3 Proposition. *Let X be a connected, locally simply connected, finite dimensional compact space. Then, for any perfect complex $F^{\cdot}$ on X, we have:*

$$W_A(X, F^{\cdot})^* = W_{A^0}(X, D_{X,A}F^{\cdot}).$$

Proof. We sketch the proof. Let $p: X^{\sim} \to X$ be the universal covering of X. By (2.4) we may assume that $F^{\cdot}$ is a bounded complex of c-soft sheaves. Let us consider the complex $D_{X,A[\Pi_1(X)]}p_!p^*F^{\cdot}$ viewed as an $A^0[\Pi_1(X)]$-complex (via the isomorphism $A^0[\Pi_1(X)] \xrightarrow{\sim} A[\Pi_1(X)]^0$). Using (3.1.1) and the definition of $p!$

(2.1), we see that the complexes $p_! D_{X^{\sim}, A} p^* F^{\cdot}$ and $D_{X, A[\Pi_1(X)]} p_! p^* F^{\cdot}$ have isomorphic injective resolutions. Moreover, since p is a covering, we have a canonical isomorphism $D_{X^{\sim}, A} p^* F^{\cdot} \simeq p^* D_{X,A} F^{\cdot}$. Therefore the complexes $\Gamma(X, p_! p^* D_{X,A} F^{\cdot})$ and $\Gamma(X, D_{X, A[\Pi_1(X)]} p_! p^* F^{\cdot})$ have isomorphic injective resolutions. Let

$$I^{\cdot}(A[\Pi_1(X)])$$

be a resolution of $A[\Pi_1(X)]$ by $A[\Pi_1(X)]$-bimodules injective on the right. It follows from the Poincaré duality theorem (3.1) that the complexes

$$\Gamma(X, D_{X, A[\Pi_1(X)]} p_! p^* F^{\cdot}) \quad \text{and} \quad \mathrm{Hom}^{\cdot}_{A[\Pi_1(X)]}(\Gamma(X, p_! p^* F^{\cdot}), I^{\cdot}(A[\Pi_1(X)]))$$

have isomorphic injective resolutions. There exist (2.2) a bounded complex of finitely generated projective $A[\Pi_1(X)]$-modules $C(X, F^{\cdot})$ and a resolution

$$C^{\cdot}(X, F^{\cdot}) \to \Gamma(X, p_! d^* F^{\cdot}).$$

We have therefore two maps of complexes

$$\mathrm{Hom}^{\cdot}_{A[\Pi_1(X)]}(C^{\cdot}(X, F^{\cdot}), A[\Pi_1(X)]) \to \mathrm{Hom}^{\cdot}_{A[\Pi_1(X)]}(C^{\cdot}(X, F^{\cdot}), I^{\cdot}(A[\Pi_1(X)]))$$

$$\mathrm{Hom}^{\cdot}_{A[\Pi_1(X)]}(\Gamma(X, p_! p^* F^{\cdot}), I^{\cdot}(A[\Pi, (X)])) \to \mathrm{Hom}^{\cdot}_{A[\Pi_1(X)]}(C^{\cdot}(X, F^{\cdot}), I^{\cdot}(A[\Pi_1(X)])),$$

which induce isomorphisms on the cohomology and the proposition follows.

3.4 Let Π be a group and A a commutative ring. We denote by $G(A[\Pi])$ the Grothendieck group of $A[\Pi]$-modules that are finitely generated and projective as A-modules. The tensor product over A of two such $A[\Pi]$-modules endowed with the diagonal action of Π is an $A[\Pi]$-module of the same kind. Hence, the tensor product defines a ring structure over G. Let M be an $A[\Pi]$-module that is finitely generated and projective as an A-module and P a finitely generated projective and projective $A[\Pi]$-module. The tensor product $A[\Pi] \otimes_A P$ with the diagonal action of Π is a finitely generated projective $A[\Pi]$-module. Hence the tensor product defines on $K(A[\Pi])$ a structure of module over $G(A[\Pi])$.

3.5 Let X be a connected and locally simply connected space and M a locally constant sheaf of finitely generated and projective A-modules. Then the stalk of M is an $A[\Pi_1(X)]$-module that is finitely generated and projective as an A-module. Its image in $G(A[\Pi_1(X)])$ is denoted by $\mathrm{cl}(M)$.

3.6 PROPOSITION. *Let X be a connected locally simply connected finite dimensional compact space, $F^{\cdot}$ a perfect complex of sheaves of A-modules, M a locally constant sheaf of finitely generated and projective A-modules. Then*

$$W_A(X, M \otimes_A F^{\cdot}) = \mathrm{cl}(M) \cdot W_A(X, F^{\cdot}).$$

PROOF. By (2.4) we may assume that $F^{\cdot}$ is a bounded complex of c-soft A-modules. Let $p: X^{\sim} \to X$ be the universal covering of X. The $A[\Pi_1(X)]$-complex $\Gamma_c(X^{\sim}, p^* F^{\cdot})$ has a resolution $C^{\cdot}(X, F^{\cdot})$ by a bounded complex of finitely generated and projective $A[\Pi_1(X)]$-modules. Since we have a canonical isomorphism $\Gamma_c(X^{\sim}, p^* M \otimes_A F^{\cdot}) \simeq \Gamma_c(X, p^* F^{\cdot}) \otimes_A M_x$, where M_x is any stalk of

M with its natural action of $\Pi_1(X)$, the complex $C^{\cdot}(X, F^{\cdot}) \otimes_A M_x$ is a resolution of the complex $\Gamma_c(X^{\sim}, p^*M \otimes_A F^{\cdot})$. q.e.d.

3.7 COROLLARY. *Let X be a compact connected n dimensional topological variety with boundary. Let ∂X_j, $1 \leqslant j \leqslant q$, be the different connected components of its boundary, $i_j : \partial X_j \to X$ the inclusion maps, and let Λ_X be the orientation $\mathbf{Z}[\Pi_1(X)]$-module. Then*

$$(-1)^n W_{\mathbf{Z}}(X, \mathbf{Z}) + (-1)^{n-1} \textstyle\sum_j i_{j*}(W_{\mathbf{Z}}(\partial X_j, \mathbf{Z})) = \mathrm{cl}(\Lambda_X) \cdot W_{\mathbf{Z}}(X, \mathbf{Z})^*.$$

PROOF. We have $W_{\mathbf{Z}}(X, \mathbf{Z})^* = W_{\mathbf{Z}}(X, T^{\cdot}_X)$ where $T^{\cdot}_X$ is the orientation complex of X (3.3). The complex $T^{\cdot}_X$ has only one zero cohomology sheaf 0_X in dimension $-n$. Hence $W_{\mathbf{Z}}(X, \mathbf{Z}) = (-1)^n{}_{W_{\mathbf{Z}}}(X, 0_X)$. The sheaf 0_X is locally constant free of rank one on $X - \partial X$ and its restriction to ∂X is zero. Let $j : X - \partial X \to X$ be the inclusion map. We have an exact sequence

$$0 \to 0_X \to j_* 0_X \to j_* 0_X / \partial X \to 0.$$

Hence (2.4) $W_{\mathbf{Z}}(X, 0_X) = W_{\mathbf{Z}}(X, j_* 0_X) + W_{\mathbf{Z}}(X, j_* 0_X/\partial X)$. The sheaf $j_* 0_X$ is locally free of rank one and is defined by the orientation module Λ_X. The formula follows from (3.6), (2.6) and trivial manipulations.

4. **Fibration.**

4.1 PROPOSITION. *Let X and Y be two connected, locally simply connected finite dimensional compact spaces, and let $F^{\cdot}$ and $G^{\cdot}$ be two perfect complexes of sheaves of A-modules on X and Y respectively (A is commutative). Assume that the stalks of the components of $F^{\cdot}$ are flat A-modules. Then*

$$W_A(X, F^{\cdot}) \cdot W_A(Y, G^{\cdot}) = W_A(X \times Y, F^{\cdot} \otimes_A G^{\cdot}),$$

where $F^{\cdot} \otimes_A G^{\cdot}$ denote the cartesian product of the two complexes of sheaves (tensor product of the two inverse images by the two projections of the product $X \times Y$).

PROOF. Immediate consequence of the Künneth formula [1 Exposé 3].

4.2 Let Π be a group and $G'(\mathbf{Z}[\Pi])$ the Grothendieck group of the $\mathbf{Z}[\Pi]$-module which are finitely generated as abelian groups. It is easy to check that the canonical map $G(\mathbf{Z}[\Pi]) \to G'(\mathbf{Z}[\Pi])$ is an isomorphism. Hence any $\mathbf{Z}[\Pi]$-module M, finitely generated as abelian group, yields an element in $G(\mathbf{Z}[\Pi])$ denoted by $\mathrm{cl}(M)$.

4.3 Let X be a connected and locally simply connected space and M a locally constant sheaf whose stalks are finitely generated abelian groups. The stalk at any point is a finitely generated abelian group on which $\Pi_1(X)$ acts hence yields an element in $G(\mathbf{Z}[\Pi_1(X)])$ denoted by $\mathrm{cl}(M)$. Let now $f : E \to X$ be a continuous map such that for any $q \in \mathbf{Z}$, $R^q f \mathbf{Z}$ is a locally constant sheaf whose stalks are of finite type, zero for q big enough.

We denote by $\mathrm{cl}(f)$ the element $\sum_q (-1)^q \mathrm{cl}(R^q f_* \mathbf{Z})$ in $G(\mathbf{Z}[\Pi_1(X)])$.

4.4 PROPOSITION. *Let $f : E \to X$ be a locally trivial fibration, where X is a connected, locally simply connected, finite dimensional compact space and where the fiber* *is a finite dimensional compact space on which the constant sheaf* $\mathbf{Z}$ *is perfect. Assume* E *is connected and locally simply connected. Then, for any* q *in* $\mathbf{Z}$, *the sheaf* $R^q f\, \mathbf{Z}$ *is a locally constant sheaf whose stalks are finitely generated abelian groups, zero for* q *big enough. Moreover, for any perfect complex* $F^\cdot$ *on* X, $f^*F^\cdot$ *is a perfect complex on* E *and*

$$f_* W_{\mathbf{Z}}(E, f^*F^\cdot) = W_{\mathbf{Z}}(X, F^\cdot) \cdot \mathrm{cl}(f).$$

PROOF. It is clear that the $R^q f_* Z$ are locally constant sheaves whose stalks are isomorphic to the cohomology of the fiber hence finitely generated abelian groups. It is also clear that $f^*F^\cdot$ is perfect whenever $F^\cdot$ is perfect. Let us prove the equality. By 2.4 we may assume that $F^\cdot$ is a bounded complex of c-soft sheaves whose stalks are torsion free. We have $f_* W_{\mathbf{Z}}(E, f^*F^\cdot) = W_{\mathbf{Z}}(X, f_* f^*F^\cdot)$ (2.6). Let $\mathbf{Z} \to \Omega^\cdot(\mathbf{Z})$ be a c-soft resolution of the constant sheaf $\mathbf{Z}$ on E. The projection formula yields a resolution $f_* f^*F^\cdot \to F^\cdot \otimes_{\mathbf{Z}} f_* \Omega^\cdot(\mathbf{Z})$ [**1**, Exposé 3]. Therefore (2.4) we have $W_{\mathbf{Z}}(X, f_* f^*F^\cdot) = W_{\mathbf{Z}}(X, F^\cdot \otimes_{\mathbf{Z}} f_* \Omega^\cdot(\mathbf{Z}))$. The cohomology sheaves of $f_* \Omega^\cdot(\mathbf{Z})$ are the $R^q f_* \mathbf{Z}$ and the complexes $F^\cdot \otimes_{\mathbf{Z}} R^q f_* Z$ are perfect. Hence, by 2.4, we have

$$W_{\mathbf{Z}}(X, F^\cdot \otimes_{\mathbf{Z}} f_* \Omega^\cdot(\mathbf{Z})) = \sum_q (-1)^q W_{\mathbf{Z}}(X, F^\cdot \otimes_{\mathbf{Z}} R^q f_* \mathbf{Z}).$$

Therefore, we are reduced to proving that, for any locally constant sheaf M whose stalks are finitely generated abelian groups, we have

$$W_{\mathbf{Z}}(X, F^\cdot \otimes_{\mathbf{Z}} M) = W_{\mathbf{Z}}(X, F^\cdot) \cdot \mathrm{cl}(M).$$

When M is locally free, this equality follows from (3.6), so that, using (2.4), we may assume that the stalks of M are torsion groups. But then M has a resolution of length two by locally free sheaves of finite rank and the equality follows from (3.6) and (2.4).

Analogous results for singular homology can be found in [**6**] and [**7**].

REFERENCES

1. Séminaire Heidelberg-Strasbourg 1966–67. Mimeographed notes.

2. A. Borel, *The Poincaré duality in generalized manifolds*, Michigan Math. J. **4** (1957), 227–239.

3. A. Borel and J. C. Moore, *Homology theory for locally compact spaces*, Michigan Math. J. **7** (1960), 137–159.

4. C. T. C. Wall, *Finiteness conditions for CW-complexes*, Ann of Math. **81** (1965), 56–69.

5. ———, *Poincaré complexes*, Ann of Math. **86** (1967), 213–245.

6. S. M. Gersten, *Product formula for Wall's obstruction*, Amer. J. Math. **88** (1966), 337–346.

7. K. Varadarajan, *On a conjecture of Milnor*, Illinois J. Math. (to appear).

COLUMBIA UNIVERSITY AND
UNIVERSITÉ DE STRASBOURG

A formula for $K_1R_\alpha[T]$[1]

F. T. Farrell and W. C. Hsiang[2]

0. **Introduction.** In this paper we obtain a formula of $K_1R_\alpha[T]$ where $R_\alpha[T]$ is the α-twisted finite Laurent series ring.[3] (See Theorem 19.) Our formula is a generalization of that obtained in [7] for $K_1R[T]$. We note that the techniques of [7] do not generalize in a straightforward fashion so as to prove our formula. As an important special case, we obtain a formula for Wh $G \times_\alpha T$ where $G \times_\alpha T$ is the semidirect product of a group G and the infinite cyclic group T with respect to an automorphism α of G. (See Theorem 21.) We obtain a geometric interpretation of this last formula in [4] and derive several applications from this interpretation. In §4 we derive a few immediate consequences of our formula. In Appendix 1, we exhibit a certain pathology as a warning to applying our formula carelessly; and in Appendix 2, we give an example to illustrate that a certain stronger generalization of the formula from [7] is false.

We wish to thank H. Bass for suggesting the possibility that the formula for $K_1R[T]$ of [7] could be generalized to $K_1R_\alpha[T]$.

1. **Notation and recollection of some facts from algebraic K-theory.** The purpose of this section is to introduce some notation to be used throughout the entire paper, and to recall those definitions and results from algebraic K-theory which will be needed in this paper.

The following is a partial list of symbols used in this paper:

R. An associative ring with identity. Ring homomorphisms are assumed to map identity to identity.

G. A group.

Z. The integers considered as a ring.

T. The infinite cyclic group (written multiplicatively).

---

[1] We announced in paragraph 2 of [18] some of the results proven in this paper.

[2] Both authors were partially supported by NSF Grant NSF-GP-6520. The second named author also held an Alfred P. Sloan Fellowship.

[3] C. T. C. Wall has independently obtained formula (31), which is the chief result of §2 of this paper.

t. A preferred generator for T.

T^+, T^-. The subsemigroups (with identity) of T generated by t and t^{-1} respectively.

$R(G)$. The R group ring of G.

α. An automorphism of R or G. If α is an automorphism of G, then the same symbol is used to denote the induced automorphism of $R(G)$ defined by

$$\alpha\Big(\sum_{g\in G} r_g g\Big) = \sum_{g\in G} r_g\alpha(g)$$

where $g \in G$ and $r_g \in R$.

$R_\alpha[T]$. Called the α-twisted finite Laurent series ring. (Also called the α-twisted R group ring of T.) Additively, $R_\alpha[T] = R[T]$. Multiplication in $R_\alpha[T]$ is defined by the following condition:

$$(rt^i)(st^j) = r\alpha^{-i}(s)t^{i+j}$$

where $r, s \in R$. There is an automorphism of $R_\alpha[T]$ induced by α and which we also denote by α, defined by the following condition:

$$\alpha(rt^i) = \alpha(r)t^i \quad \text{where} \quad r \in R.$$

$R_\alpha[t], R_\alpha[t^{-1}]$. The subrings of $R_\alpha[T]$ generated by R and t, and R and t^{-1} respectively. $R_\alpha[t]$ is called the α-twisted polynomial ring. The automorphism α of $R_\alpha[T]$ restricts to automorphisms for these two subrings which we also denote by α.

The following are ring homomorphisms which are inclusion maps.

j. $R \to R_\alpha[T]$.

k or k^+. $R \to R_\alpha[t]$.

k^-. $R \to R_\alpha[t^{-1}]$.

i or i^+. $R_\alpha[t] \to R_\alpha[T]$.

i^-. $R_\alpha[t^{-1}] \to R_\alpha[T]$.

The next two ring homomorphisms are augmentations.

ε or ε^+. $R_\alpha[t] \to R$ defined by the condition $\varepsilon(t) = 0$.

ε^-. $R_\alpha[t^{-1}] \to R$ defined by $\varepsilon^-(t^{-1}) = 0$.

$G \times_\alpha T$. The semidirect product of G and T with respect to α. Recall that if (g, t^i) and (g', t^j) are elements in $G \times_\alpha T$, then $(g, t^i)(g', t^j) = (g\alpha^{-i}(g'), t^{i+j})$.

$G \times_\alpha T^+, G \times_\alpha T^-$. The subsemigroups of $G \times_\alpha T$ generated by G and T^+, and G and T^- respectively.

The next two symbols apply to an abelian group G.

G^α. $\{g \mid \alpha(g) = g, g \in G\}$,

$I(\alpha)$. $\{g - \alpha(g) \mid g \in G\}$.

An additive map f from a right R module M_1 to a right R module M_2 is called *α linear* if $f(mr) = f(m)\alpha^{-1}(r)$ for $m \in M$ and $r \in R$.

$M(f)$. The cokernel of the α linear map f.

αM. If M is a right R module, then αM is a new right R module such that additively $\alpha M = M$ but which possesses a new scalar multiplication defined as follows: $m \cdot r = m\alpha(r)$, where $m \in M$, $r \in R$ and $\cdot$ is used to denote the scalar multiplication in αM.

c. $M \to \alpha M$, the identity map between sets "c" is an α^{-1} linear isomorphism.

r_a. $M \to M$, right multiplication by $a \in R$, i.e. $r_a(m) = ma$ for $m \in M$. If M is a right module over $R_\alpha[T]$ then r_t is an α linear endomorphism.

$M(n, m, R)$. The collection of $n \times m$ matrices with entries from R.

$\mathrm{GL}(n, R)$. The group of invertible matrices from $M(n, n, R)$.

$\alpha(A)$. If $A \in M(n, m, R)$, then $\alpha(A) \in M(n, m, R)$ such that $\alpha(A)_{ij} = \alpha(A_{ij})$. Here A_{ij} denotes the entry in the ith row and jth column of A.

f^T. If f is a ring homomorphism from R_1 to R_2 then f^T is the ring homomorphism from $R_1[T]$ to $R_2[T]$ extending f so that $f^T(t) = t$.

$\mathrm{r.dim}_R M$. The projective dimension of M, where M is a right R module. (See [**1**], p. 109.)

$\mathrm{w.dim}_R M$. Weak dimension of M over R. (See [**1**], p. 122.)

$\mathrm{r.gl.dim}\, R$. Right global dimension of R. (See [**1**], p. 111.)

$E_{pq}(r), p \neq q$. Square matrix with 1 down the diagonal and whose only off diagonal entry is $r \in R$ in the pth row, qth column. Such a matrix is said to be elementary.

(r). Square matrix whose diagonal entries are all equal to $r \in R$ and whose remaining entries are all zero.

Next, we introduce some notation and recall some simple facts concerning the relationship between matrices, α linear maps, and bases for free modules.

Let $a \in M(m, n, R)$, and let V and V' be free right R modules with ordered bases $e = (e_1, \ldots, e_n)$ and $e' = (e'_1, \ldots, e'_m)$. Then, the *$\alpha$ linear homomorphism $f: V \to V'$ associated to a with respect to e and e'* is defined by the following formula:

$$f\left(\sum_{i=1}^{n} e_i r_i\right) = \sum_{1 \le i \le n;\, 1 \le j \le m} e_j a_{ji} \alpha(r_i) \tag{1}$$

where $r_i \in R$.

With e and e' fixed, we thus obtain a 1-1 correspondence between $M(m, n, R)$ and the collection of α linear homomorphisms of V to V'. Let f' be an α' linear homomorphism from V' to a third free R module V'' corresponding to $a' \in M(k, m, R)$ with respect to e' and an ordered basis $e'' = (e''_1, \ldots, e''_k)$ for V''.

LEMMA 1. *$f'f$ is the $\alpha'\alpha$ linear homomorphism corresponding to $a'\alpha'(a)$ with respect to e and e''. (If f and g are functions, fg denotes their composition, i.e. $fg(x) = f(g(x))$ for x in the domain of g.)*

If $e = (e_1, \ldots, e_n)$ is an ordered basis for V and $a \in \mathrm{GL}(n, R)$, then $ea = (ea_1, \ldots, ea_n)$ is a new basis for V defined by the following formula:

$$ea_i = \sum_{j=1}^{n} e_j a_{ji}. \tag{2}$$

LEMMA 2. *If $a, b \in \mathrm{GL}(n, R)$, then $e(ab) = (ea)b$.*

Let $a \in M(m, n, R)$, $b \in \mathrm{GL}(n, R)$, $c \in \mathrm{GL}(m, R)$, and let V, V' be two free right R modules with ordered basis $e = (e_1, \ldots, e_n)$ and $e' = (e'_1, \ldots, e'_m)$ respectively.

LEMMA 3. *If $f: V \to V'$ is the α linear homomorphism associated to a with respect to e and e', then f is also the α linear homomorphism associated to $c^{-1}a\alpha(b)$ with respect to eb and $e'c$.*

The proofs of Lemmas 1, 2, and 3 will be left to the reader.

Next, we recall some definitions and results from algebraic K-theory. For more details, the reader is referred to [**2**] or [**3**].

$$K_1R = \operatorname*{direct\ limit}_{n\to\infty} \mathrm{GL}(n, R)/[\mathrm{GL}(n, R), \mathrm{GL}(n, R)].$$

If $a \in \mathrm{GL}(n, R)$, denote the corresponding element in K_1R by $[a]$. $J(G) =$ subgroup of $K_1Z(G)$ whose elements are $[(\pm g)]$ for $g \in G$, where $(\pm g)$ is the 1×1 matrix with single entry g or $-g$.

$\mathrm{Wh}\, G = K_1Z(G)/J(G)$ and is called the Whitehead group of G. K_1 and Wh are covariant functors. The homomorphism induced by a ring homomorphism $f: R_1 \to R_2$ (or a group homomorphism $f: G_1 \to G_2$) will be denoted by f_*.

$\mathbf{P}(R)$ denotes the category whose objects are finitely generated projective right R modules, and whose maps are R linear homomorphisms. K_0R is the Grothendieck group of $\mathbf{P}(R)$, i.e. the abelian group generated by the isomorphism classes of objects in $\mathbf{P}(R)$ modulo the relations $(P_2 - P_1 - P_3)$ for short exact sequences $0 \to P_1 \to P_2 \to P_3 \to 0$. Let $[P]$ denote the element in K_0R corresponding to $P \in \mathbf{P}(R)$. $\mathbf{F}(R)$ is the cyclic subgroup of K_0R generated by $[R]$. (Here R is considered as the free right R module of rank 1.) $\tilde{K}_0R = K_0R/\mathbf{F}(R)$ and is called the projective class group of R. Let $f: R \to R'$ be a ring homomorphism. f induces a covariant functor $f_\# : \mathbf{P}(R) \to \mathbf{P}(R')$ defined by $f_\# (P) = PR' = P \otimes_R R'$ for $P \in \mathbf{P}(R)$. $f_\#$ induces homomorphisms $K_0R \to K_0R'$ and $\tilde{K}_0R \to \tilde{K}_0R'$, both of which will again be denoted by f_*.

Now, let $f: R_1 \to R_2$ be a fixed ring homomorphism. We manufacture a category Φf whose objects are triples, $\sigma = (P, v, Q)$ where $P, Q \in \mathbf{P}(R_1)$ and $v: PR_2 \to QR_2$ is an R_2 linear isomorphism. Let $\sigma' = (P', v', Q')$ be another such triple. A morphism $\sigma \to \sigma'$ is a pair (g, h) of R_1 linear homomorphisms, $g: P \to P'$ and $h: Q \to Q'$ such that the diagram

$$\begin{array}{ccc} PR_2 & \xrightarrow{\ v\ } & QR_2 \\ \downarrow{\scriptstyle gR_2} & & \downarrow{\scriptstyle hR_2} \\ P'R_2 & \xrightarrow{\ v'\ } & Q'R_2 \end{array} \tag{3}$$

is commutative. Here gR_2 and hR_2 denote $g \otimes \mathrm{id}$ and $h \otimes \mathrm{id}$, respectively. The sequence

$$\sigma \xrightarrow{(g,h)} \sigma' \xrightarrow{(g',h')} \sigma''$$

is exact if the induced sequences

$$P \xrightarrow{g} P' \xrightarrow{g'} P'' \quad \text{and} \quad Q \xrightarrow{h} Q' \xrightarrow{h'} Q''$$

are exact sequences in $\mathbf{P}(R)$. The symbol $\sigma_1 \oplus \sigma_2$ denotes the object $(P_1 \oplus P_2,$

$v_1 \oplus v_2$, $Q_1 \oplus Q_2$) where $\sigma_1 = (P_1, v_1, P_2)$, $\sigma_2 = (P_2, v_2, Q_2)$ are objects in Φf. If $\sigma = (P, v, P')$ and $\sigma' = (P', v', P'')$, we write $\sigma'\sigma = (P, v'v, P'')$. $K_1\Phi f$ is the abelian group generated by the isomorphism classes of objects from Φf modulo the relations:

(A) (ADDITIVE). $(\sigma - \sigma' - \sigma'')$ for short exact sequences $0 \to \sigma' \to \sigma \to \sigma'' \to 0$ in Φf.

(M) (MULTIPLICATIVE). $(\sigma - \sigma' - \sigma'')$ for $\sigma = \sigma'\sigma''$.

We denote the class of $\sigma = (P, v, Q)$ in $K_1\Phi f$ by $[\sigma] = [P, v, Q]$.

Next, we define a homomorphism $q: K_1R_2 \to K_1\Phi f$. Let $a \in \mathrm{GL}(n, R_2)$ and let $\varphi: (R_2)^n \to (R_2)^n$ be the linear homomorphism associated to a with respect to the standard basis for $(R_2)^n = R_2 \oplus \cdots \oplus R_2$, where this sum has length n. By use of the standard identification of $(R_1)^n R_2$ with $(R_2)^n$, we obtain a well-defined object $\sigma = ((R_1)^n, \varphi, (R_1)^n)$ in Φf, and φ is an element in $\mathrm{Aut}_{R_2}(R_2)^n$, the automorphism group of $(R_2)^n$. For simplicity, we shall call σ an "automorphism". For each $P \in \mathbf{P}(R_1)$, let I_P denote the identity automorphism of PR_2. Then $\tau_P = (P, I_P, P)$ will be called an "identity". Clearly, the "identity" of $(R_1)^n$ is an "automorphism". Since $\tau_P\tau_P = \tau_P$, the relation (M) for $K_1\Phi f$ implies that $[\tau_P] = 0$ in $K_1\Phi f$.

Let $a \in \mathrm{GL}(n, R_2)$ and let σ be the associated "automorphism". If we replace σ by $\sigma \oplus \tau_{(R_1)m}$, we see from relation (A) that $[\sigma] = [\sigma \oplus \tau_{(R_1)m}]$ in $K_1\Phi f$. Hence by letting $q_n(a) = \sigma$ we obtain maps $q_n: \mathrm{GL}(n, R_2) \to K_1\Phi f$ which are consistant with our stabilization. By property (M) and Lemma 1, we see that each q_n is a group homomorphism, and hence they induce a homomorphism $q: K_1R_2 \to K_1\Phi f$.

Next, we give an alternate description of $K_1\Phi f$, which we will find useful in [**4**]. If $\sigma, \sigma' \in \Phi f$, we write $\sigma \sim \sigma'$ if there exist identities τ and τ' and a matrix $a \in \mathrm{GL}(n, R_2)$ with $[a] = 0$ in K_1R_2 (for some integer n) such that $(\sigma \oplus \tau)\varepsilon \cong \sigma' \oplus \tau'$, where ε is the automorphism associated to a. It can be seen that "$\sim$" is an equivalence relation.

PROPOSITION 4. *The objects of Φf under $\oplus$ modulo the relation "$\sim$" form an abelian group G. Moreover, the natural map which sends an object from Φf to its class in $K_1\Phi f$ induces an isomorphism $h: G \to K_1\Phi f$.*

For the proof of Proposition 4, see [**3**, pp. 34–35]. Finally, we define $\partial: K_1\Phi f \to K_0R$ by $\partial[P, v, Q] = [P] - [Q]$.

THEOREM 5. *Let $f: R_1 \to R_2$ be a ring homomorphism. Then, the sequence*

$$K_1R_1 \xrightarrow{f_*} K_1R_2 \xrightarrow{q} K_1\Phi f \xrightarrow{\partial} K_0R_1 \xrightarrow{f_*} K_0R_2 \tag{4}$$

is exact.

For a proof of Theorem 5 see [**3**, pp. 35–37] (See also [**5**] for a more general sequence.)

2. **The category $\mathscr{C}(R, \alpha)$ and the group $C(R, \alpha)$.** Let us first recall some terminology and results from [**6**].

$\mathscr{C}(R, \alpha)$ is the category whose objects are pairs (P, φ) where $P \in \mathbf{P}(R)$ and φ is an α linear nilpotent endomorphism of P, and whose morphisms $g:(P_1, \varphi_1) \to (P_2, \varphi_2)$ are R linear homomorphisms g from P_1 to P_2 satisfying the following commutative diagram:

$$(5) \qquad \begin{array}{ccc} P_1 & \xrightarrow{\varphi_1} & P_1 \\ \downarrow g & & \downarrow g \\ P_2 & \xrightarrow{\varphi_2} & P_2 \end{array}$$

We have the *Forgetting Functor* $\mathscr{F}: \mathscr{C}(R, \alpha) \to \mathbf{P}(R)$ by throwing away the endomorphism φ for $(P, \varphi) \in \mathscr{C}(R, \alpha)$, and the *Zero Functor* $\mathscr{T}: \mathbf{P}(R) \to \mathscr{C}(R, \alpha)$ which sends P to $(P, 0)$ for $P \in \mathbf{P}(R)$. Both $\mathscr{F}$ and $\mathscr{T}$ are covariant exact functors, and $\mathscr{F} \circ \mathscr{T}$ is the identity functor of $\mathbf{P}(R)$.

$C'(R, \alpha)$ is the abelian group generated by the isomorphism classes of objects from $\mathscr{C}(R, \alpha)$ modulo the relations $X_1 - X_0 - X_2$ for short exact sequences $0 \to X_2 \to X_1 \to X_0 \to 0$ in $\mathscr{C}(R, \alpha)$. If $X = (P, \varphi)$ is an object in $\mathscr{C}(R, \alpha)$, then the corresponding element in $C'(R, \alpha)$ is denoted by $[X]$ or $[P, \varphi]$. Let $\mathbf{F}'(R)$ denote the cyclic subgroup of $C'(R, \alpha)$ generated by $[R, 0]$; then $C(R, A) = C'(R, \alpha)/\mathbf{F}'(R)$, while $\tilde{C}(R, \alpha)$ is the subgroup of $C(R, \alpha)$ generated by $[R^n, \varphi]$ for $(R^n, \varphi) \in \mathscr{C}(R, \alpha)$. (Again, if $X = (P, \varphi) \in \mathscr{C}(R, \alpha)$, we will denote the corresponding element in any of the three groups $C'(R, \alpha)$, $C(R, \alpha)$, or $\tilde{C}(R, \alpha)$ by either $[X]$ or $[P, \varphi]$.)

PROPOSITION 6. *We have the following split exact sequences*:

$$(6) \qquad \begin{gathered} 0 \longrightarrow \tilde{C}(R, \alpha) \xrightarrow{I} C'(R, \alpha) \underset{\mathscr{T}_*}{\overset{\mathscr{F}_*}{\rightleftarrows}} K_0R \longrightarrow 0 \\ 0 \longrightarrow \tilde{C}(R, \alpha) \longrightarrow C(R, \alpha) \underset{\mathscr{T}_*}{\overset{\mathscr{F}_*}{\rightleftarrows}} \tilde{K}_0R \longrightarrow 0 \end{gathered}$$

where $\mathscr{F}_*$ and $\mathscr{T}_*$ are homomorphisms induced by $\mathscr{F}$ and $\mathscr{T}$ respectively; and $I[R^n, \varphi] = [R^n, \varphi] - [R^n, 0]$. Moreover, the splittings and I are natural with respect to homomorphisms $g:(R, \alpha) \to (R', \alpha')$, i.e. ring homomorphisms g such that $g \circ \alpha = \alpha' \circ g$. Hence, we write

$$(7) \qquad C'(R, \alpha) = \tilde{C}(R, \alpha) \oplus K_0R, \qquad C(R, \alpha) = \tilde{C}(R, \alpha) \oplus \tilde{K}_0R,$$

where in the first equation we identify $\tilde{C}(R, \alpha)$ with $I(\tilde{C}(R, \alpha))$; and K_0R and $\tilde{K}_0R$ are identified with their images under $\mathscr{T}_*$.

For the proof of this proposition, see **[6]**.

LEMMA 7. *Let $\delta_1:P_1 \to P_2$ and $\delta_2:P_2 \to P_3$ be α_1 and α_2 linear homomorphisms, where P_1, P_2, and P_3 are right R modules. If δ_2 is a monomorphism, then the following sequence is exact:*

$$(8) \qquad 0 \to \alpha_2 M(\delta_1) \to M(\delta_2\delta_1) \to M(\delta_2) \to 0.$$

PROOF. From $\delta_2\delta_1 P_1 \subset \delta_2 P_2 \subset P_3$, we obtain the following exact sequence:

$$0 \to \delta_2P_2/\delta_2\delta_1P_1 \to P_3/\delta_2\delta_1P_1 = M(\delta_2\delta_1) \to P_3/\delta_2P_2 = M(\delta_2) \to 0.$$

Since $\delta_2 : M(\delta_1) = P_2/\delta_1 P_1 \to \delta_2 P_2/\delta_2\delta_1 P_1$ is an α_2 linear isomorphsm, $\delta_2 P_2/\delta_2\delta_1 P_1 \cong \alpha_2\, M(\delta_1)$.

Next, we study the relationship of Φi to $C'(R, \alpha)$. (Recall that $i: R_\alpha[t] \to R_\alpha[T]$.) Let $S = R_\alpha[T]$ and $P \in \mathbf{P}(R_\alpha[t])$, then we identify P with $P \otimes_{R_\alpha[t]} R_\alpha[t] \subseteq PS$ by sending x to $x \otimes 1$ for $x \in P$.

THEOREM 8. *Let* $(P, v, Q) \in \Phi i$; *then*

(a) *there is an integer* $n \geq 0$ *such that* $r_{t^n} v(P) \subset Q$;

(b) $M(r_{t^n} v) \in \mathbf{P}(R)$ (*Here* $r_{t^n} v : P \to Q$ *is an* α^n *linear map between* $R_\alpha[t]$ *modules and we disregard the extra structure of* $M(r_{t^n} v)$ *as an* $R_\alpha[t]$ *module, retaining only its* R *module structure*), *and* r_t *induces an* α *linear nilpotent endomorphism on* $M(r_{t^n} v)$, *i.e.* $(M(r_{t^n} v), r_t) \in \mathscr{C}(R, \alpha)$;

(c) *and if we define* $\chi : K_1 \Phi i \to C'(R, \alpha)$ *by*

$$\chi[P, V, Q] = [M(r_{t^n} v), r_t] - [P/r_{t^n}(P), r_t],$$

then χ *is an isomorphism.*

PROOF OF (a). For any $x \in P$, we can find $n(x) \geqslant 0$ such that $r_{t^{n(x)}}(v(x)) \in Q$; and since P is finitely generated, $r_{t^n} v(P) \subset Q$ for $n \geq \max\{n(x_1), \ldots, n(x_s)\}$, where $\{x_1, \ldots, x_s\}$ are generators for P.

PROOF OF (b) By (a) applied to (Q, v^{-1}, P), there exists $m \geqslant 0$ such that $r_{t^m} v^{-1}(Q) \subset P$. From this and the fact that v is $R_\alpha[t]$ linear, we obtain that $r_{t^{n+m}}(Q) \subseteq r_{t^n} v(P)$. Hence, $r_t : M(r_{t^n} v) \to M(r_{t^n} v)$ is nilpotent. Since $M(r_{t^n} v)$ is a finitely generated $R_\alpha[T]$ module and r_t is nilpotent, $M(r_{t^n} v)$ is a finitely generated R module. Let $P' = r_{t^n} v(P)$ and $\varphi = r_{t^{n+m}}$. From $\varphi(P') \subset \varphi(Q) \subset P' \subset Q$, we obtain the following two exact sequences:

$$(9) \qquad 0 \to P'/\varphi(Q) \to Q/\varphi(Q) \to Q/P' = M(r_{t^n} v) \to 0,$$

and

$$(10) \qquad 0 \to \alpha_{n+m} M(r_{t^n} v) = \varphi(Q)/\varphi(P') \to P'/\varphi(P') \to P'/\varphi(Q) \to 0.$$

Since

$$Q/\varphi(Q),\, P'/\varphi(P') \in \mathbf{P}(R) \quad \text{and} \quad \text{r}\,.\,\dim_R M(r_{t^n} v) \leq 1$$

($0 \to P' \to Q \to M(r_{t^n} v) \to 0$ can be considered as a projective resolution of $M(r_{t^n} v)$ over R of length 1), $P'/\varphi(Q)$ is a projective R module by (9) and $\alpha_{n+m} M(r_{t^n} v)$ is a projective R module by (10). Hence $M(r_{t^n} v)$ is a projective R module, and $(M(r_{t^n} v), r_t) \in \mathscr{C}(R, \alpha)$.

PROOF OF (c). We first show that χ is well defined. Let $\chi_1 : \Phi i \to C'(R, \alpha)$ be the map defined by $\chi_1(P, v, Q) = [M(r_{t^n} v), r_t] - [P/r_{t^n}(P), r_t]$ for any n such that $r_{t^n} v(P) \subset Q$. We assert that χ_1 is independent of n. It suffices to show that $[M(r_{t^n} v), r_t] - [P/r_{t^n}(P), r_t] = [M(r_{t^{n+1}} v), r_t] - [P/r_{t^{n+1}}(P), r_t]$. We decompose

$$r_{t^{n+1}} v : P \to Q \quad \text{and} \quad r_{t^{n+1}} : P \to P$$

into

$$P \xrightarrow{r_t} P \xrightarrow{r_{t^n} v} Q \quad \text{and} \quad P \xrightarrow{r_t} P \xrightarrow{r_{t^n}} P.$$

(Observe that $r_{t^{n+1}}v = r_{t^n}vr_t$ since v is $R_\alpha[T]$ linear.) It follows from Lemma 7 that

(11) $$[M(r_{t^{n+1}}v), r_t] = [\alpha^n P/r_t P, r_t] + [M(r_{t^n}v), r_t],$$

and

(12) $$[P/r_{t^{n+1}}P, r_t] = [\alpha^n P/r_t P, r_t] + [P/r_{t^n}P, r_t].$$

Subtracting (12) from (11) we obtain the desired result, and hence χ_1 is well defined.

Next, we show that χ_1 respects relations (A) and (M). Let

$$0 \longrightarrow \sigma_1 \xrightarrow{(g_1,g_1')} \sigma_2 \xrightarrow{(g_2,g_2')} \sigma_3 \longrightarrow 0$$

be a short exact sequence in Φi where $\sigma_i = (P_i, v_i, Q_i)$ for $i = 1, 2, 3$. By (a), there exists an $n \geq 0$ such that $r_{t^n}v_2(P_2) \subseteq Q_2$. Consider the diagram

(13)
$$\begin{array}{ccccccccc} 0 & \longrightarrow & P_1 & \xrightarrow{g_1} & P_2 & \xrightarrow{g_2} & P_3 & \longrightarrow & 0 \\ & & \downarrow{\scriptstyle r_t^n v_1} & & \downarrow{\scriptstyle r_t^n v_2} & & \downarrow{\scriptstyle r_t^n v_3} & & \\ 0 & \longrightarrow & Q_1 & \xrightarrow{g_1'} & Q_2 & \xrightarrow{g_2'} & Q_3 & \longrightarrow & 0 \end{array}$$

as a short exact sequence of chain complexes. Then (13) induces the following sequence in homology:

(14) $$\cdots \to \mathrm{Ker}\, r_{t^n}v_3 \to M(r_{t^n}v_1) \to M(r_{t^n}v_2) \to M(r_{t^n}v_3) \to 0.$$

Since $\mathrm{Ker}\, r_{t^n}v_3 = 0$, we obtain from (14) the following exact sequence in $\mathscr{C}(R, \alpha)$:

(15) $$0 \to (M(r_{t^n}v_1), r_t) \to (M(r_{t^n}v_2), r_t) \to (M(r_{t^n}v_3), r_t) \to 0.$$

From (15), we obtain the following equation in $C'(R, \alpha)$:

(16) $$[M(r_{t^n}v_2), r_t] = [M(r_{t^n}v_3), r_t] + [M(r_{t^n}v_1), r_t].$$

By the same method that we obtained equation (16), we can also obtain the following equation:

(17) $$[P_2/r_{t^n}(P_2), r_t] = [P_3/r_{t^n}(P_3), r_t] + [P_1/r_{t^n}(P_1), r_t].$$

Subtracting (17) from (16) we find that $\chi_1(\sigma_2) = \chi_1(\sigma_1) + \chi_1(\sigma_3)$ and hence χ_1 respects relation (A).

Now let $\sigma_1 = (P_1, v_1, P_2)$ and $\sigma_2 = (P_2, v_2, P_3)$ be two objects from Φi. Put $\sigma_3 = (P_1, v_3, P_3)$ with $v_3 = v_2 v_1$. By (a), we can find $n, m \geq 0$ such that $r_{t^n}v_1(P_1) \subseteq P_2$, $r_{t^m}v_2(P_2) \subseteq P_3$, and hence $r_{t^{n+m}}(P_1) \subseteq P_3$. Since v_1, v_2 are $R_\alpha[T]$ linear homomorphisms, we have the following commutative diagrams:

(18)
$$\begin{array}{ccc} P_1 \xrightarrow{r_t^n v_1} P_2 & \quad & P_1 \xrightarrow{r_t^m} P_1 \\ {\scriptstyle r_t^{n+m}v_3}\searrow \downarrow{\scriptstyle r_t^m v_2} & & {\scriptstyle r_t^{n+m}}\searrow \downarrow{\scriptstyle r_t^n} \\ P_3 & & P_1 \end{array}$$

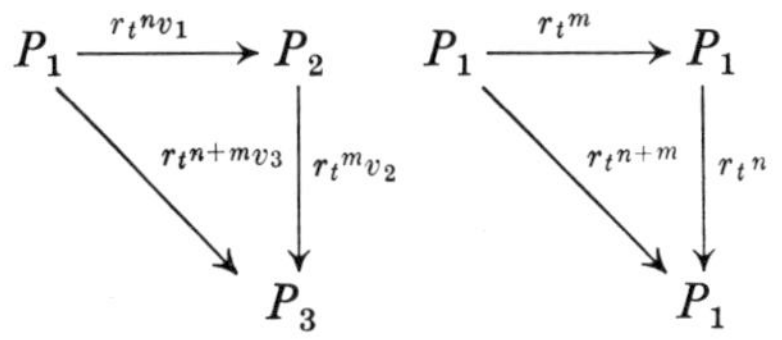

If we apply Lemma 7 to (18), we obtain the following equation in $C'(R, \alpha)$:

$$\begin{aligned}(19)\quad [M(r_{t^{n+m}}v_3), r_t] - [P_1/r_{t^{n+m}}(P_1), r_t] &= [M(r_{t^m}v_2), r_t] + [\alpha^m M(r_{t^n}v_1), r_t] \\ &\quad - [P_1/r_{t^n}(P_1), r_t] - [\alpha^n(P_1/r_{t^m}(P_1)), r_t].\end{aligned}$$

Consider the commutative diagram

$$(20)\qquad \begin{array}{ccc} P_1 & \xrightarrow{r_t{}^n v_1} & P_2 \\ \Big\downarrow{\scriptstyle r_t{}^m} & & \Big\downarrow{\scriptstyle r_t} \\ P_1 & \xrightarrow[r_t{}^n v_1]{} & P_2 \end{array}$$

By Lemma 7 applied to (20), we have

$$(21)\quad [\alpha^m M(r_{t^n}v_1), r_t] - [\alpha^n(P_1/r_{t^m}(P_1)), r_t] = [M(r_{t^n}v_1), r_t] - [P_2/r_{t^m}(P_2), r_t].$$

Substituting (21) into (19), we obtain that $\chi_1(\sigma_3) = \chi_1(\sigma_1) + \chi_1(\sigma_2)$ and hence that χ_1 preserves relation (M). Therefore χ_1 is factored through $\chi: K_1\Phi i \to C'(R, \alpha)$, i.e. χ is a well defined homomorphism.

We shall next construct χ's inverse. But first, let us obtain the "characteristic sequence" for an arbitrary right $R_\alpha[t]$ module M. Denote $r_t: M \to M$ by v. If N is a right R module and $R_\alpha[t]$ is given a left R module structure via k, then we denote $N \otimes_R R_\alpha[t]$ by NS'. Now consider M as a right R module and recall that $c: M \to \alpha M$ is an α^{-1} linear isomorphism. Hence, $\tilde{v} = vc^{-1}: \alpha M \to M$ is an R linear homomorphism. Let $l_t: R_\alpha[t] \to R_\alpha[t]$ denote left multiplication by t; then l_t is $R_\alpha[t]$ linear as a map of right $R_\alpha[t]$ modules and α^{-1} linear as a map of left R modules. Hence $c^{-1} \otimes l_t$ and $\tilde{v} \otimes \mathrm{id}: (\alpha M)S' \to MS'$ are well-defined $R_\alpha[t]$ linear homomorphisms. Let $u = c^{-1} \otimes l_t - \tilde{v} \otimes \mathrm{id}$. Let $w: MS' = M \otimes_R R_\alpha[t] \to M$ be defined by $w(m \otimes t^i) = v^i(m)$. Then u and w are both $R_\alpha[t]$ linear homomorphisms.

LEMMA 9. *The following is an exact sequence of $R_\alpha[t]$ modules, and will be referred to as the characteristic sequence of the module M:*

$$(22)\qquad 0 \longrightarrow (\alpha M)S' \xrightarrow{u} MS' \xrightarrow{w} M \longrightarrow 0.$$

The proof of Lemma 9 follows from a computation and will be left to the reader.

Now, we are in a position to construct the inverse to χ. Let $(M, v) \in \mathscr{C}(R, \alpha)$; then M can be made into a right $R_\alpha[t]$ module by defining $xt = v(x)$ for $x \in M$. Since M and $\alpha M \in \mathbf{P}(R)$, we have that MS' and $(\alpha M)S' \in \mathbf{P}(R_\alpha[t])$. If we tensor (22) with $\otimes_{R_\alpha[t]} R_\alpha[T]$ and use the facts that $R_\alpha[T]$ is a flat left $R_\alpha[t]$ module and that v is nilpotent, then we obtain that $u \otimes \mathrm{id} = uS$ is an isomorphism. Hence, $\sigma = ((\alpha M)S', uS, MS') \in \Phi i$; and, again by Lemma 9, $\chi(\sigma) = [M, v]$. Let us define $\lambda_1: \mathscr{C}(R, \alpha) \to \Phi i$ by $\lambda_1(M, v) = ((\alpha M)S', uS, MS')$. Since λ_1 is an exact functor, λ_1 induces a homomorphism $\lambda: C'(R, \alpha) \to K_1\Phi i$ such that $\chi\lambda = \mathrm{id}$.

To complete the proof of Theorem 8, it suffices to show that λ is epimorphic. To do this, we show first that $K_1\Phi i$ is generated by elements of the form $[P, v, Q]$

where $v(P) \subset Q$, and secondly that each element of the above form is in the image of λ.

Let $(P, v, Q) \in \Phi i$; then by part (a) there exists an $n \geq 0$ such that $vr_{t^n}(P) = r_{t^n}v(P) \subset Q$. Recall that $c^1:\alpha^n P \to P$ is the α^n linear identification map. Since $c' = c^{-1} \otimes \alpha^n:(\alpha^n P)S \to PS$ is an α^n linear isomorphism, $r_{t^n}c':(\alpha^n P)S \to PS$ is a linear isomorphism; and hence $(\alpha^n P, r_{t^n}c', P) \in \Phi i$. We set

$$(\alpha^n P, vr_{t^n}c', Q) = (\alpha^n P, r_{t^n}c', P)(P, v, Q),$$

and write this equation $\sigma'' = \sigma'\sigma$. Then, $[\sigma] = [\sigma''] - [\sigma']$; and therefore $K_1\Phi$ is generated by elements of the form $[P', v', Q']$ where $v'(P') \subseteq Q'$, since $[\sigma'']$ and $[\sigma']$ are both elements of this type.

Let $[P, v, Q] \in \Phi i$ be such that $v(P) \subset Q$. By part (b), $(M(v), r_t) \in \mathscr{C}(R, \alpha)$; and we wish to show that $\lambda[M(v), r_t] = [P, v, Q]$. Because of our definition of λ_1, it suffices to show that if $(P_j, v_j, Q_j) \in \Phi i$, $v_j(P_j) \subseteq Q_j$, $j = 1$ or 2, and $M(v_1) \cong M(v_2)$ as $R_\alpha[t]$ module, then $[P_1, v_1, Q_1] = [P_2, v_2, Q_2]$. From our hypothesis, we obtain the following two projective resolutions of $M(v_1) = M$:

$$(23)\qquad \begin{array}{c} 0 \longrightarrow P_1 \xrightarrow{v_1} Q_1 \xrightarrow{\varphi_1} M \longrightarrow 0 \\ 0 \longrightarrow P_2 \xrightarrow{v_2'} Q_2 \xrightarrow{\varphi_2} M \longrightarrow 0 \end{array}$$

where $v_i'S = v_i$ for $i = 1, 2$. Hence there exists a homomorphism $f: Q_2 \to Q_1$ such that $\varphi_1 f = \varphi_2$. Consider the resolution

$$(24)\qquad 0 \longrightarrow P_1 \oplus Q_2 \xrightarrow{v_1' \oplus \mathrm{id}} Q_1 \oplus Q_2 \xrightarrow{\varphi} M \longrightarrow 0$$

where φ is the composite of the projection onto Q_1 followed by φ_1. Define $g: Q_2 \to Q_1 \oplus Q_2$ to be (f, id). Since $\varphi g = \varphi_2$, there exists a unique homomorphism $g':P_2 \to P_1 \oplus Q_2$ making the following diagram commutative:

$$(25)\qquad \begin{array}{ccccccccc} 0 & \longrightarrow & P_2 & \xrightarrow{g'} & P_1 \oplus Q_2 & \longrightarrow & Y & \longrightarrow & 0 \\ & & \downarrow{\scriptstyle v_2'} & & \downarrow{\scriptstyle v_1' \oplus \mathrm{id}} & & \downarrow{\scriptstyle \psi} & & \\ 0 & \longrightarrow & Q_2 & \xrightarrow{g} & Q_1 \oplus Q_2 & \longrightarrow & X & \longrightarrow & 0 \end{array}$$

In (25), g' is monomorphic since v_2' and g are; and X, Y, ψ denote cokernel modules and induced homomorphisms, respectively. By regarding (25) as a short exact sequence of chain complexes and passing to homology, we obtain that ψ is an isomorphism. But one easily sees that $X \cong Q_1$, and hence that $(Y, \psi S, X) \in \Phi i$. Tensoring (25) with $\otimes_{R_\alpha[t]} R_\alpha[T]$ and using relation (A), we obtain the following equation in $K_1\Phi i$:

$$(26)\qquad [P_1 \oplus Q_2, v_1 \oplus \mathrm{id}, Q_1 \oplus Q_2] = [P_2, v_2, Q_2] + [Y, \psi S, X].$$

Since ψ is an isomorphism, one sees easily that $[Y, \psi S, X] = 0$; and likewise $[Q_2, \mathrm{id}, Q_2] = 0$. Using these two facts and applying (A) to the left-hand side of (26), we obtain that $[P_1, v_1, Q_1] = [P_2, v_2, Q_2]$; and hence χ is an isomorphism.

If f maps (R, α) to (R', α'), i.e. $f: R \to R'$ is a ring homorphism such that $\alpha' f = f\alpha$, then f induces an exact functor from $\mathscr{C}(R, \alpha)$ to $\mathscr{C}(R', \alpha')$ sending (P, v) to $(P \otimes_R R', v \otimes \alpha')$; and this functor induces homorphisms from $C'(R, \alpha)$ to $C'(R', \alpha')$, from $C(R, \alpha)$ to $C(R', \alpha')$, and from $\tilde{C}(R, \alpha)$ to $\tilde{C}(R', \alpha')$. All of these homomorphisms are denoted by f_*. If f happens to be α, then one easily sees that $\alpha_* [P, v] = [\alpha P, v]$.

PROPOSITION 10. $\tilde{C}(R, \alpha)^{\alpha_*} = \tilde{C}(R, \alpha)$.

PROOF. Recall that $\tilde{C}(R, \alpha)$ is generated by elements of the form $[N, v]$ where N is a free R module. Look at the characteristic sequence (22) of N regarded as a right $R_\alpha[t]$ module, and apply Lemma 7 twice, once to $r_t u$ and the second time to $u r_t$, to obtain the following two equations:

$$(27)\qquad \begin{aligned} [M(r_t u), r_t] &= [NS'/r_t(NS'), 0] + [N, v], \\ [M(r_t u), r_t] &= [(\alpha N)S'/r_t((\alpha N)S'), 0] + \alpha_* [N, v]. \end{aligned}$$

Since N is free, $[NS'/r_t(NS'), 0]$ and $[(\alpha N)S'/r_t((\alpha N)S'), 0]$ are both 0. (Recall the definitions of $\mathbf{F}'(R)$ and $C(R, \alpha)$.) Hence (27) implies that $[N, v] = \alpha_* [N, v]$.

3. **The formula for $K_1 R_\alpha[T]$.** Applying the exact sequence (4) of Theorem 5 to $i: R_\alpha[t] \to R_\alpha[T]$, we obtain

$$(28)\qquad K_1 R_\alpha[t] \xrightarrow{i_*} K_1 R_\alpha[T] \xrightarrow{q} K_1 \Phi i \xrightarrow{\partial} K_0 R_\alpha[t] \xrightarrow{i_*} K_0 R_\alpha[T].$$

Let $l: C'(R, \alpha) \to K_0 R_\alpha[t]$ be the composite map

$$(29)\qquad C'(R, \alpha) \xrightarrow{\mathscr{F}_*} K_0 R \xrightarrow{\alpha_* - \mathrm{id}} K_0 R \xrightarrow{k_*} K_0 R_\alpha[t]$$

where $\mathscr{F}_*$ is induced by the "Forgetting Functor", $\mathscr{F}$.

LEMMA 11. *The following diagram is commutative*:

$$(30)\qquad \begin{array}{ccc} K_1 \Phi i & \xrightarrow{\quad\partial\quad} & K_0 R_\alpha[t] \\ & {\scriptstyle\chi}\searrow \quad \nearrow{\scriptstyle l} & \\ & C'(R, \alpha) & \end{array}$$

where χ is the isomorphism of Theorem 8.

PROOF. Since λ is the inverse to χ, it is sufficient to prove that $\partial\lambda = l$. But this follows immediately from the definitions of ∂, λ, and l.

Now, let $p = \chi q$; then the exact sequence (28) becomes

$$(31)\qquad K_1 R_\alpha[t] \xrightarrow{i_*} K_1 R_\alpha[T] \xrightarrow{p} C'(R, \alpha) \xrightarrow{l} K_0 R_\alpha[t] \xrightarrow{i_*} K_0 R_\alpha[T].$$

Next, we wish to analyze the map p more closely.

PROPOSITION 12. Image $p = (C'(R, \alpha))^{\alpha_*}$.

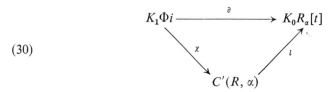

PROOF. Clearly, $l(C'(R, \alpha)^{\alpha_*}) = 0$, and hence $C'(R, \alpha)^{\alpha_*} \subseteq \text{image}\, p$. Let $[a] \in K_1R_\alpha[T]$ where $a \in \mathrm{GL}(n, R_\alpha[t])$, and let $[F, \varphi, F] = q[a]$, where $F = (R_\alpha[t])^n$. Then there exists an $m \geq 0$ such that $p[a] = [M(r_{t^m}\varphi), r_t] - [F/r_{t^m}(F), r_t]$. Applying Lemma 7 to both sides of the equation $r_t(r_{t^m}\varphi) = (r_{t^m}\varphi)r_t$ and using the fact that $(\alpha^m F, 0) \simeq (F, 0)$, we obtain that

$$\alpha_*[M(r_{t^m}\varphi), r_t] = [M(r_{t^m}\varphi), r_t].$$

But $(F/r_{t^m}(F), r_t) \simeq \alpha(F/r_{t^m}(F), r_t)$, and hence $\alpha_* p[a] = p[a]$.

Let $p' = pi_*^- : K_1R_\alpha[t^{-1}] \to C'(R, \alpha)$.

THEOREM 13. *Image* $p' = \tilde{C}(R, \alpha)$; *and the following sequence is split exact;*

$$0 \to K_1R \to K_1R_\alpha[t^{-1}] \xrightarrow{p'} \tilde{C}(R, \alpha) \to 0. \tag{32}$$

Hence, $K_1R_\alpha[t^{-1}] \simeq K_1R \oplus \tilde{C}(R, \alpha)$; *and likewise* $K_1R_\alpha[t] \simeq K_1R \oplus \tilde{C}(R, \alpha^{-1})$.

PROOF. If we consider the sequence (4) of Theorem 5 in the case where $f = k^-$ and observe that ε_*^- induces a splitting of this sequence, we obtain the following split exact sequence:

$$0 \to K_1R \underset{\varepsilon_*^-}{\overset{k_*^-}{\rightleftarrows}} K_1R_\alpha[t^{-1}] \xrightarrow{q} K_1\Phi k^- \to 0 \tag{33}$$

Next, we define a functor $\lambda_1' : \mathscr{C}(R, \alpha) \to \Phi k^-$ by $\lambda_1'(P, v) = (P, u', P)$ where $u' = \mathrm{id} - v \otimes l_{t^{-1}} : PS'' \to PS''$, $PS'' = P \otimes_R R_\alpha[t^{-1}]$. λ_1' is an exact functor, and hence induces a homomorphism $\lambda' : C'(R, \alpha) \to K_1\Phi k^-$. By use of Higman's trick (See [**8**], p. 359 and [7], Proposition 2.1.), we show that λ' is an epimorphism. By [7], Proposition 2.1, $K_1R_\alpha[t^{-1}]$ is generated by $k_*^-(K_1R)$ together with elements of the form $[1 - Nt^{-1}]$, where $N \in M(n, n, R)$ for some n and Nt^{-1} is nilpotent. Using Lemma 1 and the definition of q, we see that

$$q[1 - Nt^{-1}] = [R^n, \mathrm{id} - v \otimes l_{t^{-1}}, R^n]$$

where v is the α linear endomorphism of R^n associated to N with respect to the standard basis for R^n. Again by Lemma 1, we see that v is nilpotent, and hence that $q[1 - Nt^{-1}] = \lambda'[R^n, v]$. Therefore, λ' is an epimorphism; and in fact $\lambda' \mid \tilde{C}(R, \alpha)$ is epimorphic since $\lambda'[R^n, 0] = 0$. We define a homomorphism $\rho : K_1\Phi k' \to K_1\Phi i$ as follows: $\rho[P, \varphi, Q] = [PS', \varphi', QS']$ where φ' is defined so that the diagram

$$\begin{array}{ccc} PS'S & \xrightarrow{\varphi'} & QS'S \\ \uparrow{\scriptstyle\psi} & & \uparrow{\scriptstyle\psi} \\ PS''S''' & \xrightarrow{\varphi \otimes \mathrm{id}} & QS''S''' \end{array} \tag{34}$$

is commutative. Here, $XS''' = X \otimes_{R\alpha[t^{-1}]} R_\alpha[T]$ and ψ is the isomorphism such that $\psi(x \otimes a \otimes b) = x \otimes 1 \otimes i^-(a)b$ where $x \in P$, $a \in R_\alpha[t^{-1}]$, and $b \in R_\alpha[T]$. It can be seen easily that the following diagram is commutative:

$$(35)\qquad \begin{array}{ccc} K_1R_\alpha[t^{-1}] & \xrightarrow{\bar{i}_*} & K_1R_\alpha[T] \\ \downarrow{\scriptstyle q} & & \downarrow{\scriptstyle q} \\ K_1\Phi k^- & \xrightarrow{\rho} & K_1\Phi i \end{array}$$

We wish to calculate $\chi\rho\lambda'$. Now,

$$\rho\lambda'[P, v] = [PS', \mathrm{id} - v \otimes \alpha \otimes l_{t-1}, PS'];$$

and $r_t(\mathrm{id} - v \otimes \alpha \otimes l_{t-1}) = r_t - v \otimes \alpha \otimes \alpha$. Therefore,

$$\chi\rho\lambda'[P, v] = [M(r_t - v \otimes \alpha \otimes \alpha), r_t] - [P, 0].$$

Consider the α linear isomorphism $c \otimes \alpha \otimes \alpha : PS'S \to (\alpha P)S'S$. By composing this map with $uS = c^{-1} \otimes l_t \otimes \mathrm{id} - vc^{-1} \otimes \mathrm{id} \otimes \mathrm{id} = c^{-1} \otimes \alpha^{-1} \otimes l_t - vc^{-1} \otimes \mathrm{id} \otimes \mathrm{id}$, we obtain $r_t - v \otimes \alpha \otimes \alpha$. Since $\lambda[P, v] = [(\alpha P)S', uS, PS']$ and $c \otimes \alpha \otimes \alpha$ restricted to PS is still an isomorphism onto $(\alpha P)S'$, we have, by Lemma 7, that $\chi\rho\lambda'[P, v] = [P, v] - [P, 0]$; and hence $\chi\rho\lambda'$ restricted to $\tilde{C}(R, \alpha)$ is the identity map. From this, together with diagrams (33) and (35) and the fact that λ' restricted to $\tilde{C}(R, \alpha)$ is epimorphic, all the results stated in Theorem 13 can be immediately deduced.

The following theorem will enable us to compute the kernel of i_* (See formula (31).)

THEOREM 14. *Ker* $j_* = I(\alpha_*)$, *where* $j_* : K_1R \to K_1R_\alpha[T]$.

The proof of this theorem is motivated by the geometric arguments found in the last chapter of **[6]**.

PROOF. It is clear that $I(\alpha_*) \subseteq \ker(j_*)$; and hence we need only show that $\ker(j_*) \subseteq I(\alpha_*)$. First, let us introduce some more notation. If V is a free $R_\alpha[T]$ module and $e = (e_1, \ldots, e_n)$ is an ordered basis for V, then we define $V_e = \{\sum_{i-1}^n e_i v_i \mid v_i \in R_\alpha[t^{-1}]\}$. V_e is free both as an $R_\alpha[t^{-1}]$ module and as an R module.

LEMMA 15. *If* $a \in \mathrm{GL}(n, R_\alpha[T])$ *and each* $a_{ij} \in R_\alpha[t^{-1}]$, *then* $V_{ea} \subseteq V_e$.

The proof of this lemma is left to the reader.

Suppose that $j_*[a] = j_*[b]$. Then, after stabilizing, we may assume that $a = b\varepsilon'^1 \cdots \varepsilon'^l$ is a matrix equation in $\mathrm{GL}(n, R_\alpha[T])$ where each of ε'^i is an elementary matrix over $R_\alpha[T]$ whose off-diagonal entry is homogeneous in t. To prove that $\ker(j_*) \subseteq I(\alpha_*)$, it is sufficient to show that $[a] - [b] \in I(\alpha)$. If we conjugate a by a large enough power of t^m we obtain $\alpha^m(a) = t^{-m}b\varepsilon^1 \cdots \varepsilon^l$, where each matrix ε^i is of the form $E_{pq}(rt^{s'})t^s$, $r \in R$, $s' < s$, and $0 < s$.

Let V be a free $R_\alpha[T]$ module with ordered basis $e = (e_1, \ldots, e_n)$. Let $f: V \to V$ be the $R_\alpha[T]$ linear homomorphism associated to b with respect to bases e, e. By

Lemma 3, we see that f is also associated to $\alpha^m(a)$ with respect to $e\varepsilon^1 \cdots \varepsilon^l$ as a basis for its domain and $e(t^m)$ as a basis for its range.

Let $c^0 = (1)$, $c^1 = (t)$, $c^{i+1} = \varepsilon^1 \cdots \varepsilon^i$ for $1 \le i \le l$, and $c^{l+2} = c^{l+1}(t)$. By Lemma 15, we see that $Vec^i \subseteq Vec^{i+1}$. Consider the following two filtrations of V:

$$Ve \subset Vec^1 \subset Vec^{l+1} \subset Vec^{l+2} \subset V, \tag{36}$$

$$Ve \subset Ve(t) \subset Ve(t^m) \subset Ve(t^{m+1}) \subset V. \tag{37}$$

We see that f maps filtration (36) into filtration (37) and also f^{-1} maps (37) into (36); and hence f induces isomorphisms on the quotient modules. Let $M_1 = Vec^1/Ve$, $M_2 = Vec^{l+1}/Vec^1$, $M_3 = Vec^{l+2}/Vec^{l+1}$, $M_1' = Ve(t)/Ve$, $M_2' = Ve(t^m)/Ve(t)$, and $M_3' = Ve(t^{m+1})/Ve(t^m)$. Then f induces isomorphisms from M_i to M_i' for $i = 1, 2, 3$; we also denote these isomorphisms by f. Let $N_1 = Vec^{l+1}/Ve$, $N_2 = Vec^{l+2}/Vec^1$, $N_1' = Ve(t^m)/Ve$, and $N_2' = Ve(t^{m+1})/Ve(t)$. f induces isomorphisms (again denoted by f) from $N_i \to N_i'$ for $i = 1, 2$. Since $f\colon V \to V$ is $R_\alpha[T]$ linear, $fr_t = r_t f$; and hence the following diagram is commutative:

$$\begin{array}{ccc} N_1 & \xrightarrow{r_t} & N_2 \\ \downarrow{\scriptstyle f} & & \downarrow{\scriptstyle f} \\ N_1' & \xrightarrow{r_t} & N_2' \end{array} \tag{38}$$

where the maps denoted by r_t are induced by $r_t\colon V \to V$. The maps denoted by r_t are α linear isomorphisms.

We show next that each of the modules N_i and N_i' is a free R module. We will assign ordered bases to each of these modules and then derive from (38) a matrix equation. From this equation, we will obtain that $[a] - [b] \in I(\alpha_*)$.

A basis for a quotient module of a submodule of V will be chosen as follows. We will pick a set of elements in V and use the image of this set under the quotient map to give a basis to the sub quotient. The image of an element in V and the element itself will both be denoted by the same symbol as long as no confusion can result.

LEMMA 16. *If $a = E_{pq}(rt^{s'})(t^s)$, $s' \le s$, $0 \le s$, $r \in R$, and $e = (e_1, \dots, e_n)$ is an ordered basis for a free $R_\alpha[T]$ module V, then $(\bigcup_{i=0}^{\infty} e(t^{-i})) \cup (\bigcup_{i=1}^{s} ea^i)$ is a basis for Vea as a free R module, where $a^i = E_{pq}(rt^{s'})(t^i)$ for $i = 1, \dots, s$.*

The proof of Lemma 16 is straightforward and hence is omitted.

LEMMA 17. *Under the same hypothesis as in Lemma 16. Vea/Ve is a free R module with an ordered basis*

$$((ea^1)_1, (ea^1)_2, \dots, (ea^1)_n, (ea^2)_1, \dots, (ea^s)_1, \dots, (ea^s)_n).$$

For brevity, this basis will be denoted by the symbol $(ea^1, \dots, ea^s)$, and will be referred to as the basis for Vea/Ve associated with e (and a), or, if no confusion can result, the basis associated with e.

Lemma 17 is an immediate consequence of Lemma 16.

By Lemma 17, the modules M_i' and N_i' are free. We give the module M_1', M_2', M_3', N_1', and N_2' the bases associated with e, $e(t)$, $e(t^m)$, e, and $e(t)$ respectively. The modules M_1 and M_3 are free for the same reason; we give these modules the bases associated to e and ec^{l+1} respectively. In order to prove that the modules M_2, N_1, and N_2 are free and to assign bases to them, we need to define and examine a new collection of modules K_i. Let $K_i = Vec^i/Vec^{i-1}$ for $i = 1, \ldots, l+2$. By Lemma 17, the modules K_i are free; and we give K_i the basis associated with ec^{i-1}. Since each K_i is free, $M_2 = Vec^{l+1}/Vec' \cong \bigoplus_{i=2}^{l+1} K_i$ and hence is free. Likewise, N_1 and N_2 are free. The ordered basis assigned to M_2 is obtained by "stringing together" the ordered bases assigned to $K_2, \ldots, K_{l+1}$ in the order indicated. (That is, the basis for M_2 is the union of the bases for $K_2, \ldots, K_{l+1}$. Recall that all these elements are gotten from elements in V. If x and y are two elements in this basis, then x precedes y if either x and $y \in K_i$ and x precedes y here or $x \in K_i$ and $y \in K_j$ with $i < j$.) The ordered bases for N_1 and N_2 are obtained by stringing together the ordered bases for $K_1, \ldots, K_{l+1}$ and $K_2, \ldots, K_{l+2}$ respectively.

The number of elements in the bases assigned to N_1, N_2, N_1', and N_2' is the same and will be denoted by n'. Let A and $B \in \mathrm{GL}(n', R)$ be the matrices associated with the R linear isomorphisms $f: N_1 \to N_1'$ and $f: N_2 \to N_2'$, respectively, with respect to the bases assigned to these modules; and let C and $D \in \mathrm{GL}(n', R)$ be the matrices associated to the α linear isomorphisms $r_t: N_1 \to N_2$ and $r_t: N_1' \to N_2'$, respectively, with respect to the assigned bases. From Lemma 1 and formula (38) we obtain that $D\alpha(A) = BC$; and hence we have that

$$(39) \qquad [D] + \alpha_*[A] = [B] + [C]$$

in K_1R.

Let us now analyze the matrices A, B, C, and D more closely. D is clearly the identity matrix; hence $[D] = 1$. Consider next the following diagram:

$$(40) \qquad \begin{array}{ccccccccc} 0 & \longrightarrow & M_1 & \longrightarrow & N_1 & \longrightarrow & M_2 & \longrightarrow & 0 \\ & & \downarrow f & & \downarrow f & & \downarrow f & & \\ 0 & \longrightarrow & M_1' & \longrightarrow & N_1' & \longrightarrow & M_2' & \longrightarrow & 0 \end{array}$$

and note that the row maps in (40) respect the assigned bases. $f: M_1 \to M_1'$ is associated with b with respect to the assigned bases; we denote by d the invertible matrix associated with the R linear isomorphism $f: M_2 \to M_2'$. From (40), we see that the matrix A has the form

$$\left(\begin{array}{c|c} b & * \\ \hline 0 & d \end{array}\right)$$

and hence $[A] = [b] + [d]$ in K_1R. Likewise, by considering the following diagram:

$$\begin{array}{ccccccccc} 0 & \longrightarrow & M_2 & \longrightarrow & N_2 & \longrightarrow & M_3 & \longrightarrow & 0 \\ & & \Big\downarrow f & & \Big\downarrow f & & \Big\downarrow f & & \\ 0 & \longrightarrow & M_2' & \longrightarrow & N_2' & \longrightarrow & M_3' & \longrightarrow & 0 \end{array} \tag{41}$$

we see that B has the form

$$\left(\begin{array}{c|c} d & * \\ \hline 0 & \alpha^m(a) \end{array}\right)$$

and hence $[B] = [d] + \alpha_*^m[a]$ in K_1R. Using the above analysis of D, A, and B, equation (39) can be rewritten as

$$\alpha_*[b] - \alpha_*^m[a] = [d] - \alpha_*[d] + [C]. \tag{42}$$

By adding the quantity

$$[b] - \alpha_*[b] + \sum_{i=0}^{m-1} (\alpha_*\alpha_*^i[a] - \alpha_*^i[a])$$

to both sides of (42) we obtain

$$[b] - [a] = x - \alpha_* x + [C] \tag{43}$$

in K_1R, where $x = [d] + [b] - \sum_{i=0}^{m-1} \alpha_*^i[a]$. In view of formula (43), in order to complete the proof of Theorem 14, it suffices to show that $[C] = 0$.

LEMMA 18. $[C] = 0$ *in* K_1R.

PROOF. Let $K_i' = Vec^i(t^{-1})/Vec^{i-1}(t^{-1})$ for $i = 2, \ldots, l + 2$, and consider the following filtration of Vec^{l+1}:

$$Ve = Vec^1(t^{-1}) \subset Vec^2(t^{-1}) \subset \cdots \subset Vec^{l+2}(t^{-1}) = Vec^{l+1}. \tag{44}$$

This filtration determines a filtration of $N_1 = Vec^{l+1}/Ve$ whose successive subquotients are the modules K_i'. Since ε^i is of the form $E_{pq}(rt^{s'})(t^s)$, we see by Lemma 16 that

$$\left(\bigcup_{j=0}^{\infty} (ec^{i-1})(t^{-j})\right) \cup \left(\bigcup_{j=1}^{s-1} (ec^{i-1})a^j\right)$$

is a basis for $Vec^i(t^{-1})$ (here $a^j = E_{pq}(rt^{s'})(t^j)$); but $\bigcup_{j=1}^{\infty} (ec^{i-1}(t^{-j}))$ is a basis for $Vec^{i-1}(t^{-1})$, and hence K_1' is a free R module to which we assign the following ordered basis: $(ec^{i-1}, ec^{i-1}a^1, \ldots, ec^{i-1}a^{s-1})$. (This notation was explained in the paragraph following Lemma 17.) Note that this basis for K_i' consists of exactly those elements from the basis assigned to N_1 which lie in $Vec^i(t^{-1})$ but not in $Vec^{i-1}(t^{-1})$.

Consider the following filtration for Vec^{l+2}:

$$Vec^1 \subset Vec^2 \subset \cdots \subset Vec^{l+2}, \tag{45}$$

which induces a filtration of N_2 with subquotients the modules K_i for $i = 2, \ldots, l+2$. The ordered bases already assigned to K_i is the same as the ordered set of basis elements from the assigned ordered basis for N_2 which lie in Vec^i but not in Vec^{i-1}.

One easily sees that $r_t : V \to V$ maps filtration (44) into filtration (45) and that $r_{t^{-1}} : V \to V$ maps (45) into (44); and hence r_t determines α linear isomorphisms from K_i' to K_i, which we also denote by r_t. Let C_i be the invertible matrix associated to the α linear isomorphism $r_t : K_i' \to K_i$ with respect to the assigned bases; then $[C] = \sum_{i=2}^{l+2} [C_i]$ in $K_1 R$, and hence to prove Lemma 18, it suffices to show that $[C_i] = 0$ for $i = 2, \ldots, l+2$.

Since $\varepsilon_i = E_{pq}(rt^{s'})t^s$, the matrix C_i has one of the following three forms.

CASE I. $s' < 0$. Then C_i is the identity matrix and hence $[C_i] = 0$.

CASE II. $s' = 0$. Then,

$$C_i = \left(\begin{array}{c|c} E_{pq}(-\alpha(r)) & 0 \\ \hline 0 & I \end{array}\right)$$

and hence $[C_i] = 0$.

CASE III. $s > s' > 0$. Then,

$$C_i = \left(\begin{array}{c|c} I & * \\ \hline 0 & I \end{array}\right)$$

and hence $[C_i] = 0$.

This completes the proof of Lemma 18 and hence also of Theorem 14.

The next theorem gives the structure of $K_1 R_\alpha[T]$.

THEOREM 19. (a) *i_* and i_*^- map* ker ε_* *and* ker ε_*^-, *respectively, monomorphically into $K_1 R_\alpha[T]$ in such a way that*

$$i_*(\ker \varepsilon_*) \cap i_*^-(K_1 R_\alpha[t^{-1}]) = 0$$

and

$$i_*^-(\ker \varepsilon_*^-) \cap i_*(K_1 R_\alpha[t]) = 0.$$

(b) *$i_*(\ker \varepsilon_*) \oplus i_*^-(\ker \varepsilon_*^-)$ is a direct summand of $K_1 R_\alpha[T]$. Moreover, $i_*(\ker \varepsilon_*)$ and $i_*^-(\ker \varepsilon_*^-)$ are isomorphic to $\tilde{C}(R_1, \alpha^{-1})$ and $\tilde{C}(R, \alpha)$ respectively.*

(c) *Let $X = K_1 R_\alpha[T]/i_*(\ker \varepsilon_*) \oplus i_*^-(\ker \varepsilon_*^-)$, and hence*

$$K_1 R_\alpha[T] \simeq X \oplus \tilde{C}(R, \alpha) \oplus \tilde{C}(R, \alpha^{-1});$$

then the following sequence is exact:

$$0 \to K_1 R / I(\alpha_*) \xrightarrow{\varphi} X \xrightarrow{\psi} (K_0 R)^{\alpha_*} \to 0 \tag{46}$$

where φ is induced by $j_* : K_1 R \to K_1 R_\alpha[T]$ and ψ is induced by $\mathscr{F}_* p : K_1 R_\alpha[T] \to K_0 R$.

PROOF. (a) follows from Theorem 13. (b) follows also from Theorem 13 together with Proposition 6. In proving (c) Proposition 12 together with Proposition 6 show that image $\psi = (K_0 R)^{\alpha_*}$. Theorem 14 together with Theorem 13 shows that φ is monomorphic; and Theorems 13 shows that ker ψ = image φ.

Let $w: R \to R$ be an involution which commutes with α; that is, w is an anti-automorphism, $w^2 = id$, and $w\alpha = \alpha w$. We can extend w to an involution of $R_\alpha[T]$ which commutes with α by setting $w(rt^n) = t^{-n}w(r)$ for $r \in R$. We denote this new involution also by w, and observe that $w: R_\alpha[t] \to R_\alpha[t^{-1}]$ anti-isomorphically; w induces an involution of $\mathrm{GL}(n, R)$ (also denoted by w) as follows: $(w(a))_{ij} = w(a_{ji})$ where $a \in \mathrm{GL}(n, R)$; and hence w induces an isomorphism of period 2 of K_1R, which is again denoted by w. Likewise, w induces isomorphisms (again denoted by w) between $K_1R_\alpha[T]$ and $K_1R_\alpha[T]$, and between $K_1R_\alpha[t]$ and $K_1R_\alpha[t^{-1}]$. Since the following diagram is commutative:

$$\begin{array}{ccc} R & \underset{k}{\overset{\varepsilon}{\leftrightarrows}} & R_\alpha[t] \\ {\scriptstyle w}\downarrow & & \downarrow{\scriptstyle w} \\ R & \underset{k^-}{\overset{\varepsilon^-}{\leftrightarrows}} & R_\alpha[t^{-1}] \end{array}$$

we obtain that w induces an isomorphism between $\ker(\varepsilon_*)$ and $\ker(\varepsilon^-_*)$. From this fact together with Theorem 13 and Theorem 19, we deduce the following proposition.

PROPOSITION 20. *If there exists an involution w of R which commutes with α, then $\tilde{C}(R, \alpha) \cong \tilde{C}(R, \alpha^{-1})$. Moreover, in the direct sum decomposition of $K_1R_\alpha[T]$ given in Theorem* 19, *part* (b), *w interchanges the direct summand $i_*(\ker \varepsilon_*)$ with the direct summand $i^-_*(\ker \varepsilon^-_*)$.*

Since our main interest is integral group rings, we next specialize our results to this type of ring.

There is a standard involution w of G such that $w(g) = g^{-1}$ for $g \in G$. Let j, i, and i^- denote inclusion homomorphisms of G into $G \times_\alpha T$, $G \times_\alpha T^+$ into $G \times_\alpha T$, and $G \times_\alpha T^-$ into $G \times_\alpha T$ respectively; let k and k^- denote the inclusion homomorphisms of G into $G \times_\alpha T^+$ and G into $G \times_\alpha T^-$ respectively. If $R = Z(G)$, then there is a standard isomorphism between $R_\alpha[T]$ and $Z(G \times_\alpha T)$ which is the identity map on R and maps $t \in R_\alpha[T]$ onto $t \in Z(G \times_\alpha T)$. This isomorphism induces isomorphisms between $R_\alpha[t]$ and $Z(G \times_\alpha T^+)$ and between $R_\alpha[t^{-1}]$ and $Z(G \times_\alpha T^-)$. We denote all of these isomorphisms by γ. Then the maps j, i, i^-, k, and k^+ of §1 are converted by γ into the maps induced on the respective integral group rings by the maps j, i, i^-, k, and k^+ defined in this paragraph. Also γ will convert the augmentation maps ε and ε^- of §1 into ring homomorphisms, again denoted by ε and ε^-, of $Z(G \times_\alpha T^+)$ onto $Z(G)$ and $Z(G \times_\alpha T^-)$ onto $Z(G)$ respectively. Let $\mathrm{Wh}\, G \times_\alpha T^+ = K_1Z(G \times_\alpha T^+)/k_*J(G)$ and

$$\mathrm{Wh}\, G \times_\alpha T^- = K_1Z(G \times_\alpha T^-)/k^-_*J(G).$$

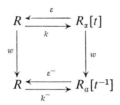

THEOREM 21. (a) *The following sequences are split exact:*

$$(48)\qquad \begin{gathered} 0 \to \mathrm{Wh}\, G \underset{\varepsilon_*}{\overset{k_*}{\rightleftarrows}} \mathrm{Wh}\, G \times_\alpha T^+ \xrightarrow{p''} \tilde{C}(Z(G), \alpha^{-1}) \to 0 \\ 0 \to \mathrm{Wh}\, G \underset{\varepsilon_*^-}{\overset{k_*^-}{\rightleftarrows}} \mathrm{Wh}\, G \times_\alpha T^- \xrightarrow{p'} \tilde{C}(Z(G), \alpha) \to 0 \end{gathered}$$

where p' is induced by p in (31) *and p'' is similarly induced; and w induces an isomorphism between $\tilde{C}(Z(G), \alpha)$ and $\tilde{C}(Z(G), \alpha^{-1})$.*

(b) *i_* and i_*^- map* $\ker \varepsilon_*$ *and* $\ker \varepsilon_*^-$, *respectively, monomorphically into* $\mathrm{Wh}\, G \times_\alpha T$ *in such a way that* $i_*(\ker \varepsilon_*) \cap i_*^-(\mathrm{Wh}\, G \times_\alpha T^-) = 0$ *and symmetrically* $i_*^-(\ker \varepsilon_*^-) \cap i_*(\mathrm{Wh}\, G \times_\alpha T^+) = 0$.

(c) *$i_*(\ker \varepsilon_*) \oplus i_*^-(\ker \varepsilon_*^-)$ is a direct summand of* $\mathrm{Wh}\, G \times_\alpha T$;

$$i_*(\ker \varepsilon_* \simeq \tilde{C}(Z(G), \alpha^{-1})$$

and $i_*^-(\ker \varepsilon_*^-) \simeq \tilde{C}(Z(G), \alpha)$.

(d) *Let* $X = \mathrm{Wh}\, G \times_\alpha T / i_*(\ker \varepsilon_*) \oplus i_*(\ker \varepsilon_*^-)$, *and hence*

$$\mathrm{Wh}\, G \times_\alpha T \simeq X \oplus \tilde{C}(Z(G), \alpha) \oplus \tilde{C}(Z(G), \alpha^{-1});$$

then the following sequence is exact:

$$(49)\qquad 0 \to \mathrm{Wh}\, G/I(\alpha_*) \xrightarrow{\varphi} X \xrightarrow{\psi} (\tilde{K}_0 Z(G))^{\alpha_*} \to 0$$

where φ and ψ are induced by the corresponding map in (46).

(e) *w is the conjugation automorphism of* $\mathrm{Wh}\, G \times_\alpha T$. (*See* [**2**], *p.* 373.) *w interchanges the summands $i_*^-(\ker \varepsilon_*^-)$ and $i_*(\ker \varepsilon_*)$.*

To prove this theorem, use Theorem 13, Theorem 19, and Proposition 20.

Next, we use Theorem 19 to obtain information about the relation of $K_0 R$ to $K_0 R_\alpha[t]$ and $K_0 R_\alpha[T]$. Let $X^+ = \ker \varepsilon_*$ where $\varepsilon_* : K_0 R_\alpha[t] \to K_0 R$; and let $X^- = \ker \varepsilon_*^-$. Note that X^+ and X^- are direct summands of $K_0 R_\alpha[t]$ and $K_0 R_\alpha[t^{-1}]$ respectively.

THEOREM 22. (a) $\operatorname{Ker} j_* = I(\alpha_*)$, *where* $j_* : K_0 R \to K_0 R_\alpha[T]$.

(b) *i_* restricted to X^+ and i_*^- restricted to X^- are both monomorphisms.*

(c) $i_*(K_0 R_\alpha[t]) \cap i_*^-(X^-) = 0$.

PROOF OF (a). From the exact sequence (31) we have that

$$(50)\qquad C'(R, \alpha) \xrightarrow{l} K_0 R_\alpha[t] \xrightarrow{i_*} K_0 R_\alpha[T]$$

is exact. But l is the composite of the maps in diagram (29), where $\mathcal{F}_*$ is epimorphic and k_* is monomorphic. Hence $(\mathrm{id} - \alpha_*)(K_0 R) = \ker(i_* k_*)$; and since $j = ik$, $\ker i_* = I(\alpha_*)$.

PROOF OF (b). Since image $(l) \cap X^+ = 0$, we see from (50) that i_* restricted to X^+ is a monomorphism; and a similar argument holds for i_*^- restricted to X^-.

PROOF OF (c). K_0R is *naturally* a direct summand of $K_1R[T]$. The maps defining this direct sum decomposition (originally given in [7]) are $\xi: K_1R[T] \to K_0R$ and $\eta: K_0R \to K_1R[T]$, where $\xi = \mathscr{F}_*p$ and η is defined as follows: Let $[P] \in K_0R$; then $\mathrm{id} \otimes r_t: P \otimes_R R[T] \to P \otimes_R R[T]$ is an automorphism of an object in $\mathbf{P}(R[T])$ and hence determines an element in $K_1R[T]$. This element is $\eta[P]$. This direct sum decomposition is natural in the following sense: If $f: R \to R'$ is a ring homomorphism, then the following two diagrams are commutative:

$$\begin{array}{ccccccc} K_0R & \xrightarrow{f_*} & K_0R' & \qquad & K_1R[T] & \xrightarrow{f_*^T} & K_1R'[T] \\ \downarrow{\scriptstyle \eta} & & \downarrow{\scriptstyle \eta} & & \downarrow{\scriptstyle \xi} & & \downarrow{\scriptstyle \xi} \\ K_1R[T] & \xrightarrow{f_*^T} & K_1R'[T] & & K_0R & \xrightarrow{f_*} & K_0R' \end{array} \tag{51}$$

Suppose $x \in X^-$ and $i_*^-(x) \in \text{image } i_*$. Then we wish to show that $x = 0$. Consider the following commutative diagram:

$$\begin{array}{ccccc} K_0R & \xleftarrow{\ \xi\ } & K_1R[T'] & \xleftrightarrow{\ \mathrm{id}\ } & K_1R[T'] \\ \downarrow{\scriptstyle k_*^-} & & \downarrow{\scriptstyle k_*^{-T'}} & & \downarrow{\scriptstyle k_*^-} \\ K_0R_\alpha[t^{-1}] & \underset{\eta}{\overset{\xi}{\rightleftarrows}} & K_1R_\alpha[t^{-1}][T'] & \xleftrightarrow{\ \varphi_1\ } & K_1R[T']_\alpha[t^{-1}] \\ \downarrow{\scriptstyle i_*^-} & & \downarrow{\scriptstyle i_*^{-T'}} & & \downarrow{\scriptstyle i_*^-} \\ K_0R_\alpha[T] & \xrightarrow[\eta]{} & K_1R_2[T][T'] & \xleftrightarrow{\ \varphi_2\ } & K_1R[T']_\alpha[T] \\ \uparrow{\scriptstyle i_*} & & \uparrow{\scriptstyle i_*^{T'}} & & \uparrow{\scriptstyle i_*} \\ K_0R_\alpha[t] & \xrightarrow[\eta]{} & K_1R_\alpha[t][T'] & \xleftrightarrow{\ \varphi_3\ } & K_1R[T']_\alpha[t] \end{array} \tag{52}$$

where φ_1, φ_2, and φ_3 are the isomorphisms induced by the ring isomorphisms obtained by sending $(t^i)t'^j \in R_\alpha[T][T']$ into $(t'^j)t^i \in R[T']_\alpha[T]$ and which are the identity map on R.

From (52), we see that $i_*^-\varphi_1\eta(x)$ is in image i_*; and hence by Theorem 19, part (a), there exists an element $x' \in K_1R[T']$ such that $k_*^-(x') = \varphi_1\eta(x)$. Applying the fact that the upper third of diagram (52) is commutative to this last equation, we obtain that $k_*^-\xi(x') = x$; and hence $x = 0$. This proves (c).

4. **Applications.** In this section we give three applications of Theorems 19, 21, and 22. Our first application is to generalize Grothendieck's Theorem (see [7], p. 545) to α-twisted polynomial and Laurent series rings. Our result in the nontwisted case is also slightly more general than that given in [**9**], since it applies to arbitrary regular rings. In order to prove our result, we prove that if R is a right regular ring then $R_\alpha[t]$ and $R_\alpha[T]$ are also right regular rings.

Our second application is that Wh $\pi_1(M)$ and $\tilde{K}_0Z(\pi_1(M))$ both vanish when M is a solvmanifold.

Our third application is that Wh $\pi_1(M) = 0$ when M is a two-dimensional manifold.

LEMMA 23 (TWISTED HILBERT SYZAGY THEOREM). *If* r . gl . dim $R = n$, *then*
(a) r . gl . dim $R_\alpha[t] = n + 1$, *and*
(b) r . gl . dim $R_\alpha[T] \le n + 1$.

PROOF OF (a). Let M be a right $R_\alpha[t]$ module. If we consider M as a right R. module, then r . $\dim_R M \le n$ and r . $\dim_R \alpha M \le n$, and hence r . $\dim_{R_\alpha[t]} MS' \le n$ and r . $\dim_{R_\alpha[t]}(\alpha M)S' \le n$. (Here, we use the fact that $R_\alpha[t]$ is free as a left R module.) Hence from the characteristic sequence of M (22) r . $\dim_{R_\alpha[t]} M \le n + 1$. (Here, we used Proposition 2.1 from Chapter VI of [**1**].) Therefore r . gl . dim $R_\alpha[t] \le n + 1$.

If M is a right R module, we can make it into a right $R_\alpha[t]$ module by defining $mt = 0$ for each $m \in M$; and we denote this module by M'. By what we have shown above, r . $\dim_{R_\alpha[t]} M' \le$ r . $\dim_{R_\alpha[t]} M + 1$. Next, we prove that the following equality is true:

$$\text{r} . \dim_{R_\alpha[t]} M' = 1 + \text{r} . \dim_R M. \tag{53}$$

Equality (53) immediately implies that r . gl . dim $R_\alpha[t] =$ r . gl . dim $R + 1$. Our proof of (53) will proceed by induction on r . $\dim_R M$. M' is never projective since $M't = 0$; therefore r . $\dim_{R_\alpha[t]} M' \ge 1$; and hence equality (53) holds when r . $\dim_R M = 0$. Now, suppose that r . $\dim_R M = 1$, then either r . $\dim_{R_\alpha[t]} M' = 1$ or 2. Suppose r . $\dim_{R_\alpha[t]} M' = 1$; then $M' \simeq P/Q$ where P and Q are projective $R_\alpha[t]$ modules. From the filtration $Qt \subset Pt \subset Q \subset P$ we obtain the following two exact sequences:

$$\begin{aligned} &0 \to Q/Pt \to P/Pt \to M \to 0 \\ &0 \to \alpha M \to Q/Qt \to Q/Pt \to 0 \end{aligned} \tag{54}$$

from which we obtain, as in the proof of Theorem 8, part (b), that r . $\dim_R M = 0$. This contradicts our assumption; hence r . $\dim_{R_\alpha[t]} M' = 2$. Now, suppose that we have proven (53) for r . $\dim_R M \le i$, $i \ge 1$; and assume that r . $\dim_R M = i + 1$. Let $0 \to N \to F \to M \to 0$ be exact as a sequence of R modules where F is free; then $0 \to N' \to F' \to M' \to 0$ is exact over $R_\alpha[t]$; $\dim_{R_\alpha[t]} F' = 1$ while $\dim_{R_\alpha[t]} N' = i + 1$; and hence it follows from [**1**], Proposition 2.1 of Chapter VI that r . $\dim_{R_\alpha[t]} M' = i + 2$. This completes the proof of (a).

PROOF OF (b). Let M be a right $R_\alpha[T]$ module; then $M \simeq M\ {}_R\otimes_{\alpha[t]} R_\alpha[T]$. By (a), M has a resolution of length $\le n + 1$ by projective $R_\alpha[t]$ modules. If we tensor this resolution with $\otimes_{R_\alpha[t]} R_\alpha[T]$, we obtain a resolution of M of length $\le n + 1$ by projective $R_\alpha[T]$ modules. (Here, we use the fact that $R_\alpha[T]$ is flat as a left $R_\alpha[t]$ module.)

LEMMA 24 (TWISTED HILBERT BASIS THEOREM). *If R is right Noetherian then both $R_\alpha[t]$ and $R_\alpha[T]$ are right Noetherian.*

The proof of Lemma 24 will be left to the reader, since it can be proven by a straightforward generalization of the proof for the untwisted case. (See, for example, **[10]**, p. 171.)

Recall that R is said to be right regular if R is right Noetherian and for any finitely generated right R module M, $\mathrm{r}\,.\,\dim_R M < \infty$.

THEOREM 25. *If R is right regular, then*
(a) $R_\alpha[t]$ is right regular; and
(b) $R_\alpha[T]$ is right regular.

PROOF OF (a). By Lemma 24, $R_\alpha[t]$ is right Noetherian. Let N be a finitely generated right $R_\alpha[t]$ module. We wish to show that $\mathrm{r}\,.\,\dim_{R_\alpha[t]} N < \infty$. Since $R_\alpha[t]$ is right Noetherian, there exists a short exact sequence of $R_\alpha[t]$ modules $0 \to M \to F \to N \to 0$ where F is free and M is finitely generated; and hence it is sufficient to show that $\mathrm{r}\,.\,\dim_{R_\alpha[t]} M < \infty$. The advantage in examining M rather than N is that $r_t: M \to M$ is a monomorphism. First, we show that $\mathrm{w}\,.\,\dim_R M < \infty$. Let $x_1, \ldots, x_n$ generate M as a $R_\alpha[t]$ module; and let M_m be the R submodule of M generated by $x_i t^j$, $1 \le i \le n$ and $0 \le j \le m$; $M = \bigcup_{i=0}^{\infty} M_i$; and hence by **[1]**, Proposition 1.3 of Chapter VI $\mathrm{w}\,.\,\dim_R M \le \sup_{0 \le i < \infty} \mathrm{w}\,.\,\dim_R M_i$. We proceed to show that $\sup_{0 \le i < \infty} \mathrm{w}\,.\,\dim_R M_i < \infty$. Let $L_i = r_{t^{i+1}}^{-1}(M_i \cap r_{t^{i+1}}(M_0))$. Then, $L_0 \subset L_1 \subset L_2 \subset \cdots$ is an ascending chain of R submodules of M_0; and since M_0 is a finitely generated module over the Noetherian ring R, there exists an integer n_0 such that $L_{n_0} = L_{n_0+1} = \cdots$. Let

$$a = 1 + \max\{\mathrm{w}\,.\,\dim_R M_i \text{ for } i = 0, 1, \ldots, n_0 \text{ and } \mathrm{w}\,.\,\dim_R L_i \text{ for } i = 0, \ldots, n_0\}.$$

We claim that the following inequality is true for all i:

$$\mathrm{w}\,.\,\dim_R M_i \le a. \tag{55}$$

The proof of (55) will proceed by induction on i. For $i = 0, \ldots, n_0$ it is part of the definition of a. Suppose (55) is true for i, $i \ge n_0$; then $M_{i+1} = M_i \cup r_{t^{i+1}}(M_0)$, and hence, by a Mayer-Vietoris type argument,

$$\mathrm{w}\,.\,\dim_R M_{i+1} \le \max\{\mathrm{w}\,.\,\dim_R M_i,\ \mathrm{w}\,.\,\dim_R r_{t^{i+1}}(M_0),\ \mathrm{w}\,.\,\dim_R v_{t^{i+1}}(L_i) + 1\}.$$

Since $r_{t^{i+1}}: M_0 \to r_{t^{i+1}}(M_0)$ and $r_{t^{i+1}}: L_{n_0} = L_i \to r_{t^{i+1}}(L_i)$ are α^{i+1} linear isomorphisms, $\mathrm{w}\,.\,\dim_R r_{t^{i+1}}(M_0) \le a - 1$ and $\mathrm{w}\,.\,\dim_R r_{t^{i+1}}(L_i) \le a - 1$. By assumption, $\mathrm{w}\,.\,\dim_R M_i \le a$; hence inequality (55) is true for all i; and therefore $\mathrm{w}\,.\,\dim_R M < \infty$. It is easily seen from this that $\mathrm{w}\,.\,\dim_{R_\alpha[t]} MS' < \infty$ and $\mathrm{w}\,.\,\dim_{R_\alpha[t]} (\alpha M)S' < \infty$. Now from the characteristic sequence for M (22), we see that $\mathrm{w}\,.\,\dim_{R_\alpha[t]} M < \infty$; but M is a finitely generated module over the Noetherian ring $R_\alpha[t]$; and hence by **[1]**, p. 122, Exercise 3(b), $\mathrm{r}\,.\,\dim_{R_\alpha[t]} M < \infty$. This proves (a).

In order to prove part (b), we need the following lemma.

LEMMA 26. *If R is a right Noetherian ring and M is a finitely generated right $R_\alpha[T]$ module, then there exists a finitely generated right $R_\alpha[t]$ module M' such that $M'S \cong M$.*

It is clear that part (b) of Theorem 25 is an immediate consequence of part (a) of the same theorem together with Lemma 26 and part (b) of Lemma 24.

PROOF OF LEMMA 26. Since $R_\alpha[T]$ is Noetherian, there exist finitely generated free $R_\alpha[T]$ modules F_1 and F_0 and a map $f: F_1 \to F_0$ such that $M(f) \cong M$. As in [7], p. 562, we may assume that $F_1 = F_1'S$, $F_0 = F_0'S$, and $f = \varphi S$, where F_0', F_1' are finitely generated free $R_\alpha[t]$ modules and $\varphi: F' \to F_1'$ is an $R_\alpha[t]$ linear map; and hence $M \cong M(\varphi)S$.

THEOREM 27 (TWISTED GROTHENDIECK THEOREM). *If R is a right regular ring, then*

(a) $k_*: K_0R \to K_0R_\alpha[t]$ *is an isomorphism; and*

(b) $j_*: K_0R \to K_0R_\alpha[T]$ *is an epimorphism whose kernel is* $I(\alpha_*)$, *i.e.* $K_0R_\alpha[T] \cong K_0R/I(\alpha_*)$.

PROOF. First, we show that $i_*: K_0R_\alpha[t] \to K_0R_\alpha[T]$ is an epimorphism. Let $[P] \in K_0R_\alpha[T], P \in \mathbf{P}(R_\alpha[T])$; then by Lemma 26 there exists a finitely generated right $R_\alpha[t]$ module M such that $MS \cong P$. By Theorem 25, part (a), there exists a resolution of M of finite length by objects from $\mathbf{P}(R_\alpha[t])$. Tensoring this resolution with $\otimes_{R_\alpha[t]} R_\alpha[T]$ and using the fact that $R_\alpha[T]$ is a flat left $R_\alpha[t]$ module, we obtain that i_* is an epimorphism. Hence by Theorem 22, part (c), $i_*^-(X^-) = 0$; and by the same theorem part (b), $X^- = 0$. Therefore, k_*^- is an isomorphism; and an analogous argument shows that k_* is an isomorphism, which completes the proof of (a). Since $j = ik$ both i_* and k_* are epimorphisms, we have that j_* is an epimorphism; and by Theorem 22, part (a), $\ker j_* = I(\alpha_*)$. This completes the proof of Theorem 27.

COROLLARY 28. *If R is a right regular ring, then*
(a) $k_*: \tilde{K}_0R \to \tilde{K}_0R_\alpha[t]$ *is an isomorphism; and*
(b) $j_*: \tilde{K}_0R \to \tilde{K}_0R_\alpha[T]$ *is an epimorphism whose kernel is* $I(\alpha_*)$.

We remark that $I(\alpha_*)$ can have in general two meanings depending on whether we consider α_* as an endomorphism of K_0R or of $\tilde{K}_0R$. In this particular case there is no ambiguity since the quotient map from K_0R onto $\tilde{K}_0R$ maps the first $I(\alpha_*)$ isomorphically onto the second. To see this, we used the fact that a right Noetherian ring has the invariant basis property (see **[11]**, Proposition 2.1), together with Lemma 24, and the fact that $F(R')$ is infinite if and only if R' has the invariant basis property. (Cf. Appendix 1.)

We leave the proof of Corollary 28 to the reader.

Any group possessing only a single element will be called a group of type 0. Inductively, we define G to be a group of type $n + 1$ if $G = H \times_\alpha T$ where G is a group of type n.

THEOREM 29. *If G is a group of type n, then both* Wh $G = 0$ *and* $\tilde{K}_0 Z(G) = 0$.

Before we prove Theorem 29, let us first obtain a corollary from it.

COROLLARY 30. *If M is a solvmanifold, then both* Wh $\pi_1(M) = 0$ *and*

$$\tilde{K}_0 Z(\pi_1(M)) = 0.$$

PROOF OF COROLLARY 30. From results of G. D. Mostow and L. Auslander, it follows that $\pi_1(M)$ is a group of type n. (See [12], Chapter 1, and [13], §§64, 66.) Hence, Corollary 30 follows directly from Theorem 29.

PROOF OF THEOREM 29. If G is a group of type n, then $Z(G)$ is a regular ring. This follows easily from Theorem 25 by induction on n.

Again, it follows by induction on n, using the above fact together with Corollary 28 Part (b), that $\tilde{K}_0 Z(G) = 0$.

Next, we show that Wh $G = 0$. Again, we proceed by induction on n. The inductive step goes as follows. Let $G = H \times_\alpha T$, where H is of type $n - 1$. By Theorem 21,

$$\text{Wh } G \cong X \oplus \tilde{C}(Z(G), \alpha) \oplus \tilde{C}(Z(G), \alpha^{-1}).$$

Now, it was proven in [6] that if R' is a regular ring, then $\tilde{C}(R', \alpha') = 0$ (see [6], Theorem 1.6); and hence $\tilde{C}(Z(G), \alpha) \cong \tilde{C}(Z(G), \alpha^{-1}) \cong 0$. Therefore, Wh $G \cong X$; but from the exact sequence (49) (in which H replaces G), we see that $X = 0$. This follows from our inductive hypothesis, that Wh $H = 0$, together with the first half of our proof in which we showed that $\tilde{K}_0 Z(H) = 0$. This completes the proof of Theorem 29.

THEOREM 31. *Let F be a free group (not necessarily finitely generated): then* Wh $F \times_\alpha T = 0$.

Before proving Theorem 31, let us first obtain a corollary from it.

COROLLARY 32. *Let M be a connected 2-dimensional manifold; then*

$$\text{Wh } \pi_1(M) = 0.$$

PROOF OF COROLLARY 32. CASE I. M is open. Then $\pi_1(M)$ is a free group. (See [14], p. 200, Problem 5.6.) A theorem due to Stallings and Gersten says that Wh $F = 0$ if F is a free group. (See [8] and [15].)

CASE II. M is the sphere or projective plane. Then $\pi_1(M) = 0$ or T_2 (cyclic) group of order 2). Wh $(0) = 0$ and Higman showed that Wh $T_2 = 0$. (See [16].)

CASE III. M is closed and $H^1(M; Z)$ has rank ≥ 1. Then $\pi_1(M) = G \times_\alpha T$ where G is the fundamental group of an infinite cyclic covering space of M and hence of an open connected 2-dimensional manifold. Therefore, as in Case I, G is a free group; and hence Wh $\pi_1(M) = 0$ by Theorem 31.

The proof of Theorem 31 will depend on the following lemma.

LEMMA 33. *Let F be a free group; then*
(a) r . gl . dim $Q(F) \leq 1$, *where Q denotes the rational numbers; and*
(b) r . gl . dim $Z(F) \leq 2$.

PROOF OF LEMMA 33. We will only prove (b), since the proof of (a) is similar and slightly simpler. Let N be a right $Z(F)$ module and let $0 \to M \to P \to N \to 0$ be exact (here P is projective). In order to prove (b), it is sufficient to show that $\mathrm{r}\,.\,\dim_{Z(F)} M \leq 1$. Since P is free as a Z module, M is also free as a Z module. Let g_i, $i \in \Omega$ be a collection of free generators for F, where Ω is some indexing set. Define a map $\varphi: P_0 = M \otimes_Z Z(F) \to M$ by $\varphi(m \otimes r) = mr$ for $m \in M$ and $r \in Z(F)$. Then φ is a $Z(F)$ linear homomorphism. Also, since M is a free Z module, P_0 is a free $Z(F)$ module. Likewise, $P_1 = \bigoplus_{i \in \Omega} M \otimes (Z(F))_i$ is a free $Z(F)$ module; here $(Z(F))_i$ denotes a copy of $Z(F)$ indexed by $i \in \Omega$. Define a map $\psi: P_1 \to P_0$ as follows: $\psi(m \otimes r_i) = m \otimes g_i r_i - m g_i \otimes r_i$ where $m \in M$ and $r_i \in (Z(F))_i$. Then ψ is a $Z(F)$ linear homomorphism, and the following sequence is exact:

$$0 \longrightarrow P_1 \xrightarrow{\psi} P_0 \xrightarrow{\varphi} M \longrightarrow 0.$$

The proof of this last statement is straightforward, but involves some tedious calculations, and hence will be left to the reader. This completes the proof of Lemma 33.

PROOF OF THEOREM 31. By Theorem 21, we have that

$$\mathrm{Wh}\, F \times_\alpha T \cong X \oplus \tilde{C}(Z(F), \alpha) \oplus \tilde{C}(Z(F), \alpha^{-1})$$

where $0 \to \mathrm{Wh}\, F/I(\alpha_*) \to X \to (\tilde{K}_0 Z(F))^{\alpha_*} \to 0$ is exact. But, by **[8]** and **[15]**, $\mathrm{Wh}\, F = 0$, while Bass has shown that $\tilde{K}_0 Z(F) = 0$ (see **[17]**); and hence $\mathrm{Wh}\, F \times_\alpha T \cong \tilde{C}(Z(F), \alpha) \oplus \tilde{C}(Z(F), \alpha^{-1})$. By Theorem 21, part (a) $\tilde{C}(Z(F), \alpha) \cong \tilde{C}(Z(F), \alpha^{-1})$; and hence in order to complete the proof of Theorem 31 it only remains to show that $\tilde{C}(Z(F), \alpha)$ vanishes.

Now, since $\mathrm{r}\,.\,\mathrm{gl}\,.\,\dim Z(F) \leq 2$ if $Z(F)$ were Noetherian, we could conclude immediately from Theorem 1.6 of **[6]** that $\tilde{C}(Z(F), \alpha) = 0$. Unfortunately, $Z(F)$ is not Noetherian if F is not infinite cyclic. But, by making strong use of Lemma 33, we are able to modify the argument used in proving Theorem 1.6 of **[6]** in such a way as to avoid use of the Noetherian condition. If we examine the proof of Theorem 1.6 of **[6]** (see also Lemma 1.2 of **[6]**), we see that an element of the form $[P, f] \in \tilde{C}(R, \alpha)$ equals zero if there exists a filtration $0 = K_0 \subset K_1 \subset \cdots \subset K_n = P$ of P by R submodules with $f(K_i) \subset K_{i-1}$ and such that each subquotient K_{i+1}/K_i has a resolution of finite length by objects from $\mathbf{P}(R)$. Using this fact, we will show that $\tilde{C}(Z(F), \alpha) = 0$. Suppose that $f^n = 0$ and let $K_i = \text{kernel } f^i$. Clearly, $0 = K_0 \subset K_1 \subset \cdots \subset K_n = P$ and $f(K_i) \subset K_{i-1}$. Also, the subquotients K_i/K_{i-1} have all resolutions of length 2 of the following form: $0 \to K_{i-1} \to K_i \to K_i/K_{i-1} \to 0$. Hence, if we can show that each $K_i \in \mathbf{P}(Z(F))$, then we will have shown that $\tilde{C}(Z(F), \alpha) = 0$. Consider the following two exact sequences:

(56) $$0 \to K_i \to P \to \text{image } f^i \to 0$$

(57) $$0 \to \text{image } f^i \to P \to \text{cokernel } f^i \to 0.$$

Since $\mathrm{r}\,.\,\mathrm{gl}\,.\,\dim Z(F) \leq 2$, we see from (56) and (57) that each K_i is projective; and hence it only remains to show that each K_i is finitely generated. Now, if we tensor sequences (56) and (57) with $\otimes_{Z(F)} Q(F)$ and use the facts that $Q(F)$ is a flat $Z(F)$ module and that r . gl . dim $Q(F) \leq 1$, we obtain that $K_i \otimes_{Z(F)} Q(F)$ is a direct summand of $P \otimes_{Z(F)} Q(F)$, and hence is finitely generated as a $Q(F)$ module. Since K_i is a projective $Z(F)$ module, $K_i \simeq K_i \otimes_{Z(F)} Z(F)$ is a $Z(F)$ submodule of $K_i \otimes_{Z(F)} Q(F)$; and hence there exists a finitely generated $Z(F)$ submodule L of K_i with the property that for any element $x \in K_i$ there exists a nonzero integer $n(x)$ such that $xn(x) \in L$.

Let N be a free $Z(F)$ module together with a map $\varphi: N \to K_i$ which is an epimorphism. Since K_i is projective, there exists a splitting $\psi: K_i \to N$ to φ. Since L is finitely generated, there exists a direct sum decomposition of N, $N = N' \oplus N''$, such that N' is finitely generated and $\psi(L) \subseteq N'$. We proceed to show that $\psi(K_i) \subseteq N'$. Let $x \in K_i$. Then $\psi(xn(x)) = \psi(x)n(x) \in N'$. But $\psi(x)$ can be uniquely written as $a + b$ where $a \in N'$ and $b \in N''$; and hence $bn(x) = 0$. But N'' is free when considered as a Z module; and therefore $b = 0$. Consequently, φ maps N' onto K_i; and hence we have shown that K_i is a finitely generated $Z(F)$ module. This completes the proof of Theorem 31.

Appendix 1. In this appendix, we give an example of a ring R possessing the I.B.P. (invariant basis property **[11]**); but such that $R_\alpha[T]$ does not have the I.B.P. Therefore, one should be careful when applying our formula, for instance, to problems concerning the torsion of a chain complex. Our example is based on a ring R' constructed by Cohn. (See **[11]**, Theorem 7.1.) R' has the I.B.P.; but in addition there exists a fixed integer n such that any finitely generated free R' module is a direct summand of $(R')^n$. Let $R = \prod_{i=-\infty}^{\infty} (R')_i$ where $(R')_i$ denotes a copy of R' indexed by i. We can consider R as the collection of functions from Z to R'; then we define α to be the shift automorphism; i.e. $\alpha(f)(i) = f(i - 1)$ where $f \in R$ and $i \in Z$.

Recall that we remarked after Corollary 28 that a ring R_1 has the I.B.P. if and only if $\mathbf{F}(R_1)$ is infinite. By **[11]**, Proposition 2.4, R possesses the I.B.P. To show that $R_\alpha[T]$ does not have the I.B.P., it suffices to show that $\mathbf{F}(R_\alpha[T])$ is finite, and hence by Theorem 22, part (a) that $I(\alpha_*) \cap \mathbf{F}(R) \neq 0$. We show, in fact that $\mathbf{F}(R) \subseteq I(\alpha_*)$ by constructing an element $x \in K_0R$ such that $x - \alpha_*(x) = [R]$. For $i \in Z$ let $R'(i) = (R')^i$ if $i \geq 0$ and equal to the zero module for $i \leq 0$. Let $P = \prod_{i=-\infty}^{\infty} R'(i)$ and $Q = \prod_{i=-\infty}^{\infty} R'(-i)$; then since each $R'(i)$ is a direct summand of $(R')^n$ we see that $P, Q \in \mathbf{P}(R)$. Now, let $x = [P] - [Q]$.

Appendix 2. Let $\tilde{p}: K_1R_\alpha[T] \to (K_0R)^{\alpha_*} \to 0$ be the composite of $p: K_1R_\alpha[T] \to C'(R, \alpha)$ and $\mathscr{F}_*: C'(R, \alpha) \to K_0R$. In this appendix, we give an example of a pair R, α such that $\tilde{p}$ does not split. Let Q_i be the adjunction space $S^{i-1} \cup_\varphi D^i$ where φ is a map of degree 2. Let $X' = Q^3 \vee Q^4$ and let $f': X' \to X'$ be a map with the following properties: $f' = \psi \vee \mathrm{id}$; $r_1\psi = \mathrm{id}$ where r_1 is the retraction map onto Q^3 and r_2 the retraction onto Q^4; and $\widetilde{r_2\psi}: S^3 = Q^3/S^2 \to S^3$ is a map of degree one, where $\widetilde{r_2\psi}$ is the map induced by $r_2\psi$. By a regular neighborhood argument, one can construct a space X and a homeomorphism f of X together with a homotopy equivalence $\lambda: X \to X'$ such that $\lambda f = f'\lambda$.

Let $R = \mathbf{C}(X)$, the ring of complex valued continuous functions defined on X, and let $\alpha(g) = gf$. That $\tilde{p}$ does not split for this pair is a consequence of the following fact. If $i: X \to X_f$ denotes the inclusion map, where X_f is the mapping torus of f (i.e. $X_f = X \times [0, 1]/(x, 1) = (f(x), 0))$, then i induces a map $i^*\tilde{K}^0(X_f) \to (\tilde{K}^0(X))^{f^*} \to 0$. If $\tilde{p}$ splits, one can show that i^* splits. But in this case, this is impossible since $f^* = \text{id}$, $\tilde{K}^0(X) = T_2$ (cyclic group of order 2) and $\tilde{K}^0(X_f) = T_4$.

ADDED IN PROOF. Theorem 21 was also discovered by L. C. Siebenmann independently in his forthcoming paper, *A total Whitehead torsion obstruction to fibering over the circle.*

REFERENCES

1. H. Cartan and S. Eilenberg, *Homological algebra*, Princeton University Press, Princeton, N.J., 1956.
2. J. Milnor, *Whitehead torsion*, Bull. Amer. Math. Soc. **72** (1966), 358–426.
3. H. Bass, *K-theory and stable algebra*, Publ. Inst. Hautes Etudes Sci. #22, 1964.
4. F. T. Farrell and W. C. Hsiang, *Manifolds with* $\pi_1 = G \times_\alpha T$, (to appear).
5. A. Heller, *Some exact sequences in algebraic K-theory*, Topology **3** (1965), 389–408.
6. F. T. Farrell, *The obstruction to fibering a manifold over a circle*, (to appear).
7. H. Bass, A. Heller, and R. Swan, *The Whitehead group of a polynomial extension*, Publ. Inst. Hautes Etudes Sci. #22, 1964.
8. J. Stallings, *Whitehead torsion of free products*, Ann. of Math. **82** (1965), 354–363.
9. H. Bass, *Algebraic K-theory*, Benjamin, New York, 1968.
10. N. Jacobson, *Lectures in abstract algebra*, vol. 1, Van Nostrand, Princeton, N.J., 1951.
11. P. M. Cohn, *Some remarks on the invariant basis property*, Topology **5** (1966), 215–228.
12. L. Auslander, L. Green, and F. Hahn, *Flows on homogeneous spaces*, Princeton University Press, Princeton, N.J., 1963.
13. A. G. Kurosh, *The theory of groups*, vol. 2, Chelsea, New York, 1956.
14. W. S. Massey, *Algebraic topology, an introduction*, Harcourt, Brace and World, New York, 1967.
15. S. Gersten, *Whitehead groups of free associative algebras*, Bull. Amer. Math. Soc. **71** (1965), 157–159.
16. G. Higman, *The units of group rings*, Proc. London Math. Soc. **46** (1940), 231–248.
17. H. Bass, *Projective modules over free groups are free*, J. Algebra **1** (1964), 367–373.
18. F. T. Farrell and W. C. Hsiang, *A geometric interpretation of the Künneth formula for algebraic K-theory*, Bull. Amer. Math. Soc. **74** (1968), 548–553.
19. L. C. Siebenmann, *A total Whitehead torsion obstruction to fibering over the circle* (to appear).

YALE UNIVERSITY

Finite flat structures

B. Mazur[1]

1. Fix a rational prime p, an algebraic closure of the p-adic numbers, $\bar{Q}_p$, and a subfield $K \subset \bar{Q}_p$, of finite degree over Q_p. Consider the category $\mathcal{G}$ of finite flat commutative group schemes over R, the ring of integers in K (call objects of $\mathcal{G}$ *group schemes*, for short). We take morphisms of $\mathcal{G}$ to be arbitrary group scheme homomorphisms over R. The category $\mathcal{G}$ is not an abelian category. One remedy for this is to imbed $\mathcal{G}$ as a full subcategory of the abelian category of sheaves of abelian groups for the *fppf* site over Spec (R) [**1**]. One thus obtains the *fppf* = flat cohomology groups over Spec (R) with coefficients in objects of $\mathcal{G}$. One has a fairly good computational understanding of this cohomology theory, (see §2 below) and it is the purpose of this paper to show how, conversely, the knowledge of flat cohomology determines the object of $\mathcal{G}$.

This is our approach: The "generic fiber" provides us with a faithful functor $M(G)$ from $\mathcal{G}$ to the abelian category $\mathcal{M}$ of finite galois modules over K. Thus $M(G) = G(\bar{K}) = G(\bar{Q}_p)$ where the galois module structure is given by the obvious formula.

It is easy to construct examples of finite galois modules M which come from no group scheme G in the above manner. There are also examples of distinct group schemes with isomorphic galois modules. In fact, after a recent result of Tate, one knows the following:

Let R contain a primitive pth root of 1. Let e denote the absolute ramification index of R. The number of nonisomorphic group schemes whose galois module is isomorphic to Z/p (taken with trivial galois action) is $e/(p-1)+1$. The problem: What information in addition to the galois module $M = M(G)$ is it necessary to give, to uniquely characterize the group scheme G? Or, what rigidification of the category $\mathcal{M}$ is needed, to allow us to refine the "generic fiber functor" $M(\ )$ to a fully faithful imbedding? Our answer, which will be presented in a moment,

[1] This research was partially supported by National Science Foundation Grant GP-6585.

provides a format for the classification of group schemes of the above sort, and may be useful in more general contexts.

If $M = M(G)$, let $h^q(M) = h^q(G)$ denote $H^q(K, M)$, galois cohomology of M over K. Let $H^q(G)$ denote the cohomology of the group scheme G for the $(fppf)$-site (e.g., $H^q(G)$ denotes *flat* cohomology over Spec R. One has the natural map $H^q(G) \to h^q(G)$ which is an isomorphism for $q = 0$, and an injection for $q = 1$ (since R is integrally closed). Let us identify $H^1(G)$ with its image in $h^1(G)$.

We shall show that the group scheme G is determined by the subgroup $H^1(G)$ in $h^1(G)$.

More explicitly, let us consider the category of pairs consisting in galois modules M together with subgroups $V \subset h^1(M)$ where morphisms $f\colon (M, V) \to (N, W)$ are those morphisms of galois modules, $f\colon M \to N$ such that $f(V) \subset W$. We have a functor $G \mapsto \tilde{G}$ from the category of group schemes to the above category, by letting $\tilde{G}$ be the pair $(M(G), H^1(G))$. We shall prove in §2

THEOREM 1. *The functor $G \mapsto \tilde{G}$ is fully faithful.*

Fix a galois module M. By a *flat structure* on M we mean a group scheme G together with an isomorphism $M(G) \underset{\approx}{\to} M$. We say $H \subset h^1(M)$ is *realizable* if $H = H^1(G)$ for some flat structure on M.

A *morphism $G_1 \to G_2$ of flat structures* is a morphism of group schemes inducing the identity map on galois modules. Since a morphism of group schemes $G_1 \to G_2$ is determined by its effect on galois modules, if a morphism of flat structures exists from G_1 to G_2, it is unique. Given two flat structures G_1, G_2 consider the diagonal imbedding of galois modules $M \hookrightarrow M(G_1 \times G_2)$. The image subgalois module determines a sub-group scheme G by "flat extension" (cf. [**1**]). By construction we have an isomorphism $M(G) \underset{\approx}{\to} M$, so we may view G as a flat structure, denoted $G_1 \wedge G_2$. The natural projections yield maps

$$\begin{array}{ccc} & G_1 \wedge G_2 & \\ \swarrow & & \searrow \\ G_1 & & G_2 \end{array}$$

expressing $G_1 \wedge G_2$ as a product of G_1, G_2 in the category of flat structures on M. Let $\hat{G}$ denote the Cartier dual of G. Define $G_1 \vee G_2 = (\hat{G}_1 \wedge \hat{G}_2)^\wedge$ and we have the natural maps,

$$\begin{array}{ccc} G_1 & & G_2 \\ \searrow & & \swarrow \\ & G_1 \vee G_2 & \end{array}$$

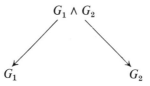

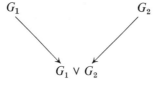

expressing $G_1 \vee G_2$ as a sum of G_1, G_2 in the category of flat structures on M. Consequently, the category of flat structures on M form a lattice. A trivial, but fundamental fact is the following: If $G_1 \to G_2$ is a morphism of flat structures and $\operatorname{disc}_R G_1 = \operatorname{disc}_R G_2$, then it is an isomorphism. Here disc_R denotes the discriminant ideal, over R. From this, and duality, one learns that there are a finite number of flat structures on M.

THEOREM 2. (a) $H^1(G_1 \wedge G_2) = H^1(G_1) \cap H^1(G_2)$.
(b) $H^1(G_1 \vee G_2) = H^1(G_1) + H^1(G_2)$.
(c) *H^1 yields an isomorphism of the finite lattice of flat structures on M onto the lattice of realizable subgroups of $h^1(M)$.*

If the galois module M admits any flat structure, it follows from the above discussion that it admits a unique "minimal" flat structure, as well as a unique "maximal" flat structure. The maximal structure may be described as that unique group scheme structure on M, admitting a group homomorphism to any other group scheme structure on M. It may also be identified as the group scheme structure on M with smallest discriminant. The minimal structure may be described dually. What are those structures for a given M?

2. **The proofs.** To prove the above theorems, we shall make use of the following results:

(2.1)
(a) $H^q(G) = 0$ for $q > 1$.

(b) $\#H^1(G) = \#(M^K) \cdot \|\operatorname{disc}_R G\|^{-1/g_0}$.

(c) (Local flat duality) There is an exact sequence,

$$0 \to H^1(G) \xrightarrow{i} h^1(G) \xrightarrow{j} H^1(\hat{G})^* \to 0,$$

where the terminology above is as follows: $M^K = G(K) = G(R)$, i.e., the subgroup of M left fixed by the galois action; $\|\operatorname{disc}_R G\|$ is the normalized absolute value of the discriminant of G; G^0 is the connected component of G; g_0 is the order of G^0, i.e., the rank of the affine algebra of G^0 regarded as an R-module; i is the natural map; $\hat{G}$ is the Cartier dual of G; $*$ denotes Pontrjagin duality; $j = i^* \cdot \tau$ where $\tau : h^1(G) \underset{\approx}{\to} h^1(\hat{G})^*$ is the isomorphism induced by cup-product (Tate duality). These results are proved in **[2]**.

We begin the proof of Theorem 2. Constructing the appropriate quotient, **[6]**, consider the exact sequence

$$0 \to G_1 \wedge G_2 \to G_1 \times G_2 \to G \to 0. \tag{2.2}$$

Applying $M(\ )$ to (2.2) one has the split exact sequence,

$$0 \to M \xrightarrow[\Delta]{} M \times M \xrightarrow[d]{} M \to 0 \tag{2.3}$$

where $d(m_1, m_2) = m_1 - m_2$.

From the above exact sequences we obtain:

$$\begin{array}{ccccccc}
0 & \longrightarrow & H^1(G_1 \wedge G_2) & \longrightarrow & H^1(G_1) \times H^1(G_2) & \longrightarrow & H^1(G) \\
 & & \downarrow \subseteq & & \downarrow \subseteq & & \downarrow \subseteq \\
0 & \longrightarrow & h^1(M) & \longrightarrow & h^1(M) \times h^1(M) & \longrightarrow & h^1(M)
\end{array}$$

from which Theorem 2(a) follows. Using local flat duality (2.1c) and Theorem 2(a) applied to $\hat{G}_1 \wedge \hat{G}_2$, Theorem 2(b) follows.

Theorem 2(c) and Theorem 1 both follow from Proposition 2, below:

PROPOSITION 1. *Let $p: G \to H$ be a morphism of flat structures such that p induces an isomorphism $H^1(G) \xrightarrow{\approx} H^1(H)$. Then p is an isomorphism.*

PROPOSITION 2. *Let G_1, G_2 be group schemes, with associated galois M_1, M_2. Let $H_i = H^1(G_i)$ $(i = 1, 2)$. Let $\varphi: M_1 \to M_2$ be a homomorphism of galois modules bringing H_1 to H_2. Then φ is induced from a morphism $f: G_1 \to G_2$ of group schemes.*

PROOF OF PROPOSITION 1. Consider the decomposition of G into connected and étale parts:

$$\begin{array}{ccccccccc}
0 & \longrightarrow & G^0 & \longrightarrow & G & \longrightarrow & G^{\text{ét}} & \longrightarrow & 0 \\
 & & & & \downarrow & & & & \\
 & & & & H & & & &
\end{array}$$

we may obtain a flat subgroup $p(G^0) = H' \subset H$ and quotient $H'' = H/H'$ to give us

$$\begin{array}{ccccccccc}
0 & \longrightarrow & G^0 & \longrightarrow & G & \longrightarrow & G^{\text{ét}} & \longrightarrow & 0 \\
 & & \downarrow p & & \downarrow p & & \downarrow p'' & & \\
0 & \longrightarrow & H' & \longrightarrow & H & \longrightarrow & H'' & \longrightarrow & 0
\end{array}$$

where all vertical maps induce isomorphisms on associated galois modules. Consider the induced cohomology maps

$$\begin{array}{ccccccc}
H^1(G^0) & \longrightarrow & H^1(G) & \longrightarrow & H^1(G^{\text{ét}}) & \longrightarrow & 0 \\
\downarrow \subset & & \downarrow = & & \downarrow \subset & & \\
H^1(H') & \longrightarrow & H^1(H) & \longrightarrow & H^1(H'') & \longrightarrow & 0.
\end{array}$$

Since the vertical maps are inclusions, and the middle vertical map an equality, we obtain all vertical maps to be equality. From (2.1b) we learn that G^0 and H' have the same discriminant (since they are both connected) and consequently p' is an isomorphism. Similarly, we conclude from (2.1b) that H'' is étale, and p'' is an isomorphism. We may then observe that G, H have equal discriminants, and so p is an isomorphism.

PROOF OF PROPOSITION 2. Let $M \subset M_1 \times M_2$ denote the graph of φ. Consider the split exact sequence of galois modules,

$$0 \to M \to M_1 \times M_2 \to M' \to 0$$

and form the finite flat subgroup $G \subset G_1 \times G_2$ obtained by "flat extension" of M. Forming the quotient, consider the exact sequence,

$$0 \to G \to G_1 \times G_2 \to G' \to 0,$$

and the two natural projections

$$\begin{array}{ccc} & G & \\ {}^{p_1}\swarrow & & \searrow^{p_2} \\ G_1 & & G_2. \end{array}$$

We get

$$\text{(2.6)} \qquad \begin{array}{ccc} & & 0 \\ & & \downarrow \\ H^1(G) & \xrightarrow{\subset} & h^1(M) \\ \downarrow & & \downarrow \\ H_1 \times H_2 & \xrightarrow{\subset} & h^1(M_1) \times h^1(M_2) \\ \downarrow & & \downarrow \\ H^1(G') & \xrightarrow{\subset} & h^1(M') \\ \downarrow & & \downarrow \\ 0 & & 0 \end{array}$$

from which it follows that p_1 induces an isomorphism,

$$p_1 : H^1(G) \underset{\approx}{\to} H^1(G_1)$$

but since p_1 also induces an isomorphism on associated galois modules, p_1 is an isomorphism of group schemes, by Proposition 1. Take $f = p_2 p_1^{-1}$. Q.E.D.

3. **Examples.** Let $M = \mu_p$, be the galois module of pth roots of 1. Kummer theory yields an isomorphism $h^1(\mu_p) \simeq K^*/K^{*p}$. A recent result of L. Roberts

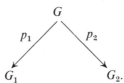

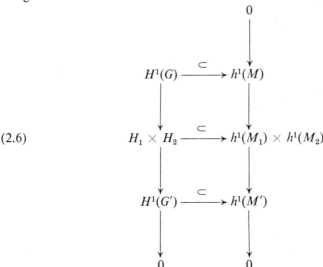

(see [**2**]) determines the lattice of realizable subgroups of K^*/K^{*p}. These turn out to be of the form

$$\frac{U^{(n)} \cdot U^p}{U^p} \subset \frac{K^*}{K^{*p}}$$

where U is the group of units in R with $U^{(n)} \subset U$ the nth subgroup of the canonical filtration. The n's that yield realizable subgroups may be easily described, using the result of Tate referred to in §1.

Now let K be absolutely unramified. Let C be an elliptic curve over K with nondegenerate reduction, and Hasse invariant zero. Let G denote the finite flat group scheme over Spec (R) obtained by taking the kernel of the pth power map on the abelian scheme induced by C over Spec (R).

By way of illustrating the use of our Theorem 2, we show that the flat group scheme G is uniquely determined by the galois module M.

One sees easily that $h^0(M) = h^2(M) = 0$, therefore Tate's Euler characteristic formula [**4**] tells us that $h^1(M)$ is a vector space of dimension two over the prime field F_p. We shall show that the lattice of admissible subspaces of $h^1(M)$ consists in exactly one subspace of dimension one. There is at least one admissible subspace, since G is a flat structure. Suppose that there were more than one. By the fact that the sum of two admissible subspaces is again admissible, and the intersection, it would follow that either $\{0\}$ or $h^1(M)$ is admissible. Using self-duality of M, it would follow in either case, that $\{0\}$ is admissible. But that implies that M admits a finite *étale* group scheme structure over Spec (R), which, in turn would imply that the action of the inertial group I of K on the space $M(\bar{K})$ is trivial. But this is not the case. In fact the action of I on $M(\bar{K})$ can be completely described ([**3**, §3]).

This suggests a definition: a galois module will be called *uni-structured over R* if it admits at most one flat group scheme structure over Spec (R). It follows from Theorem 2 that if

$$0 \to A \to B \to C \to 0$$

is an exact sequence of galois modules such that A and C are uni-structured then B is uni-structured.

Putting together what we have observed, we have: if M_n is the galois module of p^n-torsion points in the elliptic curve C above, then M_n is uni-structured.

In this very special case, then, we have come across a result for finite flat group schemes, which sharpens the general result of Tate [**5**], which asserts that a p-divisible group scheme is determined by its galois module.

References

1. M. Demazure and A. Grothendieck, *Schémas en groupes*, Seminaire Inst. Hautes Études Sci., 1963–64.

2. B. Mazur and L. Roberts, *Local Euler characteristics*, Mimeographed notes, Northeastern University.

3. J. P. Serre, *Representations abéliennes modulo l et applications*, (to appear).

4. J. Tate, *Duality theorems in Galois cohomology over number fields*, Proc. Internat. Congr. Math. 1962, Institut Mittag-Leffler, Djurrholm, 1963, pp. 288–295.

5. ———, *p-divisible groups*, Proc. Conf. on Local Fields, NUFFIC Summer School, 1966, pp. 158–183.

6. M. Raynaud, *Passage au quotient par une relation d'equivalence plate*, Proc. Conf. on Local Fields, NUFFIC Summer School, 1966, pp. 78–85.

HARVARD UNIVERSITY

Author index

Roman numbers refer to pages on which a reference is made to an author or a work of an author.

Italic numbers refer to pages on which a complete reference to a work by the author is given.

Boldface numbers indicate the first page of the articles in the book.

Subject index